NATO ASI Series

Advanced Science Institutes Series

A series presenting the results of activities sponsored by the NATO Science Committee, which aims at the dissemination of advanced scientific and technological knowledge, with a view to strengthening links between scientific communities.

The Series is published by an international board of publishers in conjunction with the NATO Scientific Affairs Division

A Life Sciences B Physics	Plenum Publishing Corporation London and New York
C Mathematical and Physical Sciences D Behavioural and Social Sciences E Applied Sciences	Kluwer Academic Publishers Dordrecht, Boston and London
F Computer and Systems Sciences G Ecological Sciences H Cell Biology I Global Environmental Change	Springer-Verlag Berlin Heidelberg New York London Paris Tokyo Hong Kong Barcelona Budapest

PARTNERSHIP SUB-SERIES

1. Disarmament Technologies	Kluwer Academic Publishers
2. Environment	Springer-Verlag/Kluwer Academic Publishers
3. High Technology	Kluwer Academic Publishers
4. Science and Technology Policy	Kluwer Academic Publishers
5. Computer Networking	Kluwer Academic Publishers

The Partnership Sub-Series incorporates activities undertaken in collaboration with NATO's Cooperation Partners, the countries of the CIS and Central and Eastern Europe, in Priority Areas of concern to those countries.

NATO-PCO DATABASE

The electronic index to the NATO ASI Series provides full bibliographical references (with keywords and/or abstracts) to about 50000 contributions from international scientists published in all sections of the NATO ASI Series. Access to the NATO-PCO DATABASE compiled by the NATO Publication Coordination Office is possible in two ways:

- via online FILE 128 (NATO-PCO DATABASE) hosted by ESRIN, Via Galileo Galilei, I-00044 Frascati, Italy.
- via CD-ROM "NATO Science & Technology Disk" with user-friendly retrieval software in English, French and German (© WTV GmbH and DATAWARE Technologies Inc. 1992).

The CD-ROM can be ordered through any member of the Board of Publishers or through NATO-PCO, Overijse, Belgium.

Series H: Cell Biology, Vol. 100

Springer
Berlin
Heidelberg
New York
Barcelona
Budapest
Hong Kong
London
Milan
Paris
Santa Clara
Singapore
Tokyo

Neurotransmitter Release and Uptake

Edited by

Şakire Pöğün

Ege University, Center for Brain Research
Ege University School of Medicine
Department of Physiology
Bornova, 35100 Izmir
Turkey

With 75 Figures

Springer

Published in cooperation with NATO Scientific Affairs Division

Proceedings of the NATO Advanced Study Institute "Neurotransmitter Release and Uptake", held at Kuşadası, Turkey, April 19–30, 1996

Library of Congress Cataloging-in-Publication Data applied for

Die Deutsche Bibliothek - CIP-Einheitsaufnahme

Neurotransmitter release and uptake : [proceedings of the NATO Advanced Study Institute "Neurotransmitter Release and Uptake", held at Kuşadası Turkey, April 19 - 30, 1996] / ed. by Şakire Pöğün. Publ. in cooperation with NATO Scientific Affairs Division. - Berlin ; Heidelberg ; New York ; Barcelona ; Budapest ; Hong Kong ; London ; Milan ; Paris ; Santa Clara ; Singapore ; Tokyo : Springer, 1997
([NATO ASI series / H] ; Vol. 100)

NE: Pöğün, Şakire [Hrsg.]; Advanced Study Institute Neurotransmitter Release and Uptake <1996, Kuşadası>; NATO: NATO ASI series / H

ISBN-13: 978-3-642-64517-4 e-ISBN-13: 978-3-642-60704-2
DOI: 10.1007/978-3-642-60704-2

Typesetting: Camera ready by authors/editor
Printed on acid-free paper
SPIN 10525531 31/3137 - 5 4 3 2 1 0

Preface

A NATO Advanced Study Institute, co-sponsored by FEBS and TUBITAK on "Neurotransmitter Release and Uptake" was held in Kuşadası, Turkey, from 19-30 April, 1996. The aim of the Institute was to bring together researchers working on the presynaptic events in transcellular communication. Even though release and uptake processes are very complementary and physically in very close proximity, researchers are usually are very focused on either one of these two aspects. This Institute provided a very good and unique opportunity to get exposure to both topics and the resulting book brings together two major presynaptic events which control transsynaptic regulation.

The topics discussed during the meeting ranged from ultrastructure and molecular control mechanisms to pathophysiology of some diseased states. Neuronal as well as non-neuronal mechanisms were covered. The course started with lectures on neurotransmitter release and continued with those on uptake and transporters. The present volume is arranged along the same lines: Part I is devoted to neurotransmitter release and Part II to neurotransmitter uptake. The concluding chapter by J.J. Furedy emphasizes the importance of interdisciplinary collaboration and provides a link between the events in the presynaptic terminal and behavior.

I thank all the lecturers and students for their participation and contributions, and the scientific and local organizing committee members for their invaluable support. Dr. Tulga Kalaycı helped with the formatting of the book with dedication and deserves special recognition; I am grateful for his excellent assistance.

Şakire Pöğün

Contents

PART I: Neurotransmitter Release

PART II: Neurotransmitter Uptake

PART I: Neurotransmitter Release

Proteins Implicated in Neurotransmitter Release and Reuptake

Heinrich Betz[1], Vincent O'Connor[1], Thomas Dresbach[1], Lorenzo Pellegrini[1], Kurt Bommert[1], William DeBello[2], James M. Hunt[2], Felix Schweizer[2], George Augustine[2], Milton P. Charlton[3], Ralph Adams[1], Kohji Sato[1], Andreas W. Püschel[1], Cyrille Sur[1] and Patrick Schloss[1]

[1]Abteilung Neurochemie, Max-Planck-Institut für Hirnforschung, D-60528 Frankfurt, Federal Republic of Germany.
[2]Department of Neurobiology, Duke University Medical Center, Durham, North Carolina 27710, USA
[3]Department of Physiology, University of Toronto, Ontario M5S 1A8, Canada.

Abstract

Neurotransmitter release and reuptake are essentiell steps in the life-cycle of neurotransmitters at the synapse. Biochemical evidence indicates that the exocytotic release process involves both evolutionarily conserved membrane proteins, the SNAREs, as well as ubiquitous cytosolic fusion proteins, NSF and SNAPs. We have tested the physiological role of both SNAREs and SNAPs at the squid giant synapse. Our data are consistent with both SNAPs and SNAREs having post-docking functions in neurotransmitter exocytosis. Analysis of the gene structure, developmental expression and pharmacology of the glycine transporter GlyT1 and the serotonin transporter SERT1 indicates a high diversity and complex regulation of mammalian neurotransmitter uptake systems.

1 Introduction

The regulated release and efficient reuptake of neurotransmitters at presynaptic nerve terminals provide the elementary processes, upon which fast synaptic transmission is based. Their high speed and efficiency are crucial for intercellular communication on a millisecond time-scale, and changes in either release or reuptake rates are thought to underlie both physiological changes in synaptic efficacy as well as pathological alterations in neurotransmission. The understanding of both processes at the molecular level therefore constitutes a major goal of our research efforts.

NATO ASI Series, Vol. H 100
Neurotransmitter Release and Uptake
Edited by Şakire Pöğün

2 Neurotransmitter Release, A Specialized Form of Subcellular Membrane Fusion

The release of neurotransmitters from presynaptic nerve terminals occurs via Ca^{2+}-regulated exocytosis from specialized storage organelles, the small synaptic vesicles. Rapid freezing of stimulated nerve terminals has allowed visualization of synaptic vesicles captured at various stages of fusion with the plasmamembrane and identified a pool of synaptic vesicles that are clustered around specialized regions of the presynaptic plasmalemma [1]. These "active zones" are thought to contain a sub-pool of readily releasable vesicles, which are "docked" close to the channels that mediate the Ca^{2+} influx that triggers release [2].

Recent biochemical and genetic studies indicate that an evolutionarily conserved protein network utilized in all eukaryotic vesicle-based membrane traffic also mediates synaptic vesicle-plasmamembrane fusion [3,4]. Experiments conducted to define the membrane receptors for ubiquitous cytosolic fusion factors, the ATPase N-ethylmaleimide sensitive fusion protein (NSF) and the α-, β- and γ-soluble NSF attachment proteins (SNAPs), identified a synaptosomal complex composed of the plasmamembrane proteins syntaxin and synaptosomal associated protein of 25 kDa (SNAP-25) and the synaptic vesicle protein synaptobrevin as the major SNAP binding component of rat brain extracts [5]. These proteins were therefore collectively named as SNAP receptors (SNAREs) and proposed to act as both synapse-specific membrane recognition molecules and acceptors for fusion catalyzing proteins [4,5].

In detergent extracts, the synaptosomal SNAREs form a complex that migrates at 7S upon density gradient centrifugation and is thought to represent the biochemical correlate of synaptic vesicles docked at the plasma membrane [6]. Under conditions that prevent ATP hydrolysis, the SNARE complex binds SNAP and NSF to form a multimeric complex that migrates at 20S. ATP hydrolysis by NSF disrupts this 20S complex in a process that has been postulated to drive membrane fusion [5,6]. Further evidence for an important role of the SNAREs in synaptic vesicle exocytosis comes from their identification as the targets of clostridial neurotoxins, the most potent inhibitors of neurotransmitter release known. The light chains (LC) of these toxins act as Zn^{2+} dependent endoproteases, with each LC specifically cleaving one of the three synaptosomal SNAREs. Synaptobrevin is cleaved by tetanus toxin and botulinum toxins B, D, F and G; syntaxin by botulinum toxin C1; and SNAP-25 by botulinum toxins A and E [7,8].

To elucidate whether the ubiquitous soluble fusion proteins NSF and SNAP are indeed involved in synaptic vesicle exocytosis, we tested the effects of SNAP at the giant synapse of the squid stellate ganglion [9]. The enormous size of this synapse allows simultaneous recording of pre- and postsynaptic potentials in a preparation amenable to microinjection of recombinant proteins, peptides and other molecular probes. The cloning of a full-length squid SNAP cDNA allowed to produce a recombinant squid SNAP protein [9]. Microinjection of either this squid SNAP or recombinant mammalian α-SNAP into the giant presynaptic terminal produced an enhancement of transmitter release as assayed by postsynaptic recording. This indicates that SNAP is a rate-limiting component of the release apparatus. In contrast, peptides corresponding to potentially surface-exposed regions of squid SNAP caused a rapid and reversible inhibition of synaptic vesicle exocytosis. [9]. Examination by imaging and electron microscopy of the effects of one of these peptides revealed that its blocking action was not related to alterations in Ca^{2+} influx, but correlated with an about 3-fold increase in the number of synaptic vesicles docked at the active zones of the presynaptic plasmamembrane. Assuming that this accumulation of docked vesicles in the peptide-inhibited terminals was the consequence of a competitive inhibition of SNAP interaction with other components of the vesicle release machinery [9, 10], our data are consistent with a post-docking role of SNAP, and by inference of NSF, in synaptic vesicle exocytosis [11].

We also have employed the squid giant synapse to test the functions of the synaptic vesicle SNARE synaptobrevin [3-6]. cDNA cloning of squid synaptobrevin revealed that the cleavage site for the clostridial Zn^{2+} proteases tetanus toxin and botulinum toxin B light chain is conserved in this invertebrate [2]. Upon microinjection into the giant terminal, both toxins produced a slow irreversible inhibition of neurotransmitter release that was prevented by coinjection of an inhibitor peptide that corresponds to the sequence of synaptobrevin that spans the cleavage site. Electron microscopy revealed that the toxin-poisoned terminals contained increased numbers of synaptic vesicles around their active zone profiles. Importantly, the number of vesicles apposed to the presynaptic membrane, i.e. the docked vesicle population, was significantly higher than in control injected terminals [2]. Thus, toxin cleavage severs synaptobrevin function at a late step of vesicle exocytosis, which lies down-stream of the vesicle-plasmamembrane recognition and/or docking process.

Biochemical data also support the conclusion that the SNAREs function in the late steps of exocytotic membrane fusion. SNAREs are accessible to proteolysis by clostridial neurotoxin only in their free non-complexed state [13,14]. Surprisingly, however, SNARE complex formation was not impaired in brain extracts, in which synaptobrevin was completely cleaved by tetanus toxin treatment [14]. Indeed, both cleavage fragments were recovered in the resultant SNARE complex. Also, assembly and disassembly of the 20S complex, i.e. binding of SNAP and NSF by the complexed SNAREs and their dissociation upon NSF-catalyzed ATP hydrolysis, was indistinguishable in tetanus toxin cleaved and control preparations [14]. Apparently, synaptobrevin cleavage does not disrupt the SNAREs' principal conformational transitions. However, a highly characteristic property of the complexed SNAREs, their ability to form a SDS-resistant trimer, was severely impaired as revealed by a different temperature sensitivity to SDS dissociation [14,15]. This shows that the stability, i.e. the low intrinsic energy, of the SNARE complex is altered upon incorporation of toxin-cleaved synaptobrevin. We therefore have proposed that clostridial neurotoxin impair SNARE function by compromising the ability of the SNARE complex to form an "activated" high-energy intermediate upon NSF-catalyzed ATP hydrolysis [13]. The conformational energy stored in this complex is proposed to ultimately drive membrane fusion upon Ca2+ triggering [16] of a release event.

3 Neurotransmitter Transporters, A Family of Na^+ Driven Plasmamembrane Co-Transport Proteins

Termination of neurotransmission requires the rapid inactivation or the removal of the released transmitter substance from the synaptic cleft. In the central nervous system, amino acid and biogenic amine transmitters are reutilized for transmission. Specialized neurotransmitter transporters are therefore crucial for their efficient reuptake into the presynaptic cytoplasm.

Our laboratory has a particular interest in synaptic inhibition mediated by glycine. Initial attempts to clone Na^+ dependent glycine transporters therefore used PCR amplification of spinal cord cDNA in order to identify gene products homologous to the previously cloned GABA transporter, GAT1 [17]. This led to the identification of the rat serotonin transporter SERT1 [17,18] and a second homologous transporter cDNA that was subsequently identified as the widely expressed creatine transporter, CREAT1 [19-21]. cDNAs encoding distinct isoforms of the mammalian glycine transporter (GlyT1 and 2) were simultaneously cloned by different laboratories [22-24].

Our work has focussed on the expression and regulation of the mouse glycine transporter, GlyT1, and on the structural and pharmacological analysis of the rat serotonin transporter, SERT1. A detailed analysis of the murine GlyT1 gene unraveled the genetic basis underlying the different splice variants of this transporter isoform [25]. Analysis of cDNA 5'-ends revealed that the GlyT1a and 1b/1c mRNAs are transcribed from two different promoters. During murine embryonic development, GlyT1 mRNAs were detectable by RNase protection assays as early as embryonic day E9 and reached maximal levels between E13 and E15 [25]. *In situ* hybridization revealed GlyT1 expression in the developing spinal cord mainly in the ventral part of the ventricular zone at E12. At later stages (E15), transcripts were also found in the lateral half of the basal and intermediate gray matter. In contrast, the second glycine transporter gene GlyT2 displayed a completely different expression pattern. At E11 it is expressed in the mantle zone, and at later stages throughout the ventral horns. In the adult rat brain and spinal cord, GlyT1 hybridization signals were found exclusively in glial cells. These data indicate that GlyT1 is an early marker of neural development and encodes glia-specific transporter proteins [25].

We also investigated a potential regulation of the recombinant GlyT1 protein by protein kinase C (PKC), an ubiquitous set of isoenzymes involved in the short-term regulation of cellular processes [26]. Treatment of human embryonic kidney cells expressing the mouse GlyT1 with the PKC activator phorbol 12-myristate 13-acetate (PMA) decreased specific [^{3}H]glycine uptake. This down-regulation resulted from a reduction of the maximal transport rate and was blocked by the PKC inhibitors 1-(5-isoquinolinylsulfonyl)-2-methylpiperazine (H7) and staurosporine. The inhibitory effect of PMA treatment was also observed after removing all five predicted phosphorylation sites for PKC in GLYT1b by site-directed mutagenesis [26]. These data indicate that glycine transport by GLYT1b is modulated by PKC activation; however, this regulation may involve indirect phosphorylation mechanisms. This interpretation is supported by analogous studies on the heterologously expressed rat GABA transporter GAT1, which also displays an about 50% down-regulation upon PKC activation [27].

Serotoninergic neurotransmission is thought to be important in the modulation of cortical functions, memory formation and control of locomotor activity, and deficiencies in the central serotonin (5-HT) system are implicated in degenerative and psychiatric disorders, such as Parkinson and depression. Drugs inhibiting 5-HT reuptake are widely used for the antidepressant treatment of depression. These antipressant drugs fall into two categories, tricyclic and non-tricyclic selective 5-HT uptake inhibitors. We found that binding of the tricyclic antidepressant [^{3}H]imipramine to the recombinant rat serotonin transporter SERT1 expressed in human embryonic kidney cells is nonhomogeneous [28]. Scatchard analysis and competition experiments revealed the existence of two distinct antidepressant binding sites (ABS). At ABS 1, [^{3}H]imipramine binding was strictly sodium-dependent with an apparent KD of A10 nM. In contrast, [^{3}H]imipramine binding to ABS 2 occurred also in the absence of sodium and exhibited a lower affinity. Binding of the non-tricyclic antidepressant [^{3}H]citalopram was observed only at ABS2. The natural substrate of this carrier, serotonin, competitively inhibited antidepressant binding at both sites; however, its affinity to site 2 was Å5-fold lower.

In uptake studies with [^{3}H]5-HT [29], both imipramine and citalopram competitively inhibited 5-HT transport mediated by recombinant SERT1. However, for citalopram the concentration producing half-maximal transport inhibition was similar to its KD value determined by equilibrium binding, whereas the inhibitory potency of imipramine was more than one order of magnitude lower than its KD value. Our data are consistent with low-affinity imipramine binding occurring at or closely to the substrate recognition site, which also binds citalopram (ABS2). Occupation of ABS1 does not affect 5-HT transport, but allosterically displaces citalopram from the substrate recognition site. Indeed, low concentrations of imipramine partially protected 5-HT transport from citalopram inhibition, and this protection was only observed in the presence of sodium ions. These date provide a molecular explanation for the distinct pharmacological actions of different antidepressants and reveal both competitive and allosteric antidepressant interactions at the neuronal serotonin transporter.

4 Perspectives

The work reported here demonstrates a high complexity of the membrane and membrane-associated proteins maintaining the synaptic life-cycle of neurotransmitters. More detailed studies of the underlying proteins should shed light on the precise molecular mechanisms of transmitter release and reuptake and help to define novel compounds that might provide a lead towards better neuropharmacologically active agents.

Acknowledgements

We thank M. Baier and H. Reitz for secretarial assistence during the preparation of this manuscript. Work in the authors′ laboratories was supported by Deutsche Forschungsgemeinschaft, Fonds der Chemischen Industrie, National Institutes of Health and Medical Research Council of Canada, and the Biomed and Health Research Program BMH1-CT93-1110.

References

1. Heuser, J.E., Reese, T.S., Dennis, M.J., Jan, Y., Jan, L. and Evans, L. (1979) J. Cell Biol. **81**, 275-300
2. Augustine, G.J., Charlton, M.P. and Smith, S.J. (1987) Annu. Rev. Neurosci. **10**, 633-693
3. Bennett, M.K. and Scheller, R.H. (1993) Proc. Natl. Acad. Sci. USA **90**, 2559-2563
4. Rothman, J.E. and Warren, G. (1994) Curr. Biol. **4**, 220-233
5. Söllner, T., Whiteheart, S.W., Brunner, M., Erdjument-Bromage, H., Geromanos, S., Tempst, P. and Rothman, J.E. (1993) Nature **362**, 318-324
6. Söllner, T., Bennett, M.K., Whiteheart, S.W., Scheller, R.H. and Rothman, J.E. (1993) Cell **75**, 409-418
7. Montecucco, C. and Schiavo, G. (1993) Trends Biol. Sci. **18**, 324-326
8. Niemann, H., Blasi, J. and Jahn, R. (1994) Trends Cell. Biol. **4**, 179-185
9. DeBello, W.M., O'Connor, V., Dresbach, T., Whiteheart, S.W., Wang, S.S-H., Schweizer, F.E., Betz, H., Rothman, J.E. and Augustine, G.J. (1995) Nature **373**, 626-630
10. Bommert, K., Charlton, M.P., DeBello, W.M., Chin, G.J., Betz, H. and Augustine, G.J. (1993) Nature **363**, 163-165
11. Schweizer, F.E., Betz, H. and Augustine, G.J. (1995) Neuron **14,** 689-696
12. Hunt, J.H., Bommert, K., Charlton, M.P., Kistner, A., Habermann, E., Augustine, G.H. and Betz, H. (1994) Neuron **12**, 1269-1279
13. Pellegrini, L.L., O'Connor, V. and Betz, H. (1994) FEBS Lett. **353**, 319-323
14. Pellegrini, L., O'Connor, V., Lottspeich, F. and Betz, H. (1995) EMBO J. **14**, 4709-4713
15. Hayashi, T., McMahon, H., Yamasaki, S., Binz, T., Hata, Y., Südhof, T.C. and Niemann, H. (1994) EMBO J. **13,** 5051-5061
16. O′Connor, V., Augustine, G.J. and Betz, H. (1994) Cell **76**, 785-787
17. Mayser, W., Betz, H. and Schloss, P. (1991) FEBS Lett. **295**, 203-206
18. Hoffmann, B.J., Mezey, E. and Brownstein, M.J. (1991) Science **254**, 579-580
19. Mayser, W., Schloss, P. and Betz, H. (1992) FEBS Lett. **305**, 31-36
20. Guimbal, C., Kilimann, M.W. (1993) J. Biol. Chem. **268**, 8418-8421

21. Schloss, P., Mayser, W. and Betz, H. (1994) Biochem. Biophys. Res. Comm. **199**, 1320-1325
22. Borowsky, B., Mezey, E. and Hoffmann, B. (1993) Neuron **10**, 851-863
23. Liu, Q.-R., Lopez-Cuerca, B., Mandiyan, S., Nelson, H. and Nelson, N. (1993) J. Biol. Chem. **268**, 22802-22808
24. Schloss, P., Püschel, A.W. and Betz, H. (1994) Curr. Opin. Cell Biol. **6**, 595-599
25. Adams, R.H., Sato, K., Shimada, S., Tohyama, M., Püschel, A.W. and Betz, H. (1995) J. Neurosci. **15**, 2524-2532
26. Sato, K., Adams, R., Betz, H. and Schloss, P. (1995) J. Neurochem. **65**, 1967-1973
27. Sato, K., Betz, H. and Schloss, P. (1995) FEBS Lett. **375**, 99-102
28. Schloss, P. and Betz, H. (1995) Biochemistry **34**, 12590-12595
29. Sur, C., Betz, H. and Schloss, P. (1996) submitted

Ion Channels in Transmitter Release

Alon Meir[1], Alexander Butkevich[1], Serdar Demirgören[2], Nilly Yakir[1] and Rami Rahamimoff[1]

[1]Department of Physiology and The Bernard Katz -Minerva Centre for Cell Biophysics Hebrew University-Hadassah Medical School, POBox 12272, Jerusalem 91120, Israel
[2]Ege University Center for Brain Research School of Medicine, Department of Physiology, Bornova, 35100 Izmir, Turkey

1 Introduction

In the nerve terminals, like in other parts of the neuron and in most other cells, ion channels play a large number of crucial roles in cell function. Some of the ion channels are involved in constitutive functions, such as the maintenance of the resting potential, while other channels are involved in the specialized function of the nerve terminal - the liberation of the neurotransmitter.

In each one of the stages leading to the release of transmitter there is an important role of ion channels. The first step in the synaptic transmission process is the *invasion* of the action potential into the nerve terminal (for a review regarding the role of ion channels in presynaptic excitability see, Jackson, 1995). Here the usual interaction among voltage sensitive channels takes place. These channels include sodium, potassium, calcium, non-selective and chloride channels.

The next step in the generation of the synaptic response is the *entry of calcium ions* into the presynaptic nerve terminal. This is achieved predominantly by selective calcium channels, but recent experiments suggest that part of the calcium entry can be accomplished by non-selective ion channels as well.

After the fusion of the synaptic vesicle with the surface membrane, there is an *incorporation* of the ion channels of the vesicle membrane into the surface membrane of the presynaptic nerve terminal. Thus the state of the nerve terminal is different before and after the release of the transmitter.

Ion channels are involved not only in the normal course of events leading to transmitter release, but also in modulation of transmitter release under different physiological and pathophysiological conditions. We show in this article how the long inactivated state of one of the potassium channels can lead to *statistical memory* and may be involved in *frequency modulation* of synaptic transmission.

Presynaptic nerve terminals possess not only voltage-gated ion channels, but also *ligand-gated ion channels* that can be involved in the regulation of transmitter release.

NATO ASI Series, Vol. H 100
Neurotransmitter Release and Uptake
Edited by Şakire Pöğün

In presynaptic nerve terminals ion channels can be studied at different levels. There are very few preparations that meet the two criteria necessary to study the ion channels with the patch clamp technique: large enough size and lack of covering cells. We describe here a preparation where some of the difficulties can be solved by *membrane fusion*.

2 Procedures

The following experimental and analytical procedures were used in the experiments done by us.

2.1 Preparation of Synaptosomes

Synaptosomes were prepared from the electric organ of the *Torpedo ocellata* and *Torpedo marmorata*, as described by Michaelson & Sokolovsky (1978). In brief, the animals were maintained in oxygen-perfused sea water at 15°C for up to 4 months prior to use. On the day of preparation, the animals were sacrificed by chilling at 4°C and the electric organs were excised. The P2 pellet was obtained from the electric organ homogenate loaded onto the density gradient. The synaptosomes were collected from the interface between 0.13 and 0.3 M-sucrose after differential centrifugation. They were washed once with three volumes of 0.8 M-glycine buffer and then with either 0.39 M KCl or NaCl and 10 mM Na-HEPES pH 7.4. The synaptosomes were pelleted by centrifugation at 17000g for 20 min.

2.2 Fusion of the Synaptosomes

Since most of the nerve terminals are too small for the various configurations of the patch-clamp technique, fusion procedures have been applied to obtain suitable structures for electrophysiological recordings.

The fusion procedure of the synaptosomes into giant structures is a modification of that described by Rahamimoff, DeRiemer, Sakmann, Stadler & Yakir, (1988). The synaptosomal pellet was suspended in 0.1 ml of fusion medium consisting of 20% dimethyl sulfoxide (DMSO), 50% polyethylene glycol 1500 (PEG) and 30% 1.3 mM KCl, 33 mM HEPES pH 7.4 and kept frozen at -70°C for up to a year. A day before the experiment, the frozen pellet was incubated at 37°C for 5 minutes, followed by two dilutions and incubations, first with 100 μl and then 200 μl of bath solution. These were done at 37°C for 15 minutes and 30 minutes, respectively. After several hours at room temperature, giant structures with diameters of up to 50 μm were observed.

2.3 Data Acquisition and Analysis

Experiments were performed at room temperature using the inside-out configuration of the patch-clamp technique described by Hamill, Marty, Neher, Sakmann & Sigworth (1981). The bath solution usually contained 390 mM of KCl or Na-Glutamate, and 10 mM Na-HEPES pH 7.4. The patch pipettes were pulled by a two-stage puller (List L/M-3P-A), coated with sylgard (Sylgard, Dow Corning 184) and fire polished. They were then filled with the appropriate solution.

In most cases, the pipette solution was composed of 390 mM NaCl, 1 mM $CaCl_2$, 1 mM $MgCl_2$, and 10 mM Na-HEPES, pH 7.4.

The patch pipette resistance before formation of the seal was between 3-10 MΩ. A seal with the resistance of several gigaohms was formed with fused synaptosomes more than 10 μm in diameter. Currents were recorded with an extracellular patch-clamp amplifier (EPC-7, List Instruments) at gains of 20 or 50 mV/pA. The data were recorded on a videotape recorder and digitized (at 20-100 kHz) by using TL-1-125 interface and the Fetchex (for steady state experiments) or Clampex (for pulses and ramps) software (part of pClamp version 5.5.1 and 6, Axon Instruments, Foster City, CA). The digitized data were analyzed using the software of pClamp, Sigmaplot (Jandel Scientific) and home-made programs for statistical analysis.

2.4 Statistical Analyses

Ion channels may open in a totally uncoordinated or random fashion or there may be some "order" in openings. Statistical analyses are used to determine whether there is an "order" in the behavior of the channel and to describe it in quantitative terms. Several selected statistical tests will be described briefly.

2.4.1 Poisson Analysis

If an event occurs randomly at a certain rate, one can estimate the number of time intervals, during which the event occurs x times (N_x). This number is given by the following equation:

$$N_x = \frac{N_T \cdot e^{-m} \cdot m^x}{x!},$$

where N_T is the total number of time intervals and m is the mean number of occurrences per interval. For an ion channel, by examining the fit of the distribution of openings to Poisson, it is possible to test for randomness. Obviously, the better the fit to Poisson, the higher the likelihood that the behavior of the channel is random.

2.4.2 Turning-Points Test

In a series of numbers the turning point is defined as a point flanked with the numbers, which are either bigger or smaller. The expected number of turning points for a completely random series (S_{exp}) is given by:

$$S_{exp} = \frac{2}{3}(n-2),$$

where n is a number of points in a series. If there is a positive interdependence between the numbers in a series, the observed number of turning points will be smaller than the expected one. The stronger the interdependence, the larger is the difference between the observed and the expected numbers of turning points.

2.4.3 Run Test

A run is defined as a series of similar values, for ion channels, a series of consecutive openings. Given the number of trials in a series (n) and the probability of opening (p), one can determine the number of runs expected from a random process: $R_{exp} = 2np(1-p)$. When the openings of a channel tend to cluster together, i.e., when there is a positive correlation between the responses, the observed number of runs will be smaller than the expected one. The departure of the behavior from randomness is determined with the use of Z statistics, given by the following equation:

$$Z = -\frac{R_{obs} - R_{exp}}{2\sqrt{np(1-p)}}.$$

When the Z value is larger than 2 the behavior is considered significantly non random, and the higher the Z value, the stronger the departure of the behavior from randomness.

2.4.4 Cross-correlation and Autocorrelation Functions

The correlation functions determine the dependence between two series. In cross-correlation, the correlation coefficient (r) is calculated between the one series and *the second series* displaced by a lag. In autocorrelation, r is calculated between the original and *the same series* displaced by a lag. Thus, this function is used to determine the dependence between the constituents in a group. The correlation coefficient (r) is given by:

$$r = \frac{Cov(x,y)}{\sigma_x \cdot \sigma_y},$$

where x and y are the two series (or original and the displaced series, respectively), σ is the standard deviation and Cov(x,y) is the covariance. The correlation coefficient may take on any value between -1 (a perfect negative correlation) and 1 (a perfect positive correlation). In autocorrelation, by definition, for lag=0, r=1. For a completely random process, r is expected to be insignificantly different from zero. Cross-correlation and autocorrelation functions have been extensively used to test for the memory and periodicity in the activity of the bursting K^+ channel (Rahamimoff, Edry-Schiller & Ginsburg, 1992; Rahamimoff, Edry-Schiller, Rubin-Fraenkel, Butkevich & Ginsburg, 1995).

3 Ion Channels and Currents in Presynaptic Nerve Terminals

3.1 Potassium Channels

The concept that the outward potassium current underlies the repolarization phase of action potential was originally proposed by Hodgkin and Huxley (1952). During the subsequent decades, potassium channels and currents were extensively investigated. In nerve terminals, several currents were identified on the basis of their biophysical and pharmacological properties. These include slow K^+ currents, Ca^{2+} activated K^+ currents, ATP sensitive K^+ current, delayed rectifier K^+ current and transient A-type current.

3.1.1 Slow K^+ Currents (IK_s)

These currents activate slowly upon depolarization (with a time constant in the order of tens of milliseconds) and do not inactivate for a long period of time (seconds) (Bartschat & Blaustein, 1985a; Bielefeldt, Rotter & Jackson, 1992). Two pharmacologically different slow currents have been identified in mouse motor nerve terminal (Bielefeldt et al., 1992; Chao & Lin-Shiau, 1995). One IK_s is blocked by 3,4 diaminopyridine (DAP) and uranyl (Bielefeldt et al., 1992), while the other is enhanced by uranyl (Chao & Lin-Shiau, 1995). Single-channel properties of the slow K^+ channel were described in posterior pituitary nerve endings (Bartschat & Blaustein, 1985a). Due to their slow activation kinetics, these currents do not seem to play a significant role in the repolarization phase of the action potential. They may be activated during long bursts of activity found in the hypothalamic-pituitary system, and thus limit the secretion towards the end of the burst (Bartschat & Blaustein, 1985a).

3.1.2 Ca^{2+} Activated K^+ Currents ($IK_{(Ca2+)}$)

Two kinetically and pharmacologically distinct Ca^{2+} activated K^+ currents are identified in different preparations: fast $IK_{(Ca2+)}$ and slow $IK_{(Ca2+)}$. (Mallart, 1985; Tabti, Bourret & Mallart, 1989; Morita & Barrett, 1990). The fast $IK_{(Ca2+)}$ is inhibited by charybdotoxin and tetraethylammonium (TEA), it activates very rapidly and accounts in part for the repolarization phase of the action potential (Morita & Barrett, 1990). In the mouse nerve terminal, this current is activated also by Sr^{2+} (Tabti et al., 1989). The slow $IK_{(Ca2+)}$ is not sensitive to charybdotoxin, it activates somewhat slower and mainly accounts for the hyperpolarizing afterpotential (Morita & Barrett, 1990; Blundon, Wright, Brodwick & Bittner, 1995).

Several Ca^{2+} activated K^+ channels were localized to the presynaptic nerve terminals by a variety of techniques. These channels differ in their single channel conductance, pharmacology, gating, kinetics and function (Bartschat & Blaustein, 1985b; Mallart, 1985; Hevron, David, Arnon & Yaari, 1986; Tabti et al., 1989; Bielefeldt et al., 1992; Wang, Thorn & Lemos, 1992; Takeuchi & Wangemann, 1993; Wangemann & Takeuchi, 1993; Issa & Hudspeth, 1994; Blundon et al., 1995; Dopico, Lemos & Treistman, 1996; Knaus, Schwarzer, Koch, Eberhart, Kaczorowski, Glossmann, Wunder, Pongs, Garcia & Sperk, 1996).

3.1.3 ATP Sensitive K^+ Current

These fast currents are activated by the decrease in intracellular ATP (Rorsman & Trube, 1985). Thus, they may serve as means of regulating the activity of the nerve terminal by its energy content. The ATP sensitive K^+ channels were found at the nerve terminals in hippocampus and the motor nerve endings (Krnjevic, 1990; Deist,

Repp & Dreyer, 1992). These channels are opened by diazoxide and some endogenous peptides (Krnjevic, 1990) and blocked by sulfonylureas and TEA in high concentrations (Deist et al., 1992). Under normal conditions, the ATP sensitive K^+ channels are closed, but when the energy content of the terminal is low, these channels become essential for maintaining resting potential (Krnjevic, 1990) and may reduce the neurotoxic release of glutamate in CNS (Deist et al., 1992).

3.1.4 Delayed Rectifier K^+ Current

This current shows a rapid, voltage-dependent activation and does not inactivate during the action potential (Augustine, 1990). In squid giant synapse, it is blocked by extracellular and intracellular application of 3,4 diaminopyridine (DAP) and intracellular application of TEA. In other preparations, it is blocked by Zn^{2+}, extracellular acetylcholine and TEA (Hevron et al., 1986; Lingdren & Moore, 1989; Miralles, Canti, Marsal, Peres & Solsona, 1994). It is proposed that the delayed rectifier current regulates the duration of the action potential and thus the magnitude of the postsynaptic response (Augustine, 1990).

3.1.5 Voltage Independent K^+ Channels

Seven different voltage-independent ion channels were found at the rat brain synaptosomes (Nelson, Roudna & Bamberg, 1983; Hirashima & Kirino, 1988). Three of these channels are TEA sensitive, slightly prefer potassium over sodium ions (permeability ratio 2:1) and differ in their single channel conductance (15.1, 28.6 and 91.0 pS) (Hirashima & Kirino, 1988).

The other four were described in rat brain synaptosomes incorporated into lipid bilayers (Nelson et al., 1983). These channels are highly selective for K^+ over Cl^- ions and differ in their conductance and kinetic properties.

These channels, being voltage independent, may play a role in maintaining the membrane resting potential of of the terminal (Nelson et al., 1983; Hirashima & Kirino, 1988).

3.1.6 A-Type K^+ Current

This current is rapidly activated by depolarization and inactivates within milliseconds (Perney & Kaczmarek, 1991). The classical channels that account for the A-type current are the members of the *Shaker* family. Many of these channels have been identified in various tissues and cells and extensively studied (for a review, see Perney & Kaczmarek, 1991). In presynaptic nerve terminals, Kv1.1, Kv1.2 and Kv1.4 channel proteins have been demonstrated. These include homo- as well as heterotetrameres (McNamara, Muniz, Wilkin & Dolly, 1993; Sheng, Liao, Jan & Jan, 1993; Wang, Kunkel, Martin, Schwartzkroin & Tempel, 1993; Sheng, Tsaur, Jan & Jan, 1994; Wang, Kunkel, Schwartzkroin & Tempel, 1994). Since it is very difficult to investigate the single-channel properties in the nerve terminals, very few of them have been studied in detail. In fused presynaptic terminals of *Torpedo* electric organ, we found the bursting potassium channel (Edry-Schiller, Ginsburg & Rahamimoff, 1991a), which will be described in the subsequent section.

The fast activation-inactivation properties of the A-type current suggest its possible function in the regulation of action potential duration and thus of the transmitter release process.

3.2 The Bursting K^+ Channel in Fused Synaptosomes

The bursting K^+ channel is the most common channel in the fused *Torpedo* synaptosomes (presynaptic nerve endings), and may also be the most common channel at the presynaptic nerve terminal *in vivo*.

The openings of the channel appear as bursts, hence the name "bursting", followed by long silent periods (figure 1A). Each burst is composed of several openings separated by short closures. This finding suggests that the channel must have at least two distinct closed states, a short one and a long one (Colquhoun & Sakmann, 1985; Edry-Schiller, Ginsburg & Rahamimoff, 1991a). With the use of Poisson analysis, it was found that the kinetic scheme of this channel is even more complex; it requires at least three distinct closed states and one open state (Edry-Schiller et al., 1991a).

The slope conductance of the bursting K^+ channel is 23.8 pS and it is 70 times more selective for potassium over sodium ions. Chloride ions do not pass significantly through the channel (Edry-Schiller et al., 1991a). A voltage step causes a rapid activation of the channel followed by inactivation, which can last for several hundreds of milliseconds to several seconds (Edry-Schiller & Rahamimoff, 1992) (figure 1B). Thus the behavior of the channel resembles that of the A-type potassium channel (Perney & Kaczmarek, 1991).

3.2.1 Statistical Memory and Oscillations

The openings of the bursting K^+ channel are not random in time, but are positively correlated, i.e., they are strongly dependent on the "history" of the channel. In response to a series of depolarizatory pulses, the probability of the channel to be opened by a given pulse is significantly augmented if the channel was opened by the previous one. We call this property a "statistical memory" of the channel (Rahamimoff et al., 1992). Upon repetitive stimulation, oscillations in the activity of the channel with a period of ~20 seconds can be observed (Rahamimoff et al., 1995)(figure 1C). These oscillations may represent an intrinsic property of the channel or be induced by the repetitive stimulation. We favor the latter possibility, although there is not enough evidence to dismiss either of them.

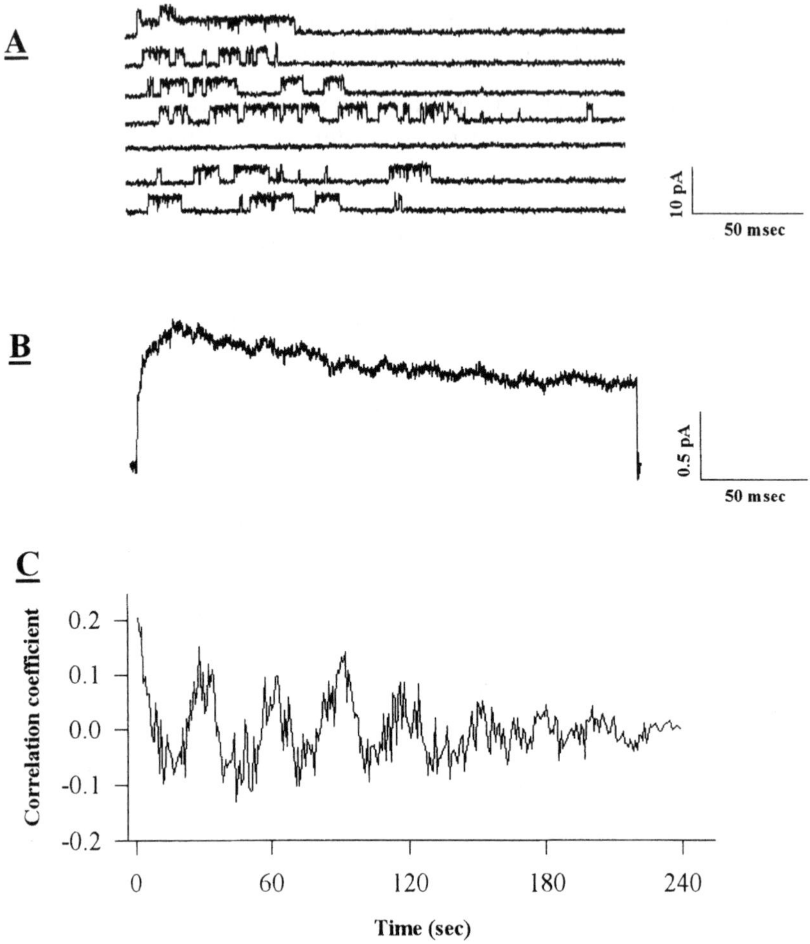

Fig 1. Activity and oscillations of the bursting K^+ channel in fused nerve terminals. A: Examples of the channel's activity after subtraction of the leak and capacitance components of the current. The membrane potential was held at -90 mV for 550 msec and stepped to +20 mV for 250 msec. See 'data acquisition and analysis' for the solutions composition. **B:** The ensemble current of the bursting K^+ channel. Experimental conditions as in A. Note rapid activation and inactivation of the current. **C:** Autocorrelogram of the 297 responses to a voltage step from -110 mV to +20 mV. The values of correlation coefficients at the maxima and minima of oscillations are significantly different from zero and from one another.

The voltage dependence, the statistical properties of the bursting K^+ channel and the fact that it is the most common channel in the fused synaptosomes, make this channel a likely candidate to be involved in the regulation of synaptic transmission.

We speculate that this channel may play a role in frequency modulation phenomena, such as paired pulse facilitation and tetanic and posttetanic potentiation.

3.3 Calcium and Calcium Permeable Currents and Channels

Calcium ions play a key role in the regulation of synaptic transmission (Locke, 1884, Feng, 1936). Their main site of action is the presynaptic nerve terminal, where they regulate the number of transmitter quanta liberated by the nerve impulse (del Castillo & Stark, 1952; del Castillo & Katz, 1954; Jenkinson, 1967; Dodge & Rahamimoff, 1967). The relation between extracellular calcium and transmitter release is highly non-linear, with a sigmoidal start, indicating an important regulatory function (Dodge & Rahamimoff, 1967; Hubbard, Jones & Landau, 1968). The high power dependence of quantal transmitter release on extracellular calcium posed the question whether the action of calcium is on extracellular sites or whether the action is intracellular and mediated by calcium channels in the presynaptic nerve terminal. In 1967, Katz and Miledi proposed the 'calcium hypothesis' of transmitter release (Katz & Miledi, 1967a; b; see also Katz, 1969), which states that the entry of calcium ions from the extracellular compartment to the presynaptic nerve terminal interior along their steep electrochemical gradient is the key step triggering quantal transmitter release. This 'calcium hypothesis' was supported by measurements done on the squid giant synapse (Bloedel, Gage, Llinas, & Quastel, 1966; Katz & Miledi, 1967a; Kusano, 1967; Llinas, Steinberg & Walton, 1981; Augustine & Eckert, 1984; Augustine, Charlton & Smith, 1985a, b), where voltage clamp is feasible. It was shown there that the size of the postsynaptic excitatory potential was strongly dependent on the calcium influx through the presynaptic membrane. The quantitative aspects of this relation are still under debate (for a review, see Augustine, Charlton & Smith, 1987).

For a number of years, the properties of the calcium currents at vertebrate nerve terminals were evaluated only by extracellular recordings (Brigant & Mallart, 1982; Gundersen, Katz & Miledi, 1982; Mallart, 1984; Lindgren & Moore, 1989) or by measurement of calcium influx into synaptosomes derived from the mammalian brain (Nachshen & Blaustein, 1982; Nelson, French & Krueger, 1984; Umbach, Gundersen & Baker, 1984) or from the neurohypophysis (Lemos & Nowycky, 1989). Recently, however, large presynaptic nerve terminals from the chick ciliary ganglion (Martin, Patel, Faille & Mallart, 1989; Stanley & Atrakchi, 1990), the goldfish retinal bipolar cells (Heidelberger & Matthews, 1992) and from the hypophysis (Wang, Treistman & Lemos, 1992) permitted the study of the calcium currents more directly.

There are clear indications for the existence of at least 5 different types of calcium channels in neurons, named L, T, N, P and Q. These channels differ from each other in their activation voltage range, inactivation voltage range, single channel conductance and pharmacology (see Miller, 1987; Uchitel, Protti, Sanchez, Cherksey, Sugimori & Llinas, 1992).

In single channel studies, Wang, Treistman & Lemos (1993) found two types of calcium channels in rat neurohypophysial nerve terminals. The first one is the N_t-type channel of 11 pS, which is High Voltage Activated (HVA) and is inactivated during a depolarization voltage pulse. The other channel is the L-type channel of 25 pS, which is Low Voltage Activated (LVA) and it is enhanced by Bay K 8644 and blocked by the dihydropyridine substance nicardipine. Similar results were reported by Lemos and Nowycky (1989). In the large presynaptic nerve terminals in the chick ciliary ganglion, single calcium channels were recorded with single channel conductance of 11-14 pS (Stanley 1991, 1993). Based on extracellular recording and on EPSP measurements to monitor transmitter release, it was suggested that P and Q type channels also mediate the process (Uchitel et al, 1992; Regher & Mintz, 1994; Wheeler, Randall & Tsien, 1994).

Of special interest is the observation that calcium channels can serve as targets for the action of neurotransmitters. For example, it was shown that noradrenaline reduces the inward calcium currents (Dunlap & Fischbach, 1981; Forscher & Oxford, 1985) by changing the voltage dependence of the activation, without much effect on the voltage dependence of the inactivation (Bean, 1989a, b). This effect of neurotransmitters is not restricted to noradrenaline, but is evident also in the action of GABA (Dolphin & Scott, 1986; Marchetti, Carbone & Lux, 1986; Deisz & Lux, 1988) and of κ-opiate agonists (MacDonald & Werz, 1986; Gross & MacDonald, 1987; Tsien, Lipscombe, Madison, Bley, & Fox, 1988; Bean, 1989a). Thus calcium channels can play an important role not only in the direct activation of transmitter release, but also in its modulation by other transmitters and drugs.

3.3.1 A large Channel in Fused Synaptosomes From Torpedo

One of the channels identified in the preparation of fused presynaptic nerve terminals of *Torpedo* electromotor nerve is a large, calcium permeable and highly voltage-dependent ion channel (Meir & Rahamimoff, 1996) (see figure 2).

The slope conductance of the channel is 846 ± 18 pS (mean ± S.E.M, n=9) as estimated by direct measurement of the single channel current amplitude at different voltages. The variance analysis of the currents gives a similar value (834 ± 23 pS, mean ± S.E.M, n=5).

This large channel is highly voltage dependent. At potentials below 0 mV, the probability of the channel to open is negligible and increases dramatically, within a very narrow voltage range, to more than 50% at +8 mV (n=8). In pulse experiments, the activation time delay is shorter as the voltage step reaches more positive values. The mean time for half activation ($T_{½}$) decreases from 15 ms at +10 mV to 4 ms at +30 mV (n=5). Ensemble currents exhibit rectification in response to voltage ramps at negative potentials (n=10).

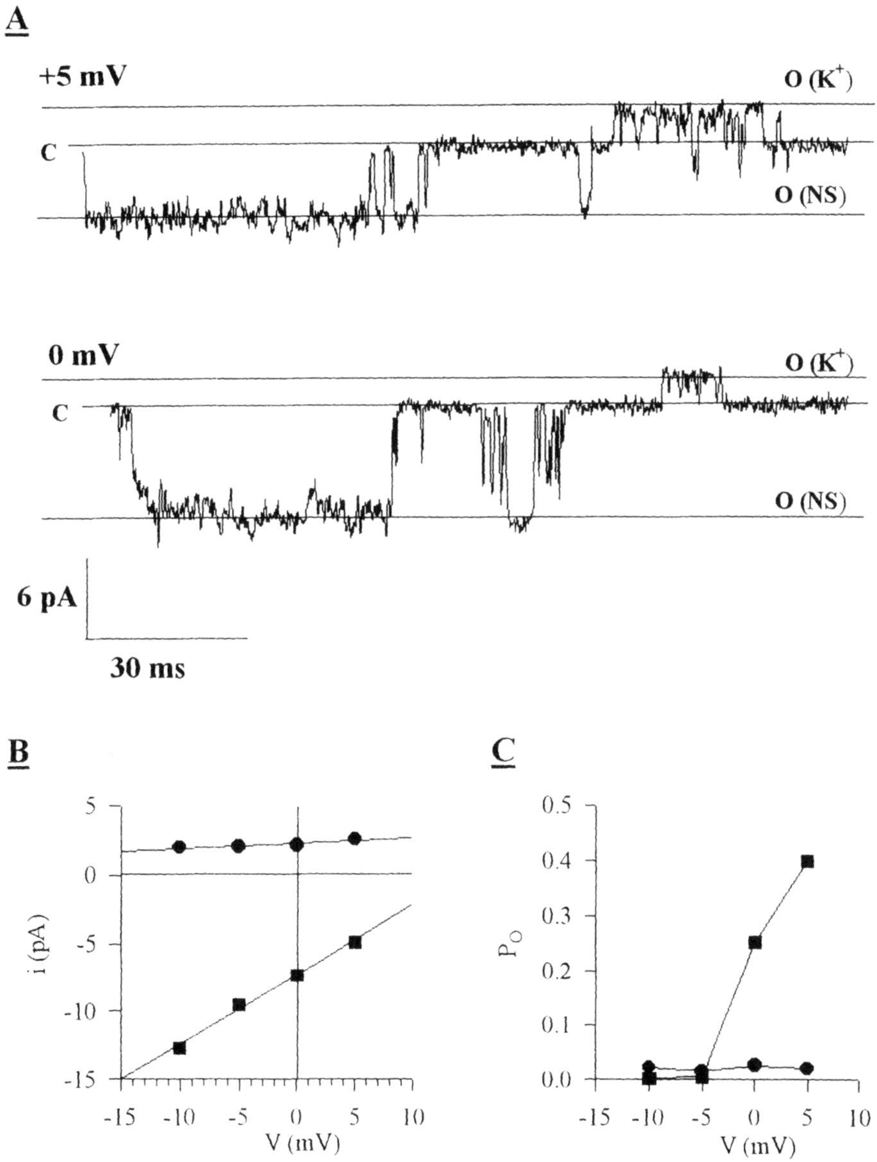

Fig 2. The activity of the bursting potassium channel and the non selective channel in the same patch from fused *Torpedo* synaptosomes. A: Two current traces in steady state voltage clamp (the holding potential is indicated on the left side of the panels), where the two channels are active. C denotes the closed state of both channels, O (K^+) the open state of the potassium channel and O (NS) the open state of the non selective channel. **B:** i-V relations obtained from the same experiment. The potassium channel (●) γ = 33 pS, V_{rev} = -60 mV. The non selective channel (■) γ = 513 pS, V_{rev} = +15 mV. The pipette contained in mM: 10 KCl, 390 Na-Glutamate, 2 $MgCl_2$ and 2 $CaCl_2$. The bath contained in mM: 390 KCl, 10 Na-HEPES and 0.01 $CaCl_2$. **C:** The open probability as a function of the holding potential of the two channels in the same experiment; symbols as in B.

In short pulse experiments from holding potential of -70 mV, we found that the channel is activated shortly after depolarization and deactivated shortly after the membrane voltage returns to the resting potential. Therefore, it may be activated and deactivated by a typical action potential invading the nerve terminal.

The channel was found to be non-selective. Its relative permeabilities (P_{ion} /P_{Na}) to the different ions, Na^+: K^+: Cl^-: Glutamate: Ba^{2+}: Ca^{2+} are : 1.00: 1.00: 1.22: 1.07: 0.85 and 0.62 respectively.

On the basis of the transport number of calcium, the calculated driving force and the mean channel open time, we estimated the number of calcium ions entering the nerve terminal upon depolarization. This calculation led to rather surprising results. For example, when the terminal membrane is depolarized to +10 mV, 1024 calcium ions will enter the terminal during the mean open time of the channel (~5 ms at +10 mV). Without taking into account buffering and extrusion, it is an elevation of $[Ca^{2+}]_{in}$ from 500 nM at rest to about 2.3 μM. For comparison, we calculated the number of calcium ions that flow through two types of calcium-selective channels. Wang et al, (1993) recorded single calcium channels from the rat neurohypophysial terminals. They found N_t type and L type channels. From their data, we calculated the number of calcium ions passing through each of these channels during a single opening. Their values correspond to the passage of 710 and 1498 calcium ions per single mean opening of these two channels respectively. Hence, the calcium entry through the non selective channel is not substantially different from that through "normal" voltage-dependent and calcium-selective channels found in other nerve terminals.

Applying the same calculation to potassium, sodium and chloride, one can conclude that the opening of the large, non-selective channel does not cause a substantial dissipation of the ion gradients across the membrane.

We speculate that this non selective ion channel may serve as a calcium entry route into the nerve terminal and hence be involved in transmitter release. The role of non selective channels in the transmitter release process was suggested also for a calcium permeable cationic channels of 50 pS found in mast cells (Matthews, Neher & Penner, 1989).

3.4 Sodium Currents and Channels

The amount of transmitter liberated from the presynaptic nerve ending depends on the amplitude and the duration of the depolarizing pulse (Katz & Miledi, 1967a). Hence sodium channels could play a very important role in the regulation of transmitter release (for a review, see Rahamimoff, Lev-Tov & Meiri, 1980).

At the frog neuromuscular junction, Katz and Miledi (1965) showed that action potentials propagate along the non-myelinated nerve terminals and that these action potentials are blocked by tetrodotoxin (TTX). Thus, the frog motor nerve terminal is excitable and apparently contains TTX-sensitive Na^+ channels. In terminals from rat neurohypophysis, several lines of evidence strongly suggest the existence of sodium channels. Optical recordings with voltage sensitive dyes point out that the action potential is propagated to the endings (Salzberg, Obaid, Senseman & Gainer, 1983). Extracellular recording indicated that the propagated action potential has sodium and calcium components (Nordmann & Stuenkel, 1986). In outside-out patches from the rat posterior pituitary nerve terminals, an inward current is blocked by TTX (Jackson & Zhang, 1995). This current is well fitted with the Hodgkin and Huxley parameters and thus contributes to the initiation and propagation of action potentials in these nerve terminals. Finally, action potentials were detected in intraterminal recordings (Borque, 1990). By measuring the release of peptides from the neurohypophysis terminals, it was shown that veratridin, a Na^+ channel agonist, causes an increase in the release and that this effect is blocked by TTX which also blocks a substantial fraction of the inward current recorded from the terminals of the crab's sinus gland (Lemos & Nordmann, 1986).

Thus, it seems reasonable to conclude that in many nerve terminals, there are sodium channels that determine the shape of the action potential triggering transmitter release.

3.5 Chloride Currents and Channels

Chloride channels were found in the fused synaptosomes of the *Torpedo* electric organ. There, they are the second most common channel after the bursting potassium channel. This channel has a conductance of 11.7 pS and is voltage dependent. The channel's open probability increases upon depolarization and decreases with increasing pH. Due to its properties, it probably participates in the determination of the membrane resting potential and in the action potential shape (Edry-Schiller, Ginsburg & Rahamimoff, 1991b).

A chloride current activated by glutamate was found in the terminals of the cone innervating bipolar cells in the retina, using the patch clamp whole cell method (Sarantis, Everett & Attwell, 1988). This current shows a dependence on $[Cl^-]_{out}$ and an inward rectification. In this synapse, the neurotransmitter is presumably glutamate and therefore, the glutamate activated Cl^- current may be involved in a positive feedback loop, contrary to a number of other autoreceptors that are involved in negative feedback loops.

Recently, GABA-activated chloride channels were found in the nerve terminals of the neurohypophysis with a single channel conductance of about 26 pS. GABA acts through $GABA_A$ receptors and causes the opening of chloride channels, a depolarization and a block of action potential spread. In this way, GABA limits the secretion presumably by interference with the spread of the action potentials into the terminal arborization (Zhang & Jackson 1993; Jackson & Zhang, 1995).

3.6 Non-Specific Cation Channels

Two types of non-specific cation channels were found in terminals obtained from the crustacean neurohaemal organ - the sinus gland. (i) The 'S' channel is Ca^{2+} activated,

with $[Ca^{2+}]_{in} > 1$ μM, has a large slope conductance of 213 pS and is equally permeable to Na^+ and K^+. (ii) The '*F*' channel is also cation selective, with equal permeability to Na^+ and K^+ ions. Its mean slope conductance is 69 pS and its open probability increases with $[Na^+]_{in}$. These two channels could be responsible for the long lasting plateau potentials which underlie bursting activity and lead to potentiation of hormone release in these terminals (Lemos, Nordmann, Cooke, & Stuenkel, 1986; Stuenkel, Ruben, Cooke & Lemos, 1990).
The channels that were characterized in different preparations of nerve terminals are summarized in Table 1

4 Ion Channels in Fused Synaptic Vesicles

Synaptic vesicles are too small for direct electrophysiological recording. This technical problem limits our knowledge regarding ionic channels in the vesicle membranes and their possible role in transmitter release. Thus, to characterize the properties of ion channels in synaptic vesicles, we fused their membranes into large structures (Rahamimoff et al., 1988). The source for cholinergic synaptic vesicles was *Torpedo ocellata* electromotor nerve endings, which innervate its electric organ. These nerve endings are a rich and homogenous origin for synaptic vesicles. The synaptic vesicles were purified by the method developed by Tashiro & Stadler (1978), with some slight modifications (Yakir & Rahamimoff, 1995). When we applied the patch clamp technique for single channel recordings, two main types of ion channels appeared consistently in the fused synaptic vesicles. The first type has a single channel conductance of 8 pS and conducts mainly chloride ions. The second type has a much higher conductance and is non-specific (figure 3A). Some of the properties of this high-conductance channel were described in detail (Yakir & Rahamimoff, 1995) and are summarized here.

4.1 The Chloride Channel

The chloride channel conductance was estimated from the slope of i-V curves. The channel had a linear i-V relation with an average single channel conductance of 8.37±0.62 pS (mean ±SEM, n=13). The channel conductance reached 12.85 pS (n=2) when the bath contained mainly (in mM): 397 KCl and the pipette (in mM): 360 NMG-Cl, 10 KCl. The selectivity of the channel was estimated from reversal potential measurements. When the reversal potentials of potassium and anions were well separated, the channel behaved as an anionic channel. The chloride channel that was found in the fused synaptic vesicles is very similar to the chloride channels that were found by Miller and others in the *Torpedo* electric organ (see Miller & Richard, 1990) and by Edry-Schiller, Ginsburg & Rahamimoff (1991b) in fused synaptosomes from nerve terminals in *Torpedo* electric organ. We speculate that the chloride channel may play a role in transmitter concentration, since the anions would counter-balance the positive charges, which are involved in the process. In addition, the channel, as any other channels in this membrane, will be involved in the determination of the vesicle membrane potential.

Table 1: Channels in nerve terminals[1]

Channel name (Ref.)	γ (pS)	Gating and modulation	Method/ preparation
L type Ca^{2+} (1)		Voltage dependent, blocked by DHP	Patch clamp rat pancreatic b cells
(2,3)	28, 25	LVA, activated by Bay-K 8644, blocked by nicardipine	Patch clamp rat neurohypophysial terminals
NPT type Ca^{2+} (4)	11-14	Voltage dependent, sensitive to ω- Conotoxin GIVA	Patch clamp Calyx synapse, chick ciliary ganglion
N_t type Ca^{2+} (2,3)	11, 10-15	HVA, inactivation	Patch clamp rat neurohypophysial terminals
P type Ca^{2+} (5,6)		Voltage dependent, blocked by FTX and ω- Agatoxin IVA	Extracellular recording mouse NMJ EPSP recordings mouse central synapse
Q type Ca^{2+} (7)		Voltage dependent, blocked by ω- Conotoxin MVIIC	EPSP recordings mouse hippocampal synapse
Na^+ (23)		Voltage dependent, blocked by TTX	Patch clamp rat posterior pituitary nerve terminals
(8)			Whole cell turtle (retina) horizontal cell axon terminals
Anion (9)	22-150		Patch clamp fused synaptosomes from rat hippocampus
Cl^- (10)	10	Voltage dependent	Patch clamp *Torpedo* fused synaptosomes
$GABA_A$ Cl^- (11)	26.3	GABA gated, blocked by picrotoxin and bicuculline	Patch clamp rat thin slice posterior pituitary nerve terminals
Cation F type (12)	69	Na^+ activated	Patch clamp crustacean sinus gland peptidergic terminals
Cation S type (12)	213	Ca^{2+} activated	Patch clamp crustacean sinus gland peptidergic terminals
Non selective (13)	846	Voltage dependent, HVA, Na^+ activated	Patch clamp *Torpedo* fused synaptosomes
Non voltage K^+ (14)	15, 28, 91	Blocked by TEA, voltage independent	Patch clamp freeze- thawed cerebellar cortex synaptosomes
(20)	12.8 - 40.8	Voltage independent	Incorporation into lipid bilayers rat brain synaptosomes
Slow K^+ (15)	27	Voltage activated, slow activation	Patch clamp rat thin slice posterior pituitary nerve terminals
ATP-sensitive K^+ (21)		Blocked by sulfonylureas	Perineural recording mouse nerve muscle preparation
Maxi K(Ca) (15)	134	Voltage and Ca^{2+} activated	Patch clamp rat thin slice posterior pituitary nerve terminals
(22)	217	Voltage, Ca^{2+} and ethanol activated	Patch clamp rat neurohypophysial terminals
(16)	221	Voltage and Ca^{2+} activated	Patch clamp rat cochlear efferent nerve terminals
(17)	231	Voltage and Ca^{2+} activated	Patch clamp rat posterior pituitary nerve terminals
A type K^+ (15)	33	Voltage dependent, rapid activation and inactivation	Patch clamp rat thin slice posterior pituitary nerve terminals
(18)	24	Rapid activation and inactivation, blocked by TEA, 4AP	Patch clamp *Torpedo* fused synaptosomes
Fusion Pore (19)	230		Capacitance recording Mast cells

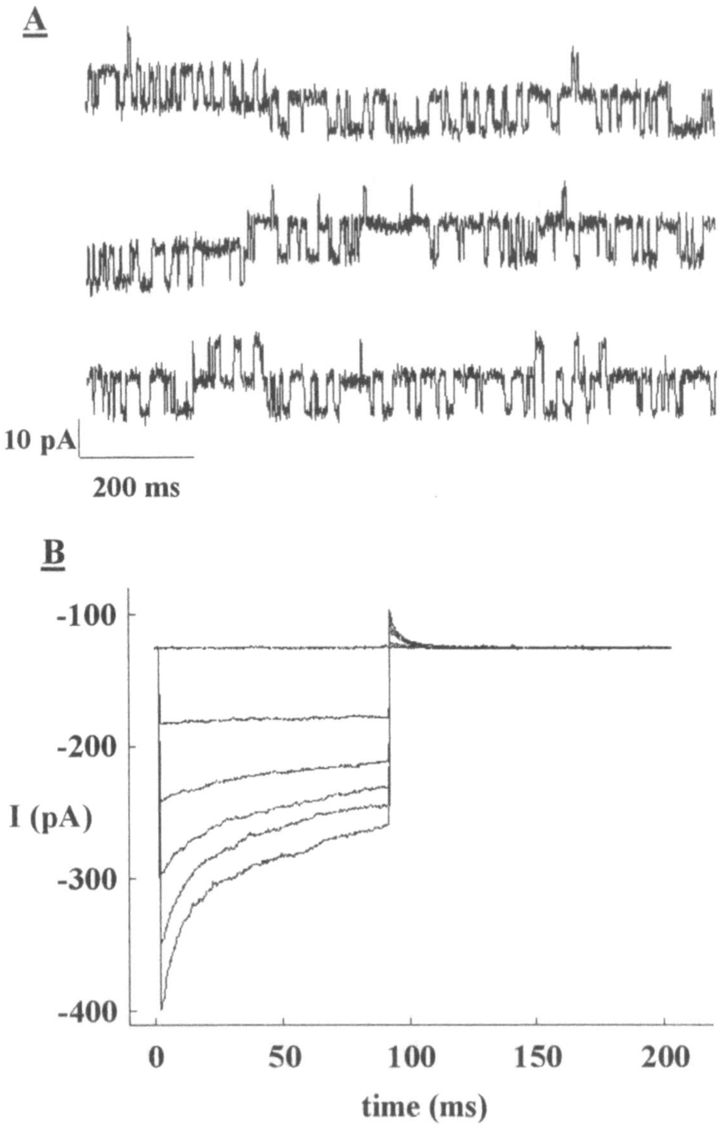

Fig 3. Activity of the non-specific channel in fused synaptic vesicles. A: A continuous record of the non-specific channel at -5 mV. Openings are downward deflections. In the records few levels of activity appear with frequent transitions. The data were filtered at 3 kHz and digitized at 20 kHz. Pipette contained (in mM): 500 Mannitol, 70 K glutamate, 10 KCl, 10 Na-HEPES, 2 $CaCl_2$; Bath (in mM): 350 K glutamate, 50 KCl, 10 Na-HEPES, 1 $CaCl_2$. **B:** Deactivation and reactivation of the non-specific channel. Ensemble current responses to voltage pulses from holding potential of -10 mV to different voltages between -60 mV to -10 mV in 10 mV steps (the lower trace is the response to a step to -60 mV). The traces are the ensemble currents of 120 pulses at each voltage given at a rate of 3.3 Hz and digitized at 10 kHz. Pipette contained (in mM): 580 Mannitol, 20 K glutamate, 20 KCl, 10 Na-HEPES, 2 $CaCl_2$; Bath (in mM): 200 K glutamate, 200 KCl, 10 Na-HEPES, 1 $CaCl_2$.

4.2 The Non-Specific Channel

1. Single channel conductance. The average single channel slope conductance of the non-specific channel was 246±15 pS (mean ±SEM, n=15), when the bath and pipette contained different combinations of KCl and KGlutamate. When glutamate was omitted from the solutions and replaced by chloride, the conductance increased and reached the value of 583± 45 pS(mean ±SEM, n=5).
2. Selectivity. The reversal potential of the channel was estimated first under the conditions of KCl gradient, where the reversal potentials of potassium and chloride were well separated. It was found, by extrapolation, that the channels do not discriminate between potassium and chloride ions. The reversal potentials were estimated also by extrapolation. The results confirmed that the current passing through the channel is carried by both cations and anions.
3. Voltage dependence in steady-state voltage conditions. The percent of open time (P) of the non-specific channel decreased dramatically when the membrane potential was clamped to negative values. Around 0 mV, the channel was open most of the time. Usually, it was impossible to estimate P reliably. In one experiment, when P was calculated, its voltage dependence was very sharp. At 0 mV, P reached the value of 0.94, while at -30 mV, it decreased to 0.02. At positive membrane potentials, the results were not as clear. In two experiments, the open probability also decreased upon polarization in this direction.
4. Voltage dependence in voltage pulse experiments. Complementary results were achieved in voltage pulse experiments (figure 3B). When the membrane potential was stepped (from holding potentials between -20 mV to +20 mV) to negative test potentials, the non-specific channel deactivated. Around 0 mV, the current stayed at high values and did not decay. As the test potential was more negative, proportionately, the current decay was larger. In addition, the decay time-$T_{1/2}$ (half of the time from the peak to the steady-state current value) decreased as the test potential increased. These voltage dependent properties of deactivation led to a strong current rectification. Thus, a short depolarization to around 0 mV would probably open the channels by increasing the open probability.

---

[1] **Abbreviations and references for the table:** 4AP- 4-aminopyridine, DHP- dihydropyridine, EPSP- excitatory post-synaptic potential, FTX- funel web toxin, GABA- g-aminobutyric acid, HVA- high voltage activated, LVA- low voltage activated, NMJ- neuromuscular junction, TEA- tetraethylammonium

(1) Bokvist, Eliasson, Ammala, Renstron & Rorsman, 1995 (2) Lemos & Nowycky, 1989 (3) Wang, Treistman & Lemos, 1993 (4) Stanley, 1991 (5) Uchitel, Protti, Sanchez, Cherksey, Sugimori & Llinás, 1992 (6) Regehr & Mintz, 1994 (7) Wheeler, Randall & Tsien, 1994 (8) Golard, Witkovsky & Tranchina ,1992 (9) Hosokawa, Sandri, Panfili & Cherubini, 1994 (10) Edry-Schiller, Ginsburg & Rahamimoff, 1991b (11) Zang & Jackson, 1995 (12) Lemos, Nordmann, Cooke & Stuenkel, 1986 (13) Meir & Rahamimoff, 1996 (14) Hirashima & Kirino, 1988 (15) Bielefeldt, Rotter & Jackson, 1992 (16) Wangemann & Takeuchi, 1993 (17) Wang, Thorn & Lemos, 1992 (18) Edry-Schiller, Ginsburg & Rahamimoff, 1991a (19) Breckenridge & Almers, 1987a (20) Nelson, Rounda & Bamberg, 1983 (21) Diest, Repp & Dreyer, 1992 (22) Dopico, Lemos & Triestmen, 1996. (23) Jackson & Zhang, 1995

5. The substate. In many cases, smaller transitions were seen in addition to the main open state. In some cases, there were indications that the smaller amplitudes represent a substate of the non-specific channel. The substate had a conductance of 44±4.4% (n=3) of the main open state. Sometimes, at negative voltages, the non-specific channel seemed to "get stuck" in the open substate and did not reach full closure.
6. Complex patterns of activity. The non-specific channel exhibits very complex patterns of activity that are very difficult to analyze, which probably reflects a highly complicated kinetic scheme. Sometimes, the change of membrane potential of a few millivolts caused a dramatic change in the pattern of activity. In addition, in some cases, the channels seemed to act in a coordinated manner.

The non-specific channel may have a role in almost every step of the synaptic vesicle life-cycle. As the chloride channel, the non-specific channel can counter-balance the concentrating positive charges in the synaptic vesicle and take part in the determination of the vesicle membrane potential. The high conductance and strong voltage dependence makes the channel a good candidate to play a role in activities that take place when the channel encounters voltage changes. The non-specific channel may be a part of a fusion complex (Breckenridge & Almers, 1987a; b). Another exciting idea about a possible role for the non-specific channel will be relevant if the channel stays intact after fusion when the synaptic vesicle membrane is polarized to the terminal resting potential. Probably, the polarization will close many of the non-specific channels, resulting in a relatively small effect on the resting nerve terminal potential. If a subsequent action potential reaches the terminal before the synaptic vesicle is recycled, the open probability would rise rapidly. It is difficult to foretell what effect this would have on the total conductance of the nerve terminal. Depending on the reversal potential of the channel under these conditions as well as on the specific charge carriers, the channel may have a depressing effect or a facilitating effect on synaptic transmission. Whichever the case, the channel could be of a modulatory significance in a variety of processes that affect synaptic efficacy.

4.3 Channels Found in Synaptic Vesicles and Secretory Granules

4.3.1 Synaptic Vesicles

Synaptophysin is one of the most abundant proteins in synaptic vesicles (Jahn, Schiebler, Ouimet & Greengard, 1985; Weidenmann & Franke, 1985). Purified synaptophysin reconstituted into black lipid bilayer acts as a voltage dependent channel (Thomas, Hartung, Langosch, Rehm, Bamberg, Franke & Betz, 1988). The mean channel conductance is 154±11 pS in symmetrical 150 mM KCl solutions. The open probability increased at positive potentials, while at negative membrane potentials no activity was detected.

Sato, Inoue & Kasai (1992) found in rat brain cerebral cortex synaptic vesicles four types of ionic channels: one anionic (94 pS) and three cationic selective channels (250, 248, 213 pS) in asymmetric KCl solutions (300 mM / 150 mM). The channels appeared infrequently. The anionic channel inactivated at negative voltages. The cationic channels were identified due to their different gating. For a summary of the channels found in synaptic vesicles, see Woodbury (1995). It should be pointed out that the different single channel conductances found in different species, do not imply necessarily the existence of different molecular species, since there is a substantial difference in the ionic composition and osmolarity between mammal and *Torpedo*.

4.3.2 Secretory Granules

1. Neurohypophysis. Two channels were identified in membranes of bovine neurohypophysial granules incorporated into planar lipid bilayers. Lee, Dayanithi, Nordmann & Lemos (1992) characterized a calcium dependent cationic channel. The channel exhibited complex patterns of activity and multiple conductance states. The single channel conductance ranged between 62-236 pS and the channel had a weak voltage dependence. The biphasic calcium dependence and some pharmacological properties are consistent with the idea that the channel is involved in the release process. The second channel is an anionic channel with a weak voltage dependence (Stanley, Ehrenstein & Russell, 1988). Lemos, Ocorr & Nordmann (1989) found that the anionic channel is not affected by calcium and under symmetric conditions of 150 mM KCl it had a single conductance of 30 pS. Apparently, the two groups characterized the same channel.
2. Adenohypophysis. Balden, Walker & Cukierman (1991) fused purified prolactin secretory granules with lipid bilayer. They found a 45 pS channel in symmetrical solutions (50 mM $CaCl_2$). Under asymmetric solutions the channel showed selectivity for calcium over chloride and a weak voltage dependence. Since prolactin release is regulated by the free cytosolic calcium concentration, this channel might be involved in this process. In addition, the channel could modulate the calcium fluxes after fusion across the plasma membrane.
3. Adrenal medulla. Arispe, Pollard & Rojas (1992) fused adrenal chromaffin granules and purified granule membrane ghosts with planar lipid bilayer membranes. A potassium channel with single channel conductance of 160 pS (in symmetrical 400 mM K-HEPES) was detected in both preparations. The channel was not affected by calcium or voltage. Arispe, De-Mazancourt & Rojas (1995) found that the channel is modulated by GTP. A different potassium channel in a similar preparation was identified by Ashley, Brown, Apps & Phillips (1994). This potassium channel had a conductance of 400 pS in symmetric 450 mM KCl solutions. The channel was insensitive to calcium and exhibited complex gating, which was voltage-dependent.
4. Peritoneal mast cells. Hirashima & Kirino (1992) isolated secretory granules from rat peritoneal mast cells and reconstituted their membranes into planar lipid bilayer. Only one channel was identified in this work: a cationic channel with single channel conductance of 80 pS (in symmetric solutions of 150 mM KCl), which had a substate. This channel was calcium and voltage dependent.

5 Epilogue

We reviewed here the different types of ion channels found in the membrane of nerve terminals and in secretory vesicles and granules. The general picture emerges that the presynaptic nerve terminal is excitable, and ion channels play a significant role in the regulation of transmitter release. Ion channels are also subject to frequency and chemical modulation, leading to plasticity in the process of synaptic transmission.

The main role of the nerve terminal is to release quanta of neurotransmitter molecules when an appropriate signal arrives. This process is not constant, but can be changed dramatically by previous activity, by the action of various neurotransmitters, hormones and drugs. Various mechanisms, such as alterations in the voltage across the membrane and in the intracellular calcium concentration are involved in this plastic behavior of the nerve terminal. The plethora of ion channels found in nerve terminals, form most probably, the targets of many of these modulatory processes.

Acknowledgments

Our work was supported by the German Israeli Foundation (GIF), the Basic research fund of the Israeli Academy of Sciences and US-Israel Binational Science Foundation (BSF). The unfailing secretarial help of Mrs. Debra Broide is very much appreciated.

The visit of Dr. S. Demirgören was supported by a frant from the Scientific and Technical Research Council of Turkey (TUBITAK).

References

Arispe, N., De-Mazancourt, P. & Rojas, E. (1995). Direct control of a large conductance K^+-selective channel by G-proteins in adrenal chromaffin granule membranes. *J. Membrane Biol.* **147(2)**: 109-119.

Arispe, N., Pollard, H. B. & Rojas, E. (1992). Calcium-independent K^+-selective channel from chromaffin granule membranes. *J. Membrane Biol.* **130**: 191-202.

Ashley, R. H., Brown, D. M., Apps, D. K. & Phillips, J. H. (1994). Evidence for a K^+ channel in bovine chromaffin granule membranes: single-channel properties and possible bioenergetic significance. *Eur. Biophys. J.* **23(4)**: 263-275.

Augustine, G. J. (1990). Regulation of transmitter release at the squid giant synapse by presynaptic delayed rectifier potassium current. *Mol. Pharmacol.* **49(1)**: 40-48.

Augustine, G. J. & Eckert, R. (1984). Divalent cations differentially support transmitter release at the squid synapse. *J. Physiol. Lond.* **364**: 257-271.

Augustine, G. J., Charlton, M. P. & Smith, S. (1985a). Calcium entry into voltage-clamped presynaptic terminals of squid. *J. Physiol. Lond.* **367:** 143- 162.

Augustine, G. J., Charlton, M. P. & Smith, S. (1985b). Calcium entry and transmitter release at voltage-clamped nerve terminals of squid. *J. Physiol. Lond.* **367**: 163-181.

Augustine, G. J., Charlton, M. P. and Smith, S. (1987) Calcium action in synaptic transmitter release. *Ann. Rev. Neurosci.* **10**: 633-693.

Balden, E., Walker, A.M. & Cukierman, S. (1991). Reconstitution of single calcium channels from secretory granules of rat adenohypophysis in planar lipid membranes. *Molecular and Cellular Endocrinology* **77**: 85-90.

Bartschat, D. K. & Blaustein, M. P. (1985a). Potassium channels in isolated presynaptic nerve terminals from rat brain. *J. Physiol. Lond.* **361**: 419-440.

Bartschat, D. K. & Blaustein, M. P. (1985b). Calcium-activated potassium channels in isolated presynaptic nerve terminals from rat brain. *J. Physiol. Lond.* **361**: 441-457.

Bean, B. P. (1989a). Classes of calcium channels in vertebrate cells. *Ann. Rev. Physiol.* **51**: 367-384.

Bean, P. (1989b). Neurotransmitter inhibition of neuronal calcium currents by changes in voltage dependence. *Nature* **340**: 153-156.

Bielefeldt, K., Rotter, J. L. & Jackson, M. B. (1992). Three potassium channels in rat posterior pituitary nerve terminals. *J. Physiol. Lond.* **458**: 41-67.

Bloedel, J. Gage, P. W., Llinas, R., Quastel, D. M. J. (1966). Transmitter release at the squid giant synapse in the presence of tetrodotoxin. *Nature* **212**: 49-50.

Blundon, J. A., Wright, S. N., Brodwick, M. S. & Bittner, G. D. (1995). Presynaptic calcium-activated potassium channels and calcium channels at a crayfish neuromuscular junction. *J. Neurophysiol.* **73**: 178-189.

Bokvist, K., Eliasson, L., Ammala, C., Renstron, E. & Rorsman, P. (1995). Colocalization of L-type Ca^{2+} channels and insulin-containing secretory granules and its significance for initiation of exocytosis in mouse pancreatic β-cells. *EMBO J.* **14**: 50-57.

Borque, C. W. (1990). Intraterminal recordings from the rat neurohypophysis *in vitro*. *J. Physiol. Lond.*, **421**: 247-262.

Breckenridge, L. J. and Almers, W. (1987a) Currents through the fusion pore that forms during exocytosis of a secretory vesicle. *Nature* **328**:814-817.

Breckenridge, L. J. and Almers, W. (1987b) Final steps in exocytosis observed in a cell with giant secretory granules. *Proc. Natl. Acad. Sci. USA* **84**:1945-1949

Brigant, J. L. & Mallart, A. (1982). Presynaptic currents in mouse motor endings. *J. Physiol. Lond.* **333**: 619-636.

Chao, K. F. & Lin-Shiau, S. Y. (1995). Enhancement of a slow potassium current component by uranyl nitrate and its relation to the antagonism on β-bungarotoxin in the mouse motor nerve terminal. *Neuropharmacol.* **34(2)**: 165-173.

Colquhoun, D. and Sakmann, B. (1985). Fast events in single-channel currents activated by acetylcholine and its analogues at the frog muscle end-plate. *J. Physiol. Lond.* **369**:501-557.

Deist, M., Repp, H. & Dreyer, F. (1992). Sulfonylurea-sensitive K^+ channels and their probable role for the membrane potential of mouse motor nerve endings. *Plfügers Arch.* **421(2-3)**: 292-294.

Deisz, R. A. & Lux, H. D. (1988). γ-aminobutyric acid-induced depression of calcium currents of chick sensory neurons. *Neurosci. Lett.* **56**: 205-210.

del Castillo, J. & Katz, B. (1954) Quantal components of the end-plate potential. *J. Physiol. Lond.* **124**: 560-573.

del Castillo, J. & Stark, L. (1952). The effect of calcium ions on the motor end-plate potentials. *J. Physiol. Lond.* **116**: 507-515.

Dodge, F. A. & Rahamimoff, R. (1967). Co-operative action of calcium ions in transmitter release at the neuromuscular junction. *J. Physiol. Lond.* **193**: 419-432.

Dolphin, A. C. & Scott, R. H.(1986). Inhibition of calcium currents in cultured rat dorsal rot ganglion neurons by baclofen. *Br. J. Pharmacol.* **88**: 213-220.

Dopico, A.M., Lemos, J. R. & Treistman, S. N. (1996). Ethanol increases the activity of large conductance, Ca^{2+}-activated K^{+} channels in isolated neurohypophysial terminals. *Mol. Pharmacol.* **49(1)**: 40-48.

Dunlap K. & Fischbach, G. D. (1981) Neurotransmitters decrease the calcium conductance activated by depolarization of embryonic chick sensory neurons. *J. Physiol. Lond.* **317**: 519-535.

Edry-Schiller, J. and Rahamimoff, R. (1992). Activation and inactivation of the bursting potassium channel from fused *Torpedo* synaptosomes. *J Physiol. Lond.* **471**: 659-678.

Edry-Schiller, J., Ginsburg, S. & Rahamimoff, R. (1991a). A bursting potassium channel in isolated cholinergic synaptosomes of *Torpedo* electric organ. *J. Physiol. Lond.* **439**: 625-647.

Edry-Schiller, J., Ginsburg, S. & Rahamimoff, R. (1991b). A chloride channel in isolated fused synaptosomes from *Torpedo* electric organ. *J. Physiol. Lond.* **438**: 231P.

Feng, T. P. (1936). Studies on the neuromuscular junction. The universal antagonism between calcium and curarizing agencies. *Clin. J. Physiol.* **10**: 513-528.

Forscher, P. & Oxford, G. S. (1985). Modulation of calcium channels by norepinephrine in internally dialyzed avian sensory neurons. *J. Gen. Physiol.* **85**: 743-763.

Golard, A., Witkovsky, P. & Tranchina, D. (1992). Membrane currents of horizontal cells isolated from turtle retina. *J. Neurophysiol.* **68**: 351-361.

Gross, R. A. & MacDonald, R. L. (1987) Dynorphin A selectively reduces a large transient (N-type) calcium current of mouse dorsal root ganglion neurons in cell culture. *Proc. Natl. Acad. Sci. USA* **84**: 5469-5473.

Gundersen, C. B., Katz, B. & Miledi, R. (1982) The antagonism between botulinum toxin and calcium in motor nerve terminals. *Proc. R. Soc. Lond [Biol]* **212**: 197-211.

Hamill, O. P., Marty, A., Neher, E., Sakmann, B. & Sigworth, F. J. (1981). Improved patch-clamp techniques for high-resolution current recording from cells and cell-free membrane patches. *Pflügers Arch.* **391**: 85-100.

Heidelberger, R. & Matthews, G. (1992). Calcium influx and calcium current in single synaptic terminals of goldfish retinal bipolar neurons. *J. Physiol. Lond.* **447**: 235-256.

Hevron, E., David, G., Arnon, A. & Yaari, Y. (1986). Acetylcholine modulates two types of presynaptic potassium channels in vertebrate motor nerve terminals. *Neurosci. Lett.* **72(1)**: 87-92.

Hirashima, N. & Kirino, Y. (1988). Potassium channels in synaptosomal membrane examined using patch-clamp techniques and reconstituted giant proteoliposomes. *Biochim. Biophys. Acta.* **946**: 209-214.

Hirashima, N. & Kirino, Y. (1992). A calcium-dependent cation channel in mast cell granule membranes reconstituted into a lipid bilayer. *Biochemical and Biophysical research communications.* **185**: 111-126.

Hodgkin, A. L. & Huxley, A. F. (1952). Na and K currents in nerve membrane. *J. Physiol. Lond.* **116**: 473-496.

Hosokawa, Y., Sandri, G., Panfili, E. & Cherubini, E. (1994). Characterization of a voltage-dependent anionic channel in fused synaptosomes isolated from rat hippocampi. *Neurosci Lett.* **169**: 167-170.

Hubbard, J. I., Jones, S. F., Landau, E. M. (1968). On the mechanism by which calcium and magnesium affect the release of transmitter by nerve impulses. *J. Physiol. Lond.* **196**: 75-86.

Issa, N. P. & Hudspeth, A. J. (1994). Clustering of Ca^{2+} channels and Ca^{2+}-activated K^+ channels at fluorescently labeled presynaptic active zones of hair cells. *Proc. Nat. Acad. Sci. USA* **91(16)**: 7578-82.

Jackson, M. B. (1995). Presynaptic excitability. *Int. Rev. Neurobiol.* **38**: 201-251.

Jackson, M. B. & Zhang, S. J. (1995). Action potential propagation and propagation block by GABA in rat posterior pituitary nerve terminals. *J. Physiol. Lond.* **483(3)**: 597-611.

Jahn, R., Schiebler, W., Ouimet, C. and Greengard, P. (1985). A 38,000-dalton membrane protein (p38) present in synaptic vesicles. *Proc. Natl. Acad. Sci. USA* **82**: 4137-4141.

Jenkinson, D. H. (1967). The nature of the antagonism between calcium and magnesium ions at the neuromuscular junction. *J. Physiol. Lond.* **138**: 434-444.

Katz, B. & Miledi, R. (1965). Propagation of electric activity in motor nerve terminals. *Proc. R. Soc. Lond. [Biol]* **161**: 453-482.

Katz, B. & Miledi, R. (1967a). A study of synaptic transmission in the absence of nerve impulses. *J. Physiol. Lond.* **192**: 407-436.

Katz, B. & Miledi, R (1967b). Tetrodotoxin-resistant electric activity in presynaptic terminals. *J. Physiol. Lond.* **207**: 789-801.

Katz, B. & Miledi, R. (1967c). The timing of calcium action during neuromuscular transmission. *J. Physiol. Lond.* **189**: 535-544.

Katz, B. (1969). The release of neural transmitter substances. **10** (The Sherrington Lecture) Liverpool University Press.

Knaus, H. G., Schwarzer, C., Koch, R. O., Eberhart, A., Kaczorowski, G. J., Glossmann, H. Wunder, F., Pongs, O., Garcia, M. L. & Sperk, G. (1996). Distribution of high-conductance Ca^{2+}-activated K^+ channels in rat brain: targeting to axons and nerve terminals. *J. Neurosci.* **16(3)**: 955-963.

Krnjevic, K. (1990). Adenosine triphosphate-sensitive potassium channels in anoxia. *Stroke.* **21**(11 Suppl): III190-193.

Kusano, K. (1967). Further study of the relationship between pre- and postsynaptic potentials in the squid giant synapse. *J. Physiol. Lond.* **52**: 326-345.

Lee, C. J., Dayanithi, G., Nordmann, J. J. & Lemos, J. R. (1992). Possible role during exocytosis of a Ca^{2+} - activated channel in neurohypophysial granules. *Neuron* **8**: 335-342.

Lemos, J. R. & Nowycky, M. C. (1989). Two types of calcium channels coexist in peptide-releasing vertebrate nerve terminals. *Neuron* **3**:1419-1426.

Lemos, J. R. Nordmann, J. J (1986). Ionic channels and hormone release from peptidergic nerve terminals. *J. Exp. Biol.* **124**: 53-72.

Lemos, J. R. Nordmann, J. J., Cooke, I. M., Stuenkel, E. L. (1986). Single channels and ionic currents in peptidergic nerve terminals. *Nature* **319**: 410-412.

Lemos, J. R., Ocorr, K. A. & Nordmann, J. J. (1989). Possible Role for Channels in Neurosecretory granules of the rat neurohypophysis. In: *Secretion and its Control*, ed. Oxford, G. S., pp. 334-347. The Rockefeller University Press, New York.

Lindgren, C.A. & Moore, J. W. (1989). Identification of ionic currents at presynaptic nerve endings of the lizard. *J. Physiol. Lond.* **414**: 201-222.

Llinas, R, Steinberg, I. Z. & Walton, K. (1981). Presynaptic calcium currents in squid giant synapse. *Biophys. J.* **33**: 289-321.

Locke, D. F. S. (1884). Notiz über Einfluß Physiologischer Kochsalz Lösung auf die Erreglarkeit von Muskel und Nerv. *Abl Physiol.* **8**: 166.

MacDonald, R. L. & Werz, M. A. (1986). Dynorphin A decreases voltage-dependent calcium conductance of mouse dorsal root ganglion neurons. *J. Physiol. Lond.* **377**: 237-249.

Mallart, A. (1985). A calcium-activated potassium current in motor nerve terminals of the mouse. *J. Physiol. Lond.* **368**: 577-591.

Mallart, A. (1984). Presynaptic currents in from motor endings. *Pflügers Arch.* **400**: 8-13.

Marchetti, C., Carbone, E. & Lux, H. D. (1986). Effects of dopamine and noradrenaline on Ca channels of cultured sensory and sympathetic neurons of chick. *Plügers Arch.*, **406**: 104-111.

Martin, A. R., Patel, V., Faille, L. & Mallart, A. (1989). Presynaptic calcium currents recorded from calcyform nerve terminals in the lizard ciliary ganglion. *Neurosci. Lett.* **105**: 14-18.

Matthews, G., Neher, E. & Penner, R. (1989). Second messenger activated calcium influx in rat peritoneal mast cells. *J. Physiol. Lond.* **418**: 105-130.

McNamara, N. M., Muniz, Z. M., Wilkin, G. P. & Dolly, J. O. (1993). Prominent location of a K^+ channel containing the alpha subunit Kv 1.2 in the basket cell nerve terminals of rat cerebellum. *Neurosci.* **57(4)**: 1039-1045.

Meir, A. and Rahamimoff, R. (1996). A voltage dependent and calcium permeable ion channel in fused presynaptic terminals of *Torpedo*. *J. Neurophysiol.* **75(5)**: 1858-1870.

Michaelson, D. M. & Sokolovsky, M. (1978). Induced acetylcholine release from active purely cholinergic *Torpedo* synaptosomes. *J. Neurochem.* **30**: 217-230.

Miller, C. and Richard, E. A. (1990) The voltage-dependent chloride channel of Torpedo electroplax: Intimations of molecular structure from quirks of single-channel function. In: *Chloride channels and Carriers in Nerve Muscle and Glial cells*. ed. Alvarez-Leemans, F. J. and Russell, J. M., pp. 383-405. Plenum Press, New York.

Miller, R. J. (1987). Multiple calcium channels and neuronal function. *Science* **235**: 46-52.

Miralles, F., Canti, C., Marsal, J., Peres, J. & Solsona, C. (1994). Zinc ions block rectifier potassium channels and calcium activated potassium channels at the frog motor nerve endings. *Brain Res.* **641**: 279-284.

Morita, K. & Barrett, E. F. (1990). Evidence for two calcium dependent potassium conductances in lizard motor nerve terminals. *J. Neurosci.* **10**: 2614-2625.

Nachshen, D. A. & Blaustein, M. P. (1982). Influx of calcium, strontium and barium in presynaptic nerve endings. *J. Gen. Physiol.* **79**: 1065-1087.

Nelson, M. T., French, R. J., Krueger, B. K., (1984). Voltage-dependent calcium channels from brain synaptosomes incorporated into planar lipid bilayers. *Nature* **308**: 440-443.

Nelson, M. T., Roudna, M. & Bamberg, E. (1983). Single K^+-channel current measurements from brain synaptosomes in lipid bilayers. *Am. J. Physiol.* **245**: C151-156.

Nordmann, J. J. & Stuenkel, E. L. (1986). Electrical properties of axons and neurohypophyseal nerve terminals and their relationship to secretion in the rat. *J. Gen. Physiol.* **380**: 521-539.

Perney, T. M. and Kaczmarek, L. K. (1991). The molecular biology of K^+ channels. *Curr. Opin. Cell. Biol.* **3**: 663-670.

Rahamimoff, R., DeRiemer, S. A., Sakmann, B., Stadler, H. and Yakir, N. (1988) Ion channels in synaptic vesicles from *Torpedo* electric organ. *Proc. Nat. Acad. Sci. USA*. **85**:5310-5314.

Rahamimoff, R., Edry-Schiller, J., and Ginsburg, S. (1992). A long closed state of the bursting potassium channel confers a statistical memory. *J. Neurophysiol.* **68(6)**: 2260-2263.

Rahamimoff, R., Edry-Schiller, J., Rubin-Fraenkel, M., Butkevich, A., and Ginsburg S. (1995). Oscillations in the activity of a potassium channel at the presynaptic nerve terminal. *J. Neurophysiol.* **74(6)**:2248-2259.

Rahamimoff, R., Lev-Tov, A. & Meiri, H. (1980). Primary and secondary regulation of quantal transmitter release: Calcium and sodium. *J. Exp. Biol.* **89**: 5-18.

Regehr, W. G. and Mintz, I. M. (1994). Participation of multiple calcium channel types in transmission at single climbing fiber to Purkinje cell synapses. *Neuron* **12(3)**: 605-613.

Rorsman, P. & Trube, G. (1985). Glucose-dependent K^+ channels in pancreatic β-cells are regulated by intracellular ATP. *Pflügers Arch.* **405**: 305-309.

Saltzberg, B. M., Obaid, A. L., Senseman D. M. & Gainer, H. (1983). Optical recording of action potentials from vertebrate nerve terminals using potentiometric probes provides evidence for sodium and calcium components. *Nature* **306**: 36-40.

Sarantis, M., Everett, K. & Attwell, D. (1988). A presynaptic action of glutamate at the cone output synapse. *Nature* **332**: 451-453.

Sato, M., Inoue, K. & Kasai, M. (1992). Ion channels on synaptic vesicle membranes studied by planar lipid bilayer method. *Biophys. J.* **259**: 1500-1505.

Sheng, M., Liao, Y. J., Jan, Y. N. & Jan, L. Y. (1993). Presynaptic A-current based on heteromultimeric K^+ channels detected *in vivo*. *Nature* **365(6441)**:72-75.

Sheng, M., Tsaur, M. L., Jan, Y. N. & Jan, L. Y. (1994). Contrasting subcellular localization of the Kv1.2 K^+ channel subunit in different neurons of rat brain. *J. Neurosci.* **14**: 2408-2417.

Stanley, E. F. (1991). Single calcium channels on a cholinergic presynaptic nerve terminal. *Neuron* **7**: 585-591.

Stanley, E. F. (1993). Single calcium channels and acetylcholine release at a presynaptic nerve terminal. *Neuron.* **11(6)**: 1007-1011.

Stanley, E. F. & Atrakchi, A. H. (1990). Calcium currents recorded from a vertebrate presynaptic nerve terminal are resistant to the dihydropyridine nifedipine. *Proc. Natl. Acad. Sci. USA* **87**: 9683-9687.

Stanley, E. F., Ehrenstein, G. & Russell, J. T. (1988). Evidence for anion channels in secretory vesicles. *Neurosci.* **25**: 1035-1039.

Stuenkel, E. L., Ruben, P., Cooke, I. M. & Lemos, J. R. (1990). Sodium-activated cation channels in peptidergic nerve terminals. *Brain Res.* **517**: 35-43.

Tabti, N., Bourret, C. & Mallart, A. (1989). Three potassium currents in mouse motor nerve terminals. *Pflügers. Arch.*, **413(4)**: 395-400.

Takeuchi, S. & Wangemann, P. (1993). Aminoglycoside antibiotics inhibit maxi K^+ channel in single isolated cochlear efferent nerve terminals. *Hear. Res.* **67(1-2)**: 13-19.

Tashiro, T. and Stadler, H. (1978) Chemical composition of cholinergic synaptic vesicles from Torpedo marmorata based on improved purification. *Eur. J. Biochem.* **90**:479-487.

Tsien, R. W., Lipscombe, D., Madison, D. V., Bley, K. R., Fox, A. P., (1988). Multiple types of neuronal calcium channels and their selective modulation. *Trends Neurosci.* **11**: 431-438.

Thomas, L., Hartung, K., Langosch, D., Rehm, H., Bamberg, E., Franke, W. W. & Betz, H. (1988). Identification of synaptophysin as a hexameric channel protein of the synaptic vesicle membrane. *Science* **242**: 1050-1053.

Uchitel, O. D., Protti, D. A. Sanchez, V., Cherksey, B. D., Sugimori, M., Llinas, R. (1992). P-type voltage-dependent calcium channel mediates presynaptic calcium influx and transmitter release in mammalian synapses. *Proc. Natl. Acad. Sci. USA* **89**: 3330-3333.

Umbach, J. A., Gundersen, CG & Baker (1984). Giant synaptosomes. *Nature* **311**: 474-477.

Wang, H., Kunkel, D. D., Martin, T. M., Schwartzkroin, P.A. & Tempel, B. L. (1993). Heteromultimeric K^+ channels in terminal and juxtaparanodal regions of neurons. *Nature* **365**: 75-79.

Wang, H., Kunkel, D. D., Schwartzkroin, P.A. & Tempel, B. L. (1994). Localization of Kv1.1 and Kv1.2, two K channel proteins, to synaptic terminals, somata, and dendrites in the mouse brain. *J. Neurosci.* **14(8)**: 4588-4599.

Wang, X., Thorn, P. & Lemos, J. R. (1992). A novel large-conductance Ca^{2+} activated potassium channel and current in nerve terminals of the rat neurohypophysis. *J Physiol. Lond.* **457**: 47-74.

Wang X., Treistman, S. N. & Lemos, J. R. (1992). Two types of high-threshold calcium currents inhibited by *ω-conotoxin* in nerve terminals of rat neurohypophysis. *J. Physiol. Lond.* **445**: 181-199.

Wang, X., Treistman, S. N. & Lemos, J. R. (1993). Single channel recordings of N_{t-} and *L*-type Ca^{2+} currents in rat neurohypophysial terminals. *J. Neurophysiol* **70**: 1617-1628.

Wangemann, P. & Takeuchi, S. (1993). Maxi-K^{+} channel in single isolated cochlear efferent nerve terminals. *Hear. Res.* **66**: 123-129.

Wheeler, D. B., Randall, A. and Tsien, R. W. (1994). Roles of N-type and Q-type Ca^{2+} channels in supporting hippocampal synaptic transmission *Science* **264**: 107-111.

Wiedenmann, B. & Franke, W. W. (1985). Identification and localization of synaptophysin, an integral membrane glycoprotein of M_r 38,000 characteristic of presynaptic vesicles. *Cell* **41**: 1017-1028.

Woodbury, D. J. (1995). Evaluation of the evidence for ion channels in synaptic vesicles. *Mol. Membr. Biol.* **12(2)**: 165-171.

Yakir, N. and Rahamimoff, R. (1995) The non-specific ion channel in *Torpedo Ocellata* fused synaptic vesicles. *J. Physiol. Lond.* **485**: 683-697.

Zhang, S. J. & Jackson, M. B. (1993). GABA-activated chloride channels in secretory nerve endings. *Science* **259**: 531-534.

Fluorescence and Confocal Microscopy

Yusuf P. Tan

Biomedical Engineering Institute, Bogaziçi University, P.K. 2, Bebek Istanbul, Turkey.

1 The Interaction of Light with the Matter

Light is an electromagnetic radiation which oscillates both in space and time, it has electric and magnetic field components which are perpendicular to each other. The dimensions of the molecules being much smaller than the wavelength of the light, the molecule will be perturbed by the latter as it's electronic charge distribution will be altered by the oscillating electric field component of the light. This interaction of the light with matter occurs via a resonance phenomenon i.e. the matter will only absorb light if the energy of the incoming photon is exactly equal to the difference of the potential energy of the lowest vibrational level of the ground state and that of one of the vibrational levels of the first excited state (fig 1). The absorbtion of light therefore occurs in discrete amounts termed quanta. The energy in a quantum of light (a photon) is given by:

$$E = h\nu = hc / \lambda$$

Where h is the Planck's constant $\nu\lambda$ are the frequency, and the wavelength, of the incoming light and c is the speed of light in vacuum. When a quantum of light is absorbed by a molecule a valence electron will be boosted into a higher energy orbit called the excited state. This phenomenon will take place in the time scale of 10^{-15} sec. resulting in the conservation of the molecular coordinates. At room temperature the electrons will be at their lowest vibrational energy level. The difference of energy between the vibrational levels being typically in the order of 10 Kcal mole $^{-1}$ there is not enough thermal energy to excite a transition to higher vibrational levels at room temperature. One might thus assume that most of the electrons will lie at the lowest vibrational level of the ground state. As the absorbtive transitions do occur to one of the vibrational levels of the excited state, had there been no interaction with the solvent molecules one could have measured the energy difference between the ground state and each of the vibrational levels of the excited state. This type of spectra can only be obtained for chemical compounds in the gaseous state. The absorbtion spectra under those circumstances would look like narrowly separated bands, however the interaction of the orbital electrons with solvent molecules will broaden those peaks, and one will get the absorbtion spectra of the more familiar form.

NATO ASI Series, Vol. H 100
Neurotransmitter Release and Uptake
Edited by Şakire Pöğün

2 The Return from the Excited State

The electrons which have been promoted to one of the vibrational levels of the excited state will loose their vibrational energy through interaction with solvent molecules by a process known as the vibrational relaxation. This process will occur in a time scale much shorter than the lifetime of the electrons in the excited state (10^{-9} to 10^{-7} sec. for aromatic molecules). The electrons which have been promoted to the excited state, will return to the ground state from the lowest lying excited vibrational state, by one of the following ways:

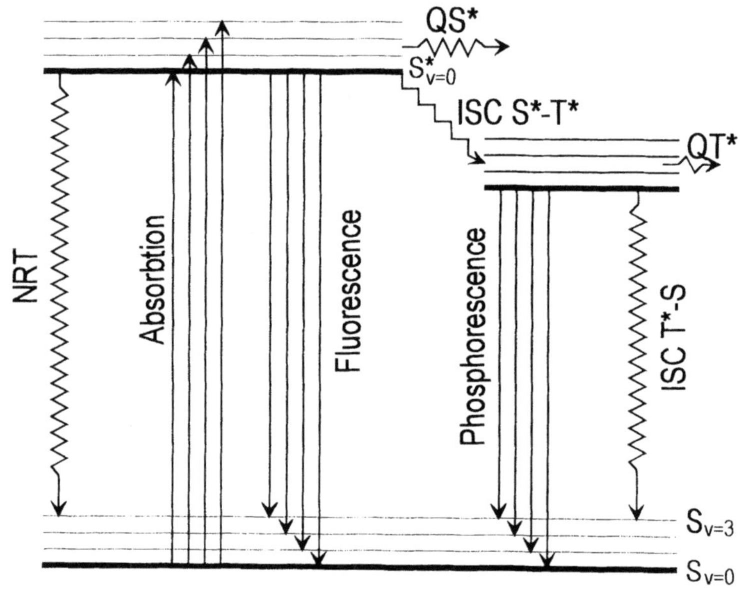

Fig 1. Jablonski diagram showing the different transitions between the ground state S and the excited singlet S* and triplet states T* (see text for details).

2.1 Fluorescence Emission

Some of the electrons at the excited state will return to one of the vibrational levels of the ground state by a radiative transition, whose frequency will be a function of the energy difference separating these levels. If one simply assumes that the energy spacing of the vibrational levels of the excited and ground states are similar one does expect the fluorescence emission spectrum to be a mirror image of the absorbtion spectrum.

A further expectation will be that the $S_{v=0}$ to $S^*_{v=0}$ absorbtion will be at the same frequency as $S^*_{v=0}$ to $S_{v=0}$ emission, this however is rarely the case, as the absorbtion process takes place in about 10^{-15} sec. the quantum coordinates of the molecule, and the orientation of the solvent molecules with respect to the electronic states will be conserved, however as the excited level lifetimes are rather long, the solvent molecules will reorient favorably about the electronic levels, stabilizing the latter in a lower energy conformation which results in a difference in the zero-zero frequencies. This difference in $S_{v=0}$ to $S^*_{v=0}$ absorbtion and $S^*_{v=0}$ to $S_{v=0}$ emission is termed as the Stoke's shift.

2.2 Non Radiative Transition

In this process the excitation energy will be lost mainly by interactions with solvent molecules, resulting in some of the electrons of the excited state returning to the ground state with a non radiative transition (denoted by NRT in the Jablonski diagram). This process will be favored by an increase in temperature, and can explain why increasing the temperature causes a decrease in fluorescence intensities.

2.3 Quenching of the Excited State

The excitation energy might be lost through interactions, in the form of collisions of quenchers with the electrons in the excited orbital. Typical quenchers like O_2, I^-, Mn^{2+} ions will quench every time they collide with an excited singlet (denoted by QS* and QT* for the quenching of the excited singlet and the excited triplet state electrons respectively).

2.4 The Inter System Crossing

The intersystem crossing is a quantum mechanically forbidden process which occurs by a spin exchange of the electron of the excited singlet state, resulting in an excited triplet state T* (ISC S* - T* on Jablonski diagram). As this process involves a forbidden transition it's probability of occurrence will be extremely low, nevertheless it will occur as the potential energy of the excited triplet is usually lower than that of the excited singlet state. The electron in the excited triplet state can then become deexcited by a non radiative transition, quenching, or by a radiative transition termed phosphorescence, the light emitted will be of longer wavelength than fluorescence because of the lower potential energy of the excited triplet . One should note that the return to the ground state necessitates a novel forbidden transition $T^*_{v=0}$ to S. For the reasons given above, the probability of this transition will be extremely low, resulting in a very long lifetime of the excited triplet state (sec. to days). This rather long lived triplet state will result in a very weak intensity of radiation, will be prone to quenching by collisions with quenchers, and the nonradiative processes will compete too well with the phosphorescence. The phosphorescence in solution will rarely be observed. In order to observe phosphorescence at all, one must very rigorously remove oxygen from the medium, and should use rigid glasses at very low temperatures.

2.5 The Delayed Fluorescence

Some of the electrons which have undergone intersystem crossing and therefore are in the T* state, may undergo a novel intersystem crossing to the S* level by the thermal energy provided by the solution, provided the energy difference between the T* and S* states are small . the return from the $S^*_{v=0}$ to S level by the fluorescence emission is called the delayed fluorescence and has the effect of lengthening the fluorescence lifetime of the molecule beyond what is expected in normal fluorescence emissions.

3 Advantages of Fluorescence Measurements

When one compares the sensitivity of fluorescence to those of the other techniques, one should note that in the absence of a chromophore, provided there is no background fluorescence of some kind, the level of the signal will be zero, so even a very small change in concentration of the chromophore can be detected by a large amplification, limited by the noise figure of the amplification chain. In the case of an absorbtion measurement however the absence of the chromophore will translate itself by zero absorbtion which will mean 100% transmission, therefore the maximum of the expected signal. In that case the amplification factor will be limited by the saturation of the amplifiers, therefore a small change in the concentration of the chromophore will result in a small percentile change over the maximum signal level, and may undergo unnoticed. One can overcome this difficulty to some extent by the use of differential detection techniques but even under these conditions the need to have extremely good match of the differential amplification chains will limit the sensitivity of the detection. As a rule of thumb one can say that with the absorbtion measurements one could detect changes in the order of 1 ppm (parts per million) and with fluorescence techniques on can easily go down to 10^{-4}-10^{-5} ppm, a sensitivity comparable to radioactive tracers.

4 The Observation of Fluorescence Emission

The fundamental principle underlying a fluorimeter is the maximization of the collection of the fluorescence emission and trying to minimize the collection of excitation light. This is usually accomplished by selecting a band of excitation wavelength which will not be present in the emission spectrum by the use of filters (interference or combination filters), or a monochromator on the excitation side, and highpass or bandpass filters on the emission side. The emission side filters will pass wavelengths longer than the excitation wavelength. For the measurements of fluorescence from individual cells one will use the epi-illuminated fluorescence microscope which is described below.

5 Epi-Luminescence Microscope

Most of the modern fluorescence microscopes will use the epi-luminescence technique, which means that both the excitation and emission lights will have a common optical path through the objective. The key element of epi-illumination is the dichroic mirror , it is an interference mirror formed by successive depositions of dielectric layers on a transparent substrate. The dichroic mirror will reflect the wavelengths below it's cutoff frequency and transmit those which are above the cutoff. This cutoff frequency should be chosen so that it will reflect all of the excitation wavelength, and transmit most of the emission wavelength. The Stoke's shift is an aid in this respect. One should note that the dichroic mirrors are far from being ideal they might have bandpass characteristics at a different wavelength, and they are very sensitive to the angle of incidence of the light beam. It is possible to find today polychroic mirrors which will allow the detection of many chromophores simultaneously.

6 Laser Scanning Confocal Microscope LSCM

No other introduction of a method has been accepted with such an excitement among researchers in biology, as that of the commercial availability of the Laser Scanning Confocal Microscope. This technique allows one to take extremely sharp thin sections of thick fluorescent specimens, penetrate deep into thick light scattering tissue, and by running parallel planar acquisitions construction of 3D view of the tissue.

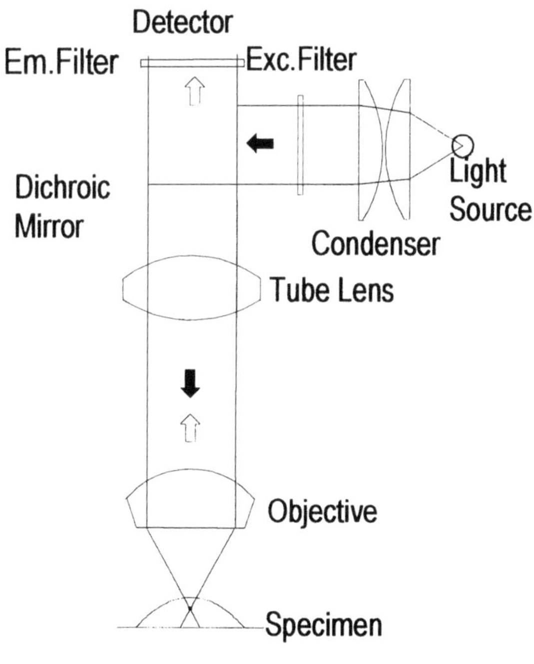

Fig 2. Epi-illumination. The light collected by the condenser from the light source is passed through interference filters to select the appropriate wavelength is than reflected by the dichroic mirror excites the chromophores in the specimen plane. The fluorescence light is collected by the objective passes through the dichroic mirror to fall on the detector.

The classical epi-illuminated fluorescence microscopy suffers a loss of detail due to blur contributed by the fluorescence of planes which are above and below the focal plane of the objective. These distortions elicited from the out of focus layers renders the epi-illuminated fluorescence technique rather useless when one is dealing with specimens of 200-300 mm thicknesses, such as work on thin brain slices. The LSCM overcomes those difficulties by forming the image of the illuminated spot on an optically confocal pinhole. The optical confocality of the detection pinhole ensures that only light emanating from the illuminated voxel (3D picture element) is fully passed to the detector and that light emanating from out of focus regions is passed with very low efficiency, thus resulting in the rejection of out of focus blur.

The original concept of confocal microscopy has been proposed by Marvin Minsky in a patent application he made in 1957, in his initial concept the specimen was illuminated throuh a pinhole, and the detection was made by a pinhole placed confocally with the illumination pinhole at the image plane and the specimen was moved by two orthogonally vibrating tuned forks. Although this technique of stage scanning microscope does not allow one to study live specimens, it provided most of the advantages of confocal microscopy for fixed specimens.

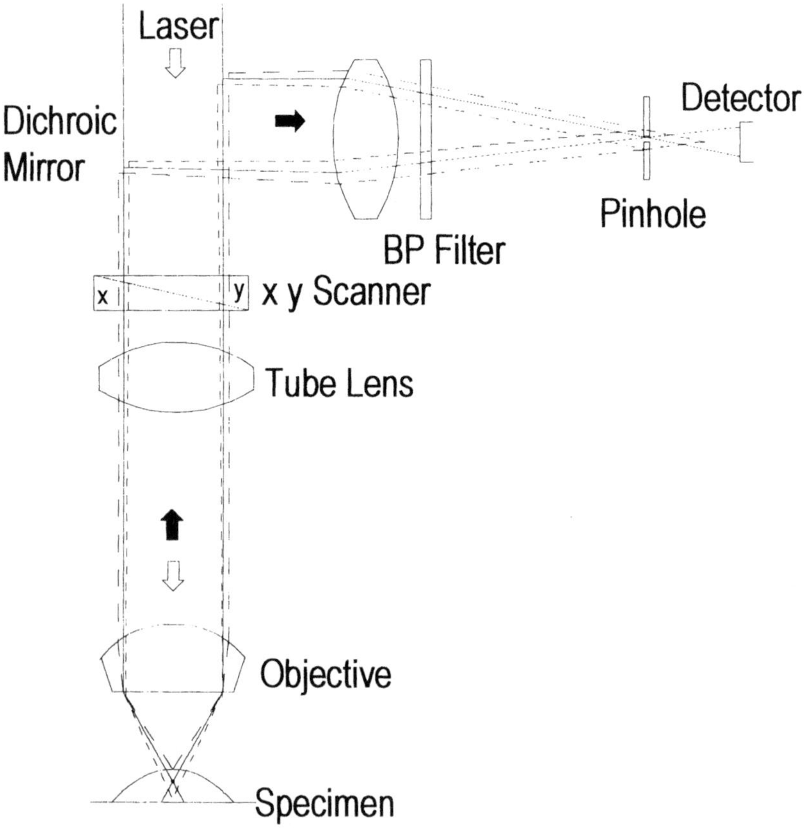

Fig 3. The confocal microscope. The solid rays form a confocal image of the excitation spot on the pinhole, the rays emanating from out of focus planes (dashed lines) are blocked by the pinhole therefore do not contribute to the image formation.

The years 1985-1987 has seen the simultaneous development from independent groups of the laser scanning confocal microscope. White and Amos developped the first commercially available LSCM whose principle of operation is depicted in Fig.3. Basically the laser light expanded to fill the objective is focused to a diffraction limited point on the specimen plane. which is scanned by moving the laser beam on the specimen plane by a set of mirrors usually driven by galvanometers. The fluorescence emission emanating from the specimen is collected by the objective and the tube lens and is deflected by the two mirrors of the scanner. this fluorescence signal is than reflected by the dichroic mirror and is focused on to the detection pinhole after passing through an interference filter. It is to be noted here that the mirrors of the scanner do form the image of the excitation spot onto the detection pinhole.

7 Wide Field Deconvolution Methods

The main advantage of confocal microscopy is the rejection of out of blur fluorescence signal emanating from out of focus planes, this however necessitates the intentional rejection of light coming from those planes. Seen in an other way a great proportion of molecules excited will not contribute to the image formation. This will result in two drawbacks. The formation of an image with a good signal to noise ratio will be rather slow and the high intensity illumination of any biological sample will have deleterious effects mainly through the toxic effects of photolysis products. An alternative way to cope with those drawbacks has been proposed by Castleman as early as in 1979. The technique consists of taking a series of 2D images along the z axis of the microscope each separated by a distance Δz and doing a 3D deconvolution on this stack of images.

Assuming that the microscope is a linear system and the shift invariance holds the image of an object can be described as the convolution of the point spread function of the microscope with the object. Linearity in this context means that the image of an object will be the sum of contributions of all the point sources which constitute the object. Shift invariance means that the imaging characteristics of the system will be uniform over the entire field. In mathematical terms,

$$g(x,y,z) = f(x,y,z) \otimes h(x,y,z)$$

where $\otimes$ denotes the convolution operator. g. f and h are the image. object. and the point spread function of the microscope respectively.

Instead of an exact solution of the above convolution which would be extremely time consuming. Castleman suggested an approximate method to recover the specimen function. Assuming that an image represents the convolution of the specimen with the in focus psf and the blur added by the convolutions of out of focus psf with the specimen. one can write at any given plane j the following relation.

$$g_j = f_j \otimes h_0 + \sum_{i=1-j}^{-1} f_{i+j} \otimes h_i + \sum_{i=1}^{N-j} f_{i+j} \otimes h_i$$

which can be rearranged to give

$$f_j \otimes h_0 = g_j - \sum_{i=1-j}^{-1} f_{i+j} \otimes h_i - \sum_{i=1}^{N-j} f_{i+j} \otimes h_i$$

which states that the object convolved with the in focus psf, is given by the image of the in focus plane minus the contributions of the of the adjacent object planes convolved with the out of focus psf. Assuming that the object of out of focus planes can be approximated by the image of out of focus planes the above equation becomes,

$$f_j \otimes h_0 = g_j - \sum_{i=1-j}^{-1} g_{i+j} \otimes h_i - \sum_{i=1}^{N-j} g_{i+j} \otimes h_i$$

ignoring the differences of the psf's above and below the plane of focus, and assuming that most of the blur will be contributed from the two adjacent planes the above equation can be simplified to

$$f_j \otimes h_0 = g_j - g_{j+1} \otimes h_1 - g_{j-1} \otimes h_1$$

rearranging and taking the Fourier transform one gets

$$F_j = \left\{G_j - c\left(G_{j-1} - G_{j+1}\right)H_j\right\} / H_0$$

where capitals denote the Fourier transforms of the variables respectively and H represents the transfer function of the microscope. In order to minimize the noise at high frequencies, Agard et all (1989). suggested the use of a Wiener filter of the form given below, instead of a filter of type $1/H_0$

$$W = H_0 / \left(H_o^2 + \alpha\right)$$

Where c and α are empirical constants. Note that the convolution becomes a multiplication in the frequency domain greatly reducing the computational load. This method is termed the nearest neighbor deconvolution.

To eliminate the need to take three time series images for Calcium measurements using FURA-2 as the reporter molecule, Monck et al. (1992) assumed that the blurred image can be used instead of neighboring images thus reducing the equation further to

$$F_j = \left\{G_j - 2cG_jH_j\right\} / W$$

The results obtained with this no neighbors method yielded images of comparable quality to that obtained with the nearest neighbors scheme.

In the case where light intensity levels and phototoxcicity are not important images obtained by confocal microscopes are unsurpassed. When light level and phototoxcicity become an issue, the method of mathematical deconvolution can become a viable alternative. For researchers not having the necessary funds to get a confocal microscope the improvement in image quality by these methods is really worth the effort of implementing them.

8 Two Photon Excitation in Laser Scanning Microscopy

As it was mentioned earlier in the laser scanning fluorescence microscope both conical volumes above and below the focused spot within the specimen are illuminated with high intensity laser light and the fluorescence emanating from those planes were rejected by the use of the detection pinhole. This approach although sufficient to form high quality images. nevertheless is prone to bleach the chromophores and cause toxic photolysis products. A new technique, introduced by Denk. et al. 1990, which holds great promise for the excitation in live tissue, is the two photon excitation technique. Two photon excitation as a single quantum event was predicted by M. Goeppert- Mayer as early as in 1931. but had to wait to be proven for couple of decades for the invention of the laser. This technique requires the simultaneous absorbtion of two photons of longer wavelength to be absorbed by a fluorophore to promote an electron to an excited state, which would then return to the ground state by one of the processes described earlier. Put in words two photons of 800 nm wavelength (the two photons need not have equal energies) can be absorbed by a fluorophore which would have normally absorbed at 400 nm. The fluorophore can than emit at a longer wavelength such as 450nm.

The two photon absorbtion probability has a quadratic dependece on the local illumination intensity, which results on the fact that this process is only possible with very high energy, ultra short pulses of the laser and will be confined to the diffraction limited spot. The typical pulses used with the mode locked Ti-sapphire laser are on the order of 10^{-13} sec. with a repetition rate of 80 MHz. One could visualize the process as such: in the cone of illumination above and below the diffraction limited spot the light will be red therefore will not be absorbed by the chromophore, therefore no photodamage and phototoxcicity effect will be seen, on the very small constrained volume at the diffraction limited spot one would have absorbtion which occurs in the blue wavelengths, and hence the photobleaching and phototoxcicity will be restrained to that small volume. The excitation occurring at a very small volume, the fluorescence emission will also be constrained to that volume and the emission of unwanted fluorescence from plans above and below the focus plane that was present in the LCSM will be absent when this technique is used for excitation. The detection pinhole of course will be unnecessary as there is no light that one wants to discriminate . The only drawback of this technique which is otherwise the ideal method, is the still high costs associated with the Ti-Sapphire laser, but as with any new technology one can expect the prices to go down within a very short time.

References.

Handbook of Confocal Microscopy 1995. J., B., Pawley ed. Plenum Press. (An excellent collection of articles on almost every aspects of confocal microscopy)

C., R. Cantor. and P., R. Schimmel. Biophysical Chemistry vol. II Techniques for the Study of Biological structure and Function 1980, W.M. Freeman and Co.

C., A. Parker. Photoluminescence of Solutions. 1968. Elsevier.

Agard D., A. 1984. Optical sectioning microscopy: cellular architecture in three dimensions. *Annu. Rev. Biophys. Bioeng.* **13:**191-219.

Agard D., A., Y. Hiaoka, P., Shaw and J., W., sedat.1989 Fluorescence microscopy in three dimensions. *Meth. Cell. Biol.* **30:** 353-377.

Denk W., J., H., Strickler, W., W., Webb. 1990. Two-photon laser scanning fluorescence microscopy. *Science* **248:** 73-76.

Monck, J., R., Oberhauser A., F., Keating T., J. and Fernandez J.,M. 1992. Thin section ratiometric Ca^{2+} images obtained by optical sectioning of Fura-2 loaded mast cells. *J. Cell Biol.* **116:** 745-759.

Calcium Confocal Microscopy of Single Synaptic Boutons

Naomi Melamed-Book and Rami Rahamimoff

Department of Physiology and the Bernard Katz Minerva Centre for Cell Biophysics, Hebrew University-Hadassah Medical School, Jerusalem, 92210, Israel

1 Preamble

Confocal microscopy is a technique of growing importance in the last decade. Its principles have been summarized in the article of Tan in this volume. The ability to observe the dynamics of important intracellular components inside the cell, without the interference of covering cells, makes it the technique of choice in the study of regulatory processes such as transmitter release.

Calcium confocal microscopy is a branch of confocal microscopy with its own problems and solutions. Its increasing popularity can be assessed from the *Medline* search on calcium confocal microscopy for the years 1990 to 1995, presented in Figure 1.

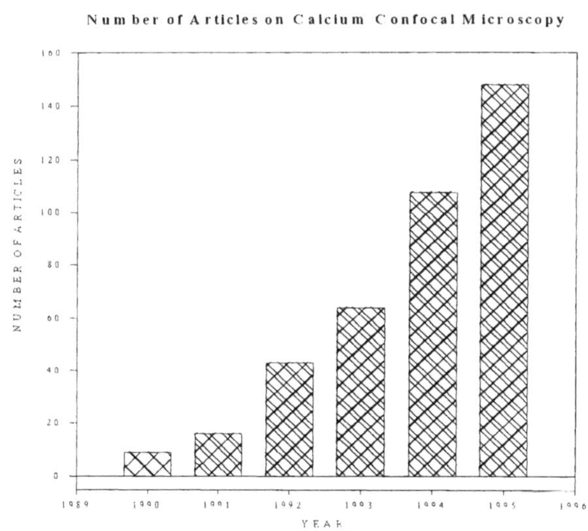

Fig 1. The number of articles listed in the Medline dealing with calcium confocal microscopy. No attempt was made to fit a curve.

The number of articles increased from 9 in 1990 to 148 in 1995. In spite of its increasing popularity, calcium confocal microscopy is not without major problems.

NATO ASI Series, Vol. H 100
Neurotransmitter Release and Uptake
Edited by Şakire Pöğün

In this article, we would like to summarize the applicability of confocal microscopy to the study of calcium dynamics in small synaptic structures, such as synaptic boutons, responsible for transmitter release. We would like to dwell on the inherent difficulties and on the use of stochastic methods in solving some of these difficulties. The methods have been published previously (Melamed, Helm and Rahamimoff 1993; Rahamimoff and Melamed 1993).

2 Calcium and Transmitter Release

It is well known that calcium ions play a major role in transmitter release (Feng 1936; del Castillo and Stark 1952; Jenkinson 1957; Dodge and Rahamimoff 1967; Hubbard, Jones and Landau 1968; Katz 1969; Rahamimoff, Lev-Tov and Meiri 1980; Llinas, Steinberg and Walton 1981; Augustine, Charlton and Smith 1987; Bertolino and Llinas 1992; Llinas, Sugimori and Silver 1992; Zucker 1993; Heidelberger, Heinemann, Neher and Matthews 1994; Burgoyne and Morgon 1995; Markram, Helm and Sakmann 1995; Mintz, Sabatini and Regehr 1995; Randall and Tsien 1995; Reuter 1995). Initially, it was not known whether calcium ions act on extracellular sites or on intracellular ones. However, after the findings that the entry of calcium ions through the plasma membrane to the interior of the nerve terminal is well correlated with transmitter release (Katz and Miledi 1967c; Katz 1969), it became increasingly clear that the intracellular calcium concentration is of key importance in the regulation of transmitter release. The role of calcium in transmitter release is a very large topic, which is outside the scope of this chapter. It can be found in other chapters of this book and in a large number of articles (Rahamimoff et al. 1980; Augustine, Charlton and Smith 1985b; Augustine et al. 1987; Bean 1989b; Robitaille, Adler and Charlton 1990; Komuro and Rakic 1992; van den Pol, Finkbeiner and Cornell Bell 1992; Friedman and Haddad 1993; Rahamimoff and Melamed 1993; Dailey and Smith 1994; Silinsky, Watanabe, Redman, Qiu, Hirsh, Hunt, Solsona, Alford and MacDonald 1995).

3 Intracellular Calcium Concentration

Confocal microscopy can be used to study the dynamics of intracellular calcium at rest and after alteration of the intracellular calcium concentration by various means.

4 Methods to Change Intracellular Calcium Concentration

There are several different methods employed to change the intracellular calcium concentration

4.1 Presynaptic Nerve Stimulation

The most physiological method to change the intracellular calcium concentration is to use nerve stimulation either by direct current injection, or by other means of depolarization (such as high$[K]_o$. The advantage of this method is that calcium ions enter into the nerve terminal by their natural route. The disadvantage is that there is no direct control on the amount of calcium entering the nerve terminal nor on its time course. The entry of calcium may be achieved through different types of calcium channels (Reuter 1996).

4.2 Injection of Calcium by Liposomes.

An alternative way to introduce calcium ions into the nerve terminals is by ion containing liposomes. This method relies on the fusion of the liposomes with the surface membrane of the nerve terminal and the 'injection' of the content of the liposome into the terminal. This method was introduced many years ago (Rahamimoff, Meiri, Erulkar and Barenholtz 1978) and is still used in various contexts (Kato, Akiyoshi, Furuno, Nakanishi, Kikuchi, Kataoka and Sunamoto 1994; Silinsky et al. 1995).

4.3 Perfusion

The intracellular Ca^{++} concentration can be changed by a perfusion of solutions containing different Ca^{++} concentrations or by injecting Ca^{++} buffers, such as EGTA or BAPTA (Tanabe and Kijima 1988; Augustine, Adler, Charlton, Hans, Swandulla and Zipser 1992). However, this is possible only in large terminals, such as the terminals of the squid (Augustine et al. 1992; Smith, Buchanan, Osses, Charlton and Augustine 1993).

4.4 Caged Calcium

In recent years, substances have been developed where calcium can be released by an optical method. The principle is rather simple. In the darkness, calcium ions are tightly bound to the chelator. In a sense, they are in a 'cage'. After shining a light at a specific wavelength, usually by a laser, the bond between the calcium and the chelator weakens and ionized calcium is released into the intracellular space (Kaplan 1990; Fricker, Gilroy, Read and Trewavas 1991; Parker and Ivorra 1993; Augustine 1994; Ilyin and Parker 1994; Parker and Yao 1994; Davis, Chang and Trump 1995; Kubitscheck, Pratsch, Passow and Peters 1995; Shiraishi, Okada, Shirakawa, Nakanishi, Mikoshiba and Miyazaki 1995; Stricker 1995; Wang and Augustine 1995; Yamaguchi, Kajita and Madison 1995).

Intracellular calcium concentration is an important regulator of many cellular functions (Casteels and Droogmans 1981; Rink, Smith and Tsien 1982; Tsien, Pozzan and Rink 1982; Woods, Cuthbertson and Cobbold 1986; Ambler, Poenie, Tsien and Taylor 1987; Berridge, Cobbold and Cuthbertson 1988; Harootunian, Kao and Tsien 1988; Jacob, Merritt, Hallam and Rink 1988; Meldolesi, Volpe and Pozzan 1988; Meyer and Stryer 1988; Prentki, Glennon, Thomas, Morris, Matschinsky and Corkey 1988; Yule and Gallacher 1988; Rink and Hallam 1989; Sage, Adams and van Breemen 1989). For further details, see (Lester and Nerbonne 1982; Nerbonne 1986; Gurney and Lester 1987; Kaplan and Somlyo 1989; Adams and Tsien 1993; Ellis-Davies and Kaplan 1994; Heidelberger et al. 1994; Currie, Wootton and Scott 1995; Lamb and Stephenson 1995; Nerbonne 1996).

This method can be used not only for calcium, but also for other substances important in synaptic transmission, such as NO (Makings and Tsien 1994; Pou, Anderson, Surichamorn, Keaton and Tod 1994; Lev-Ram, Makings, Keitz, Kao and Tsien 1995), ATP (Kaplan, Forbush and Hoffman 1978; Thirlwell, Corrie, Reid, Trentham and Ferenczi 1994; Stapleton, Bell, Wootton and Scott 1995), ADP-ribose (Lee 1994; Aarhus, Gee and Lee 1995), GABA (Denk, Delaney, Gelperin, Kleinfeld, Strowbridge, Tank and Yuste 1994), glutamate (Callaway and Katz 1993; Callaway 1994; Dalva and Katz 1994) and adrenergic modulators (Muralidharan, Mayer, Boyle and Nerbonne 1993; Walker, Martin, Schmitt and Barsotti 1993; Muralidharan and Nerbonne 1995).

5 Monitoring Intracellular Calcium Concentration by Confocal Microscopy

To monitor intracellular calcium concentration one needs an adequate 'reporter'. In recent years many calcium sensitive fluorescent probes have been developed, starting with the pioneering work of Roger Tsien and his collaborators -for a review see (Grynkiewicz, Poenie and Tsien 1985). The principle of all these probes is rather similar. The calcium probe bounds calcium, and the affinity for calcium is different for various probes, hence permitting the estimation of calcium in different concentration ranges.

When calcium ions bind to the fluorescent probe, the optical properties of the probe change. It is this change in the optical properties that is monitored and it permits the estimation of the intracellular calcium concentration.

Among the various methods employed to monitor intracellular calcium, we would like to dwell only on confocal microscopy. By using confocal microscopy and Ca^{++} sensitive fluorescent probes, it is possible to monitor the dynamics of $[Ca^{++}]_{in}$ in both good space and time resolution. An additional advantage of confocal microscopy is that it produces optical sections in a predetermined level, thus avoiding signals from adjacent cells such as the Schwann cell in the case of peripheral nerve terminals. This consideration is important because it has been realized that there are substantial calcium transients in these cells (Cornell-Bell, Finkbeiner, Cooper and Smith 1990; Charles, Merrill, Dirksen and Sanderson 1991; Dani, Chernjavsky and Smith 1992; Lieberman and Sanzenbacher 1992; Reist and Smith 1992; Charles, Dirksen, Merrill and Sanderson 1993; Chiu and Krigler 1994; Kim, Rioult and Cornell-Bell 1994; Lev Ram and Ellisman 1995). There are even indications for interactions between glial cells and neurons in their calcium signaling (Boitano, Dirksen and Sanderson 1992; Jahromi, Robitaille and Charlton 1992; van den Pol et al. 1992; Finkbeiner 1993; Murphy, Blatter, Wier and Baraban 1993; Charles 1994; Parpura, Basarsky, Liu, Jeftinija, Jeftinija and Haydon 1994).

There are two main ways to perform calcium confocal microscopy on living cells. The first is a *whole cell calcium confocal microscopy*. Using this variant the photons coming from the entire cell at a given confocal depth is measured by the detector. The alternative method to obtain the optical signal from a limited portion of the cell. According to the size of the scan, this subcellular scanning can be a *point scan*, *a line scan*, or several lines, which are nicknamed a *street scan*. We shall see the advantages and the disadvantages of the various scanning procedures in view of the limitations of confocal microscopy.

6 Limitations of Confocal Microscopy

Confocal microscopy is not without problems. Some of the problems are inherent within the method while others are a consequence of the biological reality.

6.1 The Number of Calcium Ions

The first problem we would like to deal with is the number of calcium ions in a typical optical section used in confocal microscopy. If we assume that $[Ca^{++}]_{in}$ is 300 nM (see, for example (Delaney, Zucker and Tank 1989; Clapham 1995), we can calculate the number of Ca^{++} ions in a typical voxel (3D unit) of $0.5\mu m^3$. Taking this concentration (3×10^{-7} moles of calcium in a liter), then in a volume of $0.5\mu m^3$ we can expect 22.5 calcium ions on the average, taking the Avogadro number into account. Figure 2 gives the number of calcium ions according to the size of a pixel in a confocal section. It is quite clear that reducing the size of the voxel to improve spatial resolution causes a reduction in the number of calcium ions, and thus, an increase in the noise.

6.2 The Concentration of the Fluorophore

In order to obtain a photon, the calcium ion has to meet the fluorophore. All the calcium-sensitive fluorophores are calcium buffers in nature, and calcium buffers affect the intracellular calcium concentration. Thus, to avoid large perturbations of the intracellular calcium concentration, we must limit the concentration of the fluorophore to suboptimal levels.

6.3 The Number of Photons Reaching the Detector

The next limitation is the number of photons which reach the detector (the photomultiplier, in our case). Confocal microscopy uses a small pinhole to produce optical sections, thus severely limiting the number of photons reaching the detector at any given time.

Due to these and other limitations, the number of photons which reach the detector is small. There are, however, several ways to overcome this inherent difficulty. One option is to use larger areas of measurement. This, of course, severely reduces the spatial resolution, so we opted for an alternative approach - namely, to use a time-series analysis to extract the signal from the inherent noise.

7 Time Series Analysis of Confocal Calcium Signals

In the time-series analysis, we used six different types of analysis:

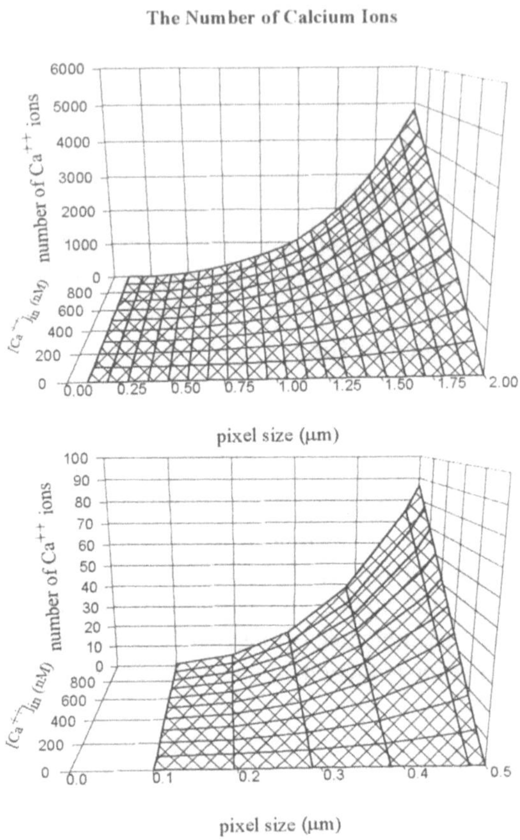

Fig 2. The calculated number of calcium ions as a function of pixel size and calcium concentration.

7.1 Space Averaging of Pixels

This is the simplest solution for the inherent noise problem of calcium confocal microscopy. One can analyze several adjacent pixels, and thus reduce the noise. The problem is that such a procedure greatly reduces the spatial resolution.

7.2 Time Averaging of Pixels

During homogeneous periods (for example, the synaptic bouton at rest), the time averaging reduces the noise, at the obvious expense of time resolution. This procedure was nevertheless routinely used for **boundary detection**. At the neuromuscular junction, the synaptic bouton sits on top of a muscle, contraction of the muscle usually causes an irreversible change of the scanning coordinates, and thus invalidates the results. Hence, the progressive time averaging of the pixels is a useful subroutine in boundary detection. In addition, it permits one to obtain the averaged $[Ca^{++}]_{in}$ and the ability to see whether, at a given time, the $[Ca^{++}]_{in}$ is above or below the average at a given pixel.

7.3 Variance Analysis

The average is the first moment of statistical analysis. The fluctuations around the mean are also important. Hence, we performed a variance analysis of each pixel - the confocal scan.

7.4 Regression Analysis

The next step in the time series analysis is to perform a regression on the original data. This is rather important to estimate whether the series is stationary in time. There are methodological reasons for a slight non-stationarity, such as bleaching of the fluorophore. Small non-stationarity can easily be corrected by the regression analysis. When a substantial systematic non-linearity was detected, the results were discarded.

7.5 Autocorrelation

After obtaining the primary data for each pixel, we proceeded with the examination, whether the fluctuations at each individual location are random in the time domain. The simplest way to achieve this task is to perform an autocorrelation on the original data. The principle of autocorrelation is rather simple. As a first step, the correlation coefficient of adjacent time points was estimated. Usually, data points were collected every 10ms or 20ms. If the intracellular calcium concentration is a completely random process in the time domain, then the signals 10ms apart will be independent of each other and the correlation coefficient will be 0. The correlation coefficient can obtain values between 1 (for a perfect positive correlation) and -1 (for a perfect negative correlation).

After obtaining the correlation coefficient for a lag of 1 (adjacent time points), one proceeds with longer lags. Lag of 2 means the concentration of the correlation between the fluorescent signal at a given time and the signal after two intervals (20 or 40ms). Subsequently, a computer program calculates the correlation coefficients with lags up to one half of the points (minus 1). The result of this procedure is the **autocorrelogram**, which is a very effective way to unravel whether there are oscillation in the intracellular calcium signals in the time domain. Figure 3 shows the results of such a procedure, illustrating the oscillations in $[Ca^{++}]_{in}$ in a single pixel in a synaptic bouton.

7.6 Crosscorrelation

The autocorrelation procedure described in the previous section examines the change in the time domain. The same correlation procedure can be applied on different pixels to check whether there are interactions in the $[Ca^{++}]_{in}$ between different locations in the same bouton in the space domain. This technique shows the existence of strong interaction in $[Ca^{++}]_{in}$ between different locations in the same bouton.

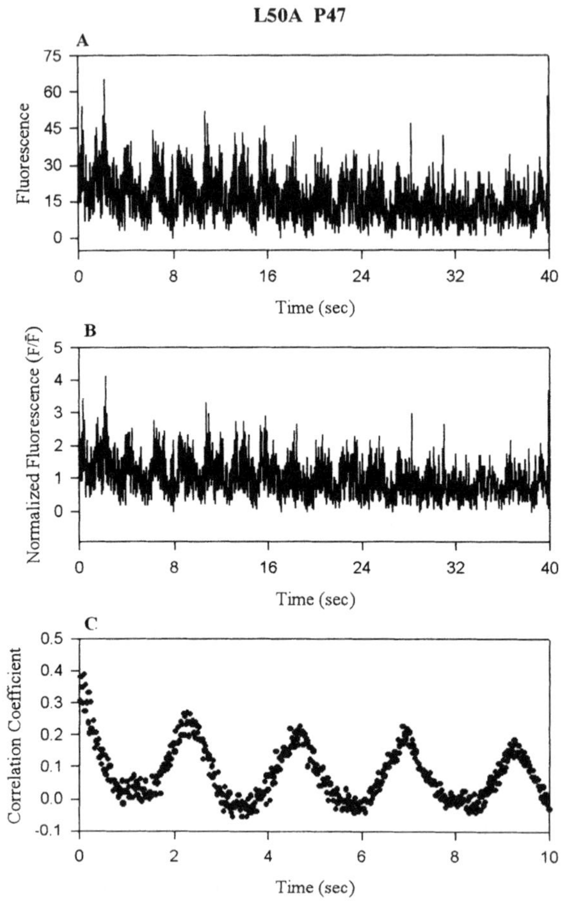

Fig 3. A) Fluorescence values (arbitrary units) of a single pixel. Pixels were scanned once every 20 msec. **B)** Normalized (F/F_{mean}) fluorescence values (same pixel as in A). **C)** Autocorrelation analysis reveals calcium oscillations once every 2 seconds.

8 Typical Results

Using these methods on Rhod-2 stained, single-synaptic boutons of the resting lizard neuromuscular junction, we get the following results:

8.1. A single bouton, which was divided on the average to X pixels, frequently behaves as one unit. At a given time, the $[Ca^{++}]_{in}$ in the whole bouton is either above or below the averaged $[Ca^{++}]_{in}$.

8.2. Two types of $[Ca^{++}]_{in}$ **oscillations** are found in the resting synaptic boutons. The more common one is once every two seconds, and the second is once every 20 seconds. Sometimes, both oscillations appear together.

8.3. Most of the time, two synaptic boutons correlate to one another, but sometimes, they behave in a different manner from one another.

9 Calcium Oscillations

The most interesting result in our means of averaging these presented here is the existence of calcium oscillation in individual synaptic boutons (Melamed et al. 1993; Rahamimoff and Melamed 1993). This result is of interest in two aspects. First, it puts the synaptic bouton in the growing family of cells and subcellular organelles that have an oscillating $[Ca^{++}]_{in}$. To many investigators, this property seems to be of importance in intracellular regulation, although the mechanism is not yet completely understood (Cuthbertson and Cobbold 1985; Merritt, Taylor, Rubin and Putney 1986; Ueda, Oiki and Okada 1986; Schlegel, Winger, Mollard, Vacher, Wuarin, Zahnd, Wollheim and Dufy 1987; Berridge and Gallione 1988; Harootunian et al. 1988; Holl, Thorner, Mandell, Sullivan, Sinha and Leong 1988; Jacob et al. 1988; Jaconi, Rivest, Schlegel, Wollheim, Pittet and Lew 1988; Meyer and Stryer 1988; Prentki et al. 1988; Quinn, Williams and Tillotson 1988; Gray 1989; Kurtz and Penner 1989; Sage et al. 1989; Weissberg, Little and Bobik 1989; Berridge and Potter 1990; Brooker, Seki, Croll and Wahlestedt 1990; Cornell-Bell et al. 1990; Malgaroli, Fesce and Meldolesi 1990; Milani, Malgaroli, Guidolin, Fasolato, Skaper, Meldolesi and Pozzan 1990; Parker and Ivorra 1990; Rooney, Sass and Thomas 1990; Takamatsu and Wier 1990; Tsunoda, Stuenkel and Williams 1990a; Delisle 1991; Harootunian, Kao, Paranjape and Tsien 1991; Tsunoda 1991; Tsunoda, Williams and DelValle 1991; Wier and Blatter 1991; Friel 1995). The second aspect is that the release of transmitter from presynaptic nerve endings is an oscillatory process (Erulkar and Rahamimoff 1976; Rahamimoff, Meiri and Erulkar 1976; Meiri and Rahamimoff 1978; Grinnell and Pawson 1988; Pawson and Grinnell 1989). Are these two phenomena related? It is not possible at present to answer this question affirmatively with confidence, since the two sets of observations were shown in different preparations. Until a simultaneous recording of the quantal release and calcium oscillation will be achieved, this remains an attractive working hypothesis.

Acknowledgments

We thank our colleagues Alon Meir, Serdar Demirgören, Alexander Butkevich and Ronit Ahdut for reading the manuscript and for helpful comments. The unfailing administrative help of Mrs. Debra Broide is greatly appreciated.

This work was supported by the German Israeli Foundation (GIF), the Basic research fund of the Israeli Academy of Sciences and US-Israel Binational Science Foundation (BSF).

References

Aarhus, R., Gee, K. and Lee, H.C. (1995). Caged cyclic ADP-ribose. Synthesis and use. *Journal of Biological Chemistry* **270:** 7745-7749.

Adams, S.R. and Tsien, R.Y. (1993). Controlling cell chemistry with caged compounds. *Annual Review of Physiology* **55:** 755-784.

Ambler, S.K., Poenie, M., Tsien, R.Y. and Taylor, P. (1987). Agonist-stimulated oscillations and cycling of intracellular free calcium in individual cultured muscle cells. *Journal of Biological Chemistry* **263:** 1952-1958.

Augustine, G.J. (1994). Combining patch-clamp and optical methods in brain slices. *J Neurosci Methods* **54(2):** 163-169.

Augustine, G.J., Adler, E.M., Charlton, M.P., Hans, M., Swandulla, D. and Zipser, K. (1992). Presynaptic calcium signals during neurotransmitter release: detection with fluorescent indicators and other calcium chelators. *Journal of Physiology (Paris)* **86:** 129-134.

Augustine, G.J., Charlton, M.P. and Smith, S. (1985b). Calcium entry and transmitter release at voltage-clamped nerve terminals of squid. *Journal of Physiology, London* **367:** 163-181.

Augustine, G.J., Charlton, M.P. and Smith, S. (1987). Calcium action in synaptic transmitter release. *Annual Review of Neuroscience* **10:** 633-693.

Bean, B.P. (1989b). Neurotransmitter inhibition of neuronal calcium currents by changes in voltage dependence. *Nature* **340:** 153-156.

Berridge, M.J., Cobbold, P.H. and Cuthbertson, K.S.R. (1988). Spatial and temporal aspects of cell signalling. *Philos. Trans. R. Soc. Lond. (Biol)* **320:** 325-343.

Berridge, M.J. and Gallione, A. (1988). Cytosolic calcium oscillators. *FASEB Journal* **2:** 2074-3082.

Berridge, M.J. and Potter, B.V.L. (1990). Inositol trisphosphate analogues induce different oscillatory patterns in Xenopus oocytes. *Cell Regulation* **1:** 675-681.

Bertolino, M. and Llinas, R.R. (1992). The central role of voltage-activated and receptor-operated calcium channels in neuronal cells. *Annual Review of Pharmacology and Toxicology* **32:** 399-421.

Boitano, S., Dirksen, E.R. and Sanderson, M.J. (1992). Intercellular propagation of calcium waves mediated by inositol triphosphate. *Science* **258:** 292-295.

Brooker, G., Seki, T., Croll, D. and Wahlestedt, C. (1990). Calcium wave evoked by activation of endogenous or exogenously expressed receptors in Xenopus oocytes. *Proceedings of the National Academy of Science, USA* **87:** 2813-2817.

Burgoyne, R.D. and Morgon, A. (1995). Ca^{2+} and secretory-vesicle dynamic. *Trends in Neuroscience* **18:** 191-196.

Callaway, E.M. (1994). Caged neurotransmitters. Shedding light on neural circuits. *Current Biology* **4:** 1010-1012.

Callaway, E.M. and Katz, L.C. (1993). Photostimulation using caged glutamate reveals functional circuitry in living brain slices. *Proceedings of the National Academy of Science USA* **90:** 7661-7665.

Casteels, R. and Droogmans, G. (1981). Exchange characteristics of the nonadrenaline-sensitive calcium store in vascular smooth muscle cells of rabbit ear artery. *Journal of Physiology* **317:** 263-279.

Charles, A.C. (1994). Glia-neuron intercellular calcium signaling. *Development in Neuroscience* **16:** 196-206.

Charles, A.C., Dirksen, E.R., Merrill, J.E. and Sanderson, M.J. (1993). Mechanisms of intercellular calcium signaling in glial cells studied with dantrolene and thapsigargin. *Glia* **7:** 134-145.

Charles, A.C., Merrill, J.E., Dirksen, E.R. and Sanderson, M.J. (1991). Intercellular signaling in glial cells: Calcium waves and oscillations in response to mechanical stimulation and glutamate. *Neuron* **6:** 983-992.

Chiu, S.Y. and Krigler, S. (1994). Neurotransmitter-mediated signaling between axons and glial cells. *Glia* **11(2):** 191-200.

Clapham, D.E. (1995). Calcium signaling. *Cell* **80:** 259-268.

Cornell-Bell, A.H., Finkbeiner, S.M., Cooper, M.S. and Smith, S.J. (1990). Glutamate induces calcium waves in cultured astrocytes: Long-range glial signaling. *Science* **247:** 470-473.

Currie, K.P., Wootton, J.F. and Scott, R.H. (1995). Activation of Ca^{2+} -dependent Cl-currents in cultured rat sensory neurones by flash photolysis of DM-nitrophen. *Journal of Physiology, London* **482:** 291-307.

Cuthbertson, K.S.R. and Cobbold, P.H. (1985). Phorbol ester and sperm activate mouse oocytes by inducing sustained oscillations in cell Ca^{2+}. *Nature* **316:** 541-542.

Dailey, M.E. and Smith, S.J. (1994). Spontaneous Ca^{2+} transients in developing hippocampal pyramidal cells. *J Neurobiol* **25(3):** 243-251.

Dalva, M.B. and Katz, L.C. (1994). Rearrangements of synaptic connections in visula cortex revealed by lase photostimulation. *Science* **265:** 255-258.

Dani, J.W., Chernjavsky, A. and Smith, S.J. (1992). Neuronal activity triggers calcium waves in hippocampal astrocyte networks. *Neuron* **8:** 429-440.

Davis, M.A., Chang, S.H. and Trump, B.F. (1995). IP3-mediated cytosolic and nuclear calcium elevation in NRK-52E cells using 'caged' GPIP2. *Cell Calcium* **17(6):** 453-458.

del Castillo, J. and Stark, L. (1952). The effect of calcium ions on the motor end-plate potentials. *Journal of Physiology, London* **116:** 507-515.

Delaney, K.R., Zucker, R.S. and Tank, D.W. (1989). Calcium in motor nerve terminals associated with post tetanic potentiation. *Journal of Neuroscience* **9(10):** 3558-3567.

Delisle, S. (1991). The four dimensions of calcium signalling in Xenopus oocytes. *Cell Calcium* **12:** 217-227.

Denk, W., Delaney, K.R., Gelperin, A., Kleinfeld, D., Strowbridge, B.W., Tank, D.W. and Yuste, R. (1994). Anatomical and functional imaging of neurons using 2-photon laser scanning microscopy. *Journal of Neuroscience Methods* **54:** 151-162.

Dodge, F.A. and Rahamimoff, R. (1967). Co-operative action of calcium ions in transmitter release at the neuromuscular junction. *Journal of Physiology, London* **193:** 419-432.

Ellis-Davies, G.C. and Kaplan, J.H. (1994). Nitrophenyl-EGTA, a photolabile chelator that selectively binds Ca^{2+} with high affinity and releases it rapidly upon photolysis. *Proceedings of the National Academy of Science, USA* **91:** 187-191.

Erulkar, S.D. and Rahamimoff, R. (1976). Periodic fluctuations in transmitter release at the frog neuromuscular junction. *Journal of Physiology, London* **256:** 20-21p.

Feng, T.P. (1936). Studies on the neuromuscular junction. The universla antagonism between calcium and curarizing agents. *Chin. J. Physiol.* **16:** 341-372.

Finkbeiner, S.M. (1993). Glial calcium. *Glia* **9:** 83-104.

Fricker, M.D., Gilroy, S., Read, N.D. and Trewavas, A.J. (1991). Visualisation and measurement of the calcium message in guard cells. *Symp Soc Exp Biol* **45:** 177-190.

Friedman, J.E. and Haddad, G.G. (1993). Major differences in $Ca^{2+}{}_{i}$ response to anoxia between neonatal and adult rat CA1 neurons: role of $Ca^{2+}{}_{o}$ and $Na^{+}{}_{o}$. *J Neurosci* **13(1):** 63-72.

Friel, D.D. (1995). $[Ca^{2+}]_i$ oscillations in sympathetic neurons: an experimental test of a theoretical model. *Biophysical Journal* **68:** 1752-66.

Gray, P.T.A. (1989). Oscillations of free cytosolic calcium evoked by cholinergic and catecholaminergic agonists in rat parotid acinar cells. *Journal of Physiology, London* **406:** 35-53.

Grinnell, A.D. and Pawson, P.A. (1988). Oscillation period of mEPP frequency at frog neuromuscular junctions is inversely correlated with release efficacy and independent of acute Ca^{++}-loading. *Society of Neuroscience Abstract* **14:** 68.

Grynkiewicz, C., Poenie, M. and Tsien, R.Y. (1985). A new generation of Ca^{2+} indicators with greatly improved fluorescence properties. *Journal of Biological Chemistry* **260:** 3440-3450.

Gurney, A.M. and Lester, H.A. (1987). Light flash physiology with synthetic photosensitive compounds. *Physiological Review* **67:** 583-617.

Harootunian, A.T., Kao, J.P., Paranjape, S. and Tsien, R.Y. (1991). Generation of calcium oscillations in fibroblasts by positive feedback between calcium and IP3. *Science* **251:** 75-78.

Harootunian, A.T., Kao, J.P.Y. and Tsien, R.Y. (1988). Agonist-induced calcium oscillations in depolarized fibroblasts and their manipulation by photoreleased Ins (1,4,5,) P_3, Ca^{2+} and Ca^{2+} buffer. *Cold Spring Harbor Symp. Quant. Biol.* 53.

Heidelberger, R., Heinemann, C., Neher, E. and Matthews, G. (1994). Calcium dependence of the rate of exocytosis in a synaptic terminal. *Nature* **371:** 513-515.

Holl, R.W., Thorner, M.O., Mandell, G.L., Sullivan, J.A., Sinha, Y. and Leong, D.A. (1988). Spontaneous oscillations of intracellular calcium and growth hormone secretion. *Journal of Biological Chemistry* **263:** 9682-9685.

Hubbard, J.I., Jones, S.F. and Landau, E.M. (1968). On the mechanism by which calcium and magnesium affect the release of transmitter by nerve impulses. *Journal of Physiology, London* **196:** 75-86.

Ilyin, V. and Parker, I. (1994). Role of cytosolic Ca2+ in inhibition of InsP3-evoked Ca2+ release in Xenopus oocytes. *J Physiol Lond* **477(Pt 3):** 503-509.

Jacob, R., Merritt, J.M., Hallam, T.J. and Rink, T.J. (1988). Repetitive spikes in cytoplasmic calcium evoked by histamine in human endothelial cells. *Nature* **335:** 40-45.

Jaconi, M.E.E., Rivest, R.W., Schlegel, W., Wollheim, C.B., Pittet, D. and Lew, P.D. (1988). Spontaneous and chemoattractant-induced oscillations of cytosolic free calcium in single adherent human neutrophils. *Journal of Biological Chemistry* **263:** 10557-10560.

Jahromi, B.S., Robitaille, R. and Charlton, M.P. (1992). Transmitter release increases intracellular calcium in perisynaptic Schwann cells in situ. *Neuron* **8:** 1069-1077.

Jenkinson, D.H. (1957). The nature of the antagonism between calcium and magnesium ions at the neuromuscular junction. *Journal of Physiology* **138:** 434-444.

Kaplan, J.H. (1990). Photochemical manipulation of divalent cation levels. *Annual Review of Physiology* **52:** 897-914.

Kaplan, J.H., Forbush, B. and Hoffman, J.F. (1978). Rapid photolytic release of adenosine 5'-triphosphate from a protected analogue: utilization of the Na:K pump of human red cell ghosts. *Biochemistry* **17:** 1929-1935.

Kaplan, J.H. and Somlyo, A.P. (1989). Flash photolysis of caged compounds: new tools for cellular physiology. *Trends in Neuroscience* **12:** 54-59.

Kato, E., Akiyoshi, K., Furuno, T., Nakanishi, M., Kikuchi, A., Kataoka, K. and Sunamoto, J. (1994). Interaction between ganglioside-containing liposome and rat T-lymphocyte: confocal fluorescence microscopic study. *Biochem Biophys Res Commun* **203(3):** 1750-1755.

Katz, B. (1969). The Release of Neural Transmitter Substances. *The Sherrington Lectures No. X. The Liverpool University Press, Liverpool* .

Katz, B. and Miledi, R. (1967c). The timing of calcium action during neuromuscular transmission. *Journal of Physiology, London* **189:** 535-544.

Kim, W.T., Rioult, M.G. and Cornell-Bell, A.H. (1994). Glutamate-induced calcium signaling in astrocytes. *Glia* **11:** 173-184.

Komuro, H. and Rakic, P. (1992). Selective role of N-type calcium channels in neuronal migration. *Science* **257(5071):** 806-809.

Kubitscheck, U., Pratsch, L., Passow, H. and Peters, R. (1995). Calcium pump kinetics determined in single erythrocyte ghosts by microphotolysis and confocal imaging. *Biophys J* **69(1):** 30-41.

Kurtz, A. and Penner, R. (1989). Angiotensin II induces oscillations of intracellular calcium and blocks anomalous inward rectifying potassium current in mouse renal juxtaglomerular cells. *Proceedings of the National Academy of Science, USA* **86:** 3423-3427.

Lamb, G.D. and Stephenson, D.G. (1995). Activation of ryanodine receptors by flash photolysis of caged Ca^{2+}. *Biophysical Journal* **68:** 423-430.

Lee, H.C. (1994). Cyclic ADP-ribose: a calcium mobilizing metabolite of NAD+. *Molecular and Cellular Biochemistry* **138:** 229-235.

Lester, H.A. and Nerbonne, J.M. (1982). Physiological and pharmacological manipulations with light flashes. *Annual Review of Biophysical Bioengineering* **11:** 151-175.

Lev Ram, V. and Ellisman, M.H. (1995). Axonal activation-induced calcium transients in myelinating Schwann cells, sources, and mechanisms. *J Neurosci* **15(4):** 2628-37.

Lev-Ram, V., Makings, L.R., Keitz, P.F., Kao, J.P.Y. and Tsien, R.Y. (1995). Long-term depression in cerebellar Purkinje neurons results from coincidence of nitric oxide and depolarization-induced Ca^{2+} transients. *Neuron* **15:** 407-415.

Lieberman, E.M. and Sanzenbacher, E. (1992). Mechanisms of glutamate activation of axon to Schwann cell signaling in the squid. *Neuroscience* **47(4):** 931-939.

Llinas, R., Steinberg, I.Z. and Walton, K. (1981). Relationship between presynaptic calcium current and postsynaptic potential in squid giant synapse. *Biophysical Journal* **33:** 323-352.

Llinas, R., Sugimori, M. and Silver, R.B. (1992). Microdomains of high calcium concentration in a presynaptic terminal. *Science* **256:** 677-679.

Makings, L.R. and Tsien, R.Y. (1994). Caged nitric oxide. Stable organic molecules from which nitric oxide can be photoreleased. *Journal of Biological Chemistry* **269:** 6282-6285.

Malgaroli, A., Fesce, R. and Meldolesi, J. (1990). Spontaneous $[Ca^{2+}]_1$ fluctuations in rat chromaffin cells do not require inositol 1, 4, 5-trisphosphate elevations but are generated by a caffeine- and ryanodine-sensitive intracellular Ca^{2+} store. *Journal of Biological Chemistry* **265:** 3005-3008.

Markram, H., Helm, P.J. and Sakmann, B. (1995). Dendritic calcium transients evoked by single back-propogating action potentials in rat neocortical pyramidal neurons. *Journal of Physiology, London* **485:** 1-20.

Meiri, H. and Rahamimoff, R. (1978). Clumping and oscillations in evoked transmitter release at the frog neuromuscular junction. *Journal of Physiology, London* **278:** 513-523.

Melamed, N., Helm, P.J. and Rahamimoff, R. (1993). Confocal microscopy reveals coordinated calcium fluctuations and oscillations in synaptic boutons. *J Neurosci* **13(2):** 632-649.

Meldolesi, J., Volpe, P. and Pozzan, T. (1988). The intracellular distribution of calcium. *Trends in Neuroscience* **11:** 449-452.

Merritt, J.E., Taylor, C.W., Rubin, R. and Putney, J.W.J. (1986). Isomers of inositol trisphosphate in exocrine pancreas. *Biochemical Journal* **238:** 825-829.

Meyer, T. and Stryer, L. (1988). Molecular model for receptor-stimulated calcium spiking. *Proceedings of the National Academy of Science, USA* **85:** 5051-5055.

Milani, D., Malgaroli, A., Guidolin, D., Fasolato, C., Skaper, S.D., Meldolesi, J. and Pozzan, T. (1990). Ca^{2+} channels and intracellular Ca^{2+} stores in neuronal and neuroendocrine cells. *Cell Calcium* **11(2-3):** 191-199.

Mintz, I.M., Sabatini, B.L. and Regehr, W.G. (1995). Calcium control of transmitter release at a cerebellar synapse. *Neuron* **15:** 675-688.

Muralidharan, S. and Nerbonne, J.M. (1995). Photolabile 'caged' adrenergic receptor agonists and related model compounds. *Journal of Photochemistry and Photobiology (B)* **27:** 123-137.

Muralidharan, S.M., Mayer, G.M., Boyle, W.B. and Nerbonne, J.M. (1993). 'Caged' phenylephrine development and application to probe the mechanism of alpha-receptor mediated vasoconstriction. *Proceedings of the National Academy of Science, USA* **90:** 5199-5203.

Murphy, T.H., Blatter, L.A., Wier, W.G. and Baraban, J.M. (1993). Rapid communication between neurons and astrocytes in primary cortical cultures. *Journal of Neuroscience* **13:** 2672-2679.

Nerbonne, J.M. (1986). Design and application of photolabile intracellular probes. **Optical Methods in Cell Physiology**. New York, Wiley-Interscience: 417-445.

Nerbonne, J.M. (1996). Caged compounds: tools for illuminating neuronal responses and connections. *Current Opinion in Neurobiology* **6(3):** 379-386.

Parker, I. and Ivorra, I. (1990). Inhibition by a Ca^{2+} liberation: A possible mechanism for oscillatory release of Ca^{2+}. *Proceedings of the National Academy of Science, USA* **87:** 260-264.

Parker, I. and Ivorra, I. (1993). Confocal microfluorimetry of Ca^{2+} signals evoked in Xenopus oocytes by photoreleased inositol trisphosphate. *J Physiol Lond* **461:** 133-165.

Parker, I. and Yao, Y. (1994). Relation between intracellular Ca^{2+} signals and Ca Ca^{2+}-activated Cl- current in Xenopus oocytes. *Cell Calcium* **15(4):** 276-288.

Parpura, V., Basarsky, T., Liu, F., Jeftinija, K., Jeftinija, S. and Haydon, P. (1994). Glutamate-mediated astrocyte-neuron signalling. *Nature* **369:** 744-747.

Pawson, P.A. and Grinnell, A.D. (1989). Oscillation period of MEPP frequency at frog neuromuscular junctions is inversely correlated with release efficacy and independent of acute Ca^{2+} loading. *Proceedings of the Royal Society of London (Biol)* **237:** 489-499.

Pou, S.J., Anderson, D.E., Surichamorn, W., Keaton, L.L. and Tod, M.L. (1994). Biological studies of a nitroso compound that releases nitric oxide upon illumination. *Molecular Pharmacology* **46:** 709-715.

Prentki, M., Glennon, M.C., Thomas, A.P., Morris, R.L., Matschinsky, F.M. and Corkey, B.E. (1988). Cell-specific patterns of oscillating free Ca^{2+} in carbamylcholine-stimulated insulinoma cells. *Journal of Biological Chemistry* **263:** 11044-11047.

Quinn, S.J., Williams, G.H. and Tillotson, D.L. (1988). Calcium oscillations in single adrenal glomerulosa cells stimulated by angiotensin II. *Proceedings of the National Academy of Science, USA* **85:** 5754-5758.

Rahamimoff, R., Lev-Tov, A. and Meiri, H. (1980). Primary and secondary regulation of quantal transmitter release: Calcium and sodium. *Journal of Experimental Biology* **89:** 5-18.

Rahamimoff, R., Meiri, H., Erulkar, S. and Barenholtz, Y. (1978). Changes in transmitter release induced by ion-containing liposomes. *Proceedings of the National Academy of Science, USA* **75:** 5214-5216.

Rahamimoff, R., Meiri, H. and Erulkar, S.D. (1976). Periodic oscillations in transmitter release at the frog neuromuscular junction. *Israeli Journal of Medical Science* **12:** 1511-1512.

Rahamimoff, R. and Melamed, N. (1993). Visualization of synaptic structure and function with confocal microscopy: calcium fluctuations and oscillations. *Neurosci Res* **16(3):** 173-80.

Randall, A. and Tsien, R.W. (1995). Pharmacological dissection of multiple types of Ca^{2+} channel currents in rat cerebellar granule neurons. *Journal of Neuroscience* **15:** 2995-3012.

Reist, N.E. and Smith, S.J. (1992). Neurally evoked calcium transients in terminal Schwann cells at the neuromuscular junction. *Proc Natl Acad Sci U S A* **89(16):** 7625-7629.

Reuter, H. (1995). Measurements of exocytosis from singlew presynaptic nerve terminals reveal heterogeneous inhibition by Ca^{2+} channel blockers. *Neuron* **14:** 773-779.

Reuter, H. (1996). Diversity and function of presynaptic calcium channels in the brain. *Current Opinion in Neurobiology* **6:** 331-337.

Rink, T.J. and Hallam, T.J. (1989). Calcium signalling in non-excitable cells: Notes on oscillations and store refilling. *Cell Calcium* **10:** 385-395.

Rink, T.J., Smith, S.W. and Tsien, R.Y. (1982). Cytoplasmic free Ca^{2+} in human platelets: Ca^{2+} thresholds and Ca-independent activation for shape-change in secretion. *FEBS Letters* **148:** 21-26.

Robitaille, R., Adler, E.M. and Charlton, M.P. (1990). Strategic location of calcium channels at transmitter release sites of frog neuromuscular synapses. *Neuron* **5(6):** 773-779.

Rooney, T.A., Sass, E.J. and Thomas, A.P. (1990). Agonist-induced cytosolic calcium oscillations originate from a specific locus in single hepatocytes. *Journal of Biological Chemistry* **265:** 10792-10796.

Sage, S.O., Adams, D.J. and van Breemen, C. (1989). Synchronized oscillations in cytoplasmic free calcium concentration in confluent bradykinin-stimulated bovine pulmonary artery endothelial cell monolayers. *Journal of Biological Chemistry* **264:** 6-9.

Schlegel, W., Winger, B.P., Mollard, P., Vacher, P., Wuarin, F., Zahnd, G.R., Wollheim, C.B. and Dufy, B. (1987). Oscillations of cytosolic Ca^{2+} in pituitary cells due to action potentials. *Nature* **329:** 719-721.

Shiraishi, K., Okada, A., Shirakawa, H., Nakanishi, S., Mikoshiba, K. and Miyazaki, S. (1995). Developmental changes in the distribution of the endoplasmic reticulum and inositol 1,4,5-trisphosphate receptors and the spatial pattern of Ca^{2+} release during maturation of hamster oocytes. *Dev Biol* **170(2):** 594-606.

Silinsky, E.M., Watanabe, M., Redman, R.S., Qiu, R., Hirsh, J.K., Hunt, J.M., Solsona, C.S., Alford, S. and MacDonald, R.C. (1995). Neurotransmitter release evoked by nerve impulses without Ca^{2+} entry through Ca^{2+} channels in frog motor nerve endings. *J Physiol Lond* **482(Pt 3):** 511-520.

Smith, S.J., Buchanan, J., Osses, L.R., Charlton, M.P. and Augustine, G.J. (1993). The spatial distribution of calcium signals in squid presynaptic terminals. *Journal of Physiology, (London)* **472:** 573-593.

Stapleton, S.R., Bell, B.A., Wootton, J.F. and Scott, R.H. (1995). Modulation of Ca^{2+} -dependent currents in metabolically stressed cultured sensory neurones by intracellular photorelease of ATP. *British Journal of Pharmacology* **114:** 544-550.

Stricker, S.A. (1995). Time-lapse confocal imaging of calcium dynamics in starfish embryos. *Dev Biol* **170(2):** 496-518.

Takamatsu, T. and Wier, G. (1990). Calcium waves in mammalian heart: qualification of origin, magnitude, waveform and velocity. *FASEB Journal* **4:** 1529-1525.

Tanabe, N. and Kijima, H. (1988). Transmitter release at frog end-plate loaded with a Ca^{2+} chelator, BAPTA: hypertonicity and erythrosin B augment the release independently of internal Ca^{2+}. *Neuroscience Letters* **92(1):** 52-57.

Thirlwell, H., Corrie, J.E., Reid, G.P., Trentham, D.R. and Ferenczi, M.A. (1994). Kinetics of relaxation from rigor of permeabilized fast-twitch skeletal fibers from the rabbit using a novel caged ATP and apyrase. *Biophysical Journal* **67:** 2436-2447.

Tsien, R.Y., Pozzan, T. and Rink, T.J. (1982). Calcium homeostasis in intact lymphocytes: cytoplasmic free calcium monitored with a new, intracelluarly trapped fluorescent indicator. *Journal of Cell Biology* **94:** 325-334.

Tsunoda, Y. (1991). Oscillatory Ca^{2+} signaling and its cellular function. *New Biologist* **3(1):** 3-17.

Tsunoda, Y., Stuenkel, E.L. and Williams, J.A. (1990a). Oscillatory mode of calcium signaling in rat pancreatic acinar cells. *American Journal of Physiology* **258:** C147-C155.

Tsunoda, Y., Williams, J.A. and DelValle, J. (1991). Secretagogue-induced Ca^{2+} oscillations in isolated canine gastric chief cells. *Biochim. Biophys. Acta.* **1091(2):** 251-254.

Ueda, S., Oiki, S. and Okada, Y. (1986). Oscillations of cytoplasmic concentrations of Ca^{2+} and K^{+} in fused L cells. *Journal of Membrane Biology* **91:** 65-72.

van den Pol, A.N., Finkbeiner, S.M. and Cornell Bell, A.H. (1992). Calcium excitability and oscillations in suprachiasmatic nucleus neurons and glia in vitro. *J Neurosci* **12(7):** 2648-2664.

Walker, J.W., Martin, H., Schmitt, F.R. and Barsotti, R.J. (1993). Rapid release of alpha-adrenergic receptor ligand from photolabile analogues. *Biochemistry* **32:** 1338-1345.

Wang, S.S. and Augustine, G.J. (1995). Confocal imaging and local photolysis of caged compounds: dual probes of synaptic function. *Neuron* **15(4):** 755-760.

Weissberg, P.L., Little, P.J. and Bobik, A. (1989). Spontaneous oscillations in cytoplasmic calcium concentration in vascular smooth muscle. *American Journal of Physiology* **256:** C951-C957.

Wier, W.G. and Blatter, L.A. (1991). Ca^{2+} -oscillations and Ca^{2+}-waves in mammalian cardiac and vascular smooth muscle cells. *Cell Calcium* **12(2-3):** 241-254.

Woods, N.M., Cuthbertson, K.S.R. and Cobbold, P.H. (1986). Repetitive transient rises in cytoplasmic free calcium in hormone-stimulated hepatocytes. *Nature* **319:** 600-602.

Yamaguchi, H., Kajita, J. and Madison, J.M. (1995). Isoproterenol increases peripheral $[Ca^{2+}]_i$ and decreases inner $[Ca^{2+}]_i$ in single airway smooth muscle cells. *Am J Physiol* **268(3 Pt 1):** C771-779.

Yule, D.I. and Gallacher, D.V. (1988). Oscillations of cytosolic calcium in single pancreatic acinar cells stimulated by acetylcholine. *FEBS Letters* **239:** 358-362.

Zucker, R.S. (1993). Calcium and transmitter release. *Journal of Physiology, Paris* **87:** 25-36.

Practical Aspects of Near-Field Optical Microscopy

Silvio P. Marchese-Ragona and Philip G. Haydon

Laboratory of Cellular Signaling, Dept. of Zoology and Genetics, rm. 339 Science II, Iowa State University, Ames, IA 50011, USA.

1 Introduction

The study of synaptic transmission has been significantly facilitated by the use of optical techniques to monitor calcium influx and vesicle cycling. However, the optical microscope is limited by diffraction putting a lower limit of 250 nm on spatial resolution. Near-field Scanning Optical Microscopy (NSOM) is a scanning probe technique that utilizes a sub wavelength light source in close proximity to the surface of a sample to generate optical images with a lateral optical resolution below the diffraction limit (about 20-50 nm). In this chapter the NSOM technique is discussed together with its initial application to biological samples. The concept of NSOM predates its more familiar cousins of Atomic Force Microscopy (AFM) and Scanning Tunneling Microscopy (STM) by six decades. It was first considered theoretically in 1928 by Synge[1] and first demonstrated with microwave radiation in 1972 by Ash and Nicholls [2]. However, due to experimental difficulties it was not until 1984, and the advent of STM, that it was possible to demonstrate the technique using visible light [3].

The essential elements for a NSOM are: **1.** A sub wavelength light source, which is generally a single mode optical fiber that is coated with aluminum except for a small 20-100 nm aperture at the apex., **2.** A feedback mechanism that allows the tip to remain a fixed distance from the surface of the specimen, as the specimen is scanned relative to the probe. **3.** A raster mechanism for scanning the sample in x and y for image formation, and **4.** A photosensor (usually a PMT or photon counter) with associated filters and analyzers.

2 Instrumentation

The system used in our laboratory is shown in Figure 1. The NSOM is mounted on a Nikon Diaphot inverted microscope equipped with epi-illumination. The sample is scanned on a piezo driven xyz stage with a maximum xy scan range of 18 microns and a z-range of 3.2 microns. We use a shear force feedback mechanism to allow the probe to maintain a fixed distance above the sample, while the sample is being scanned. Thus, by controlling the sample height the near-field fiber can remain stationary and in the focal plane of the objective lens. This mechanism of sample scanning produces a near-field image and a corresponding topographic image of the same area.

NATO ASI Series, Vol. H 100
Neurotransmitter Release and Uptake
Edited by Şakire Pöğün

3 Near-Field Probes

Sub wavelength apertures are produced at the apex of a drawn single mode fiber. The fabrication of the aperture is essentially a two step procedure. First, the optical fiber is heated and drawn down to a fine taper using a laser based micropipette puller. Second, aluminum is evaporated onto the sides of the optical fiber to prevent leakage of light through the sides of the fiber. The aluminum is angle evaporated on to fiber , while the fiber is rotated about its long axis with the tip of the fiber facing away from the evaporation source. This configuration allows the sides of the fiber to be coated while allowing the apex of the tip to be uncoated, and function as the sub wavelength aperture.

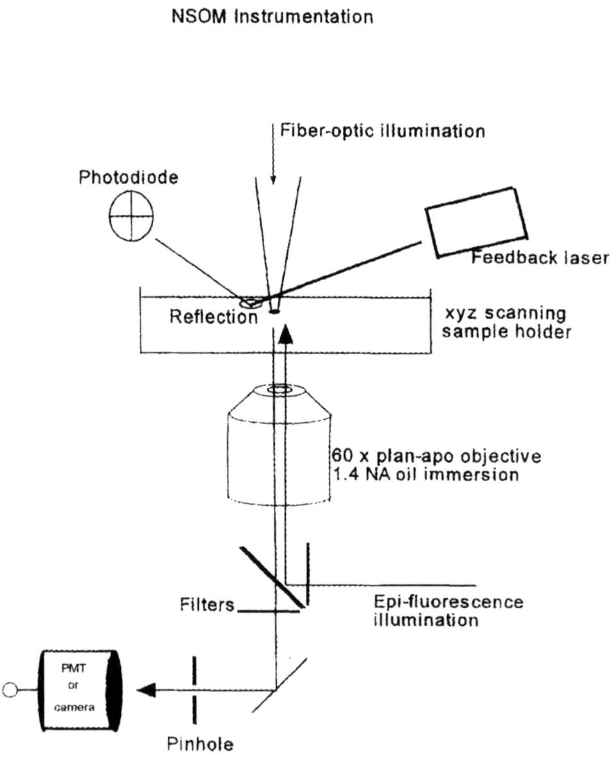

Fig 1. Schematic diagram of equipment configuration

Before the pulled fibers are coated with aluminum, each fiber is coupled to a laser and inspected under an optical microscope. Only fibers that show a single light source at the end are selected for coating. Fibers that show multiple light sources or a single light source distal to the tip of the fiber are discarded. Fibers are glued into capillary mounts (Fig. 2a) and are stored until used. Fig. 2b shows primary taper of the pulled fiber, whereas Fig. 2c shows a 50 nm aperture at the apex of an aluminum coated pulled optical fiber.

NEAR-FIELD TIP FABRICATION

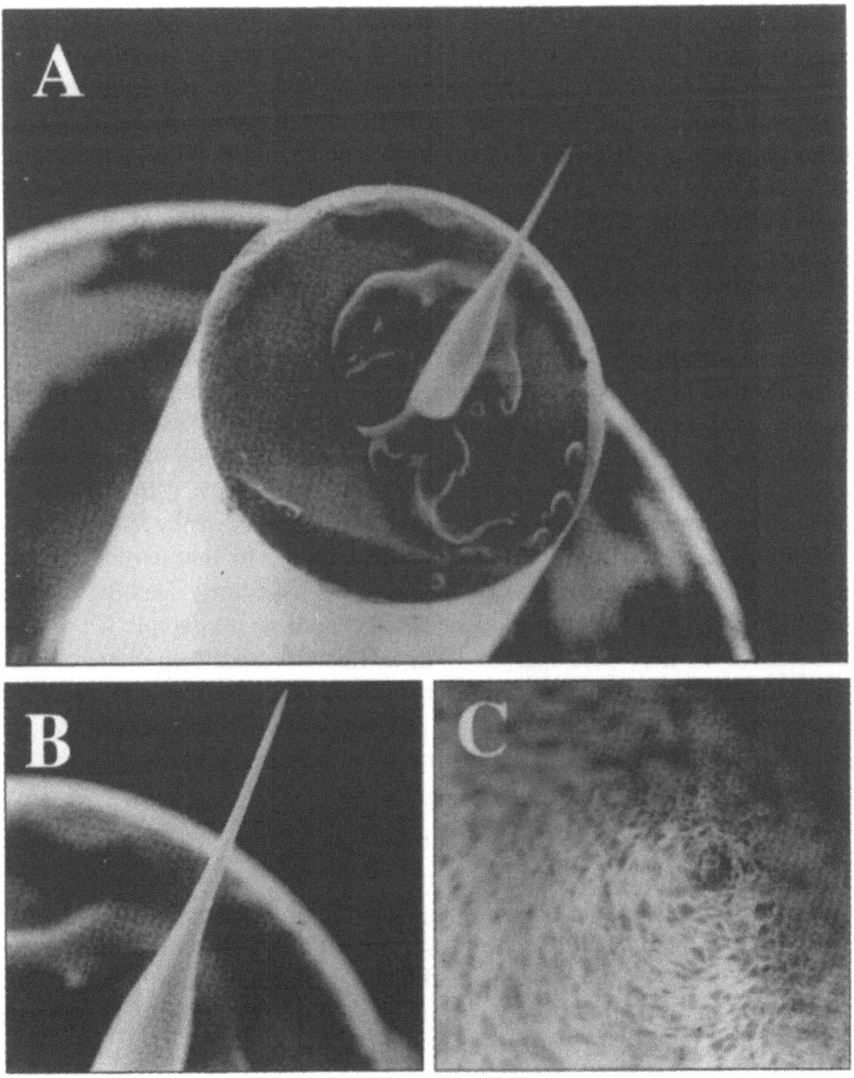

Fig 2. The top panel shows a near-field probe glued into a capillary mount and ready for use. The scale bar represents 1.0 mm. Figure 2c shows the primary taper of the pulled fiber, whereas, Figure 2c shows the sub-wavelength aperture at the apex of an aluminum coated near-field probe. The scale bar represents 200 nm.

4 Shear Force Feedback

The shear force feedback mechanism consists of a fiber oscillating at its resonant frequency using a dither piezo that is coupled to the fiber mount. The resonant frequency is generally between 50-200 kHz depending on the length of the fiber, shorter fibers generally having the higher frequency. As the resonating probe approaches to within a few nanometers of the surface of the sample, shear forces act on the probe causing a drop in dither amplitude and a corresponding phase lag in the dither motion with respect to the signal generator. The shear force sensing optics are also contained in the NSOM head. The sensing optics consist of a laser diode (wavelength 670 nm) that is focused to within 5-30 μm from the apex of the fiber at a shallow angle to produce a shadow of the probe that bisects the laser light on the sample. The reflection of this bisected illumination falls on to a split photodiode that is sensitive to perturbations in the dither amplitude and phase shift of the resonating fiber. By locking in to a specific percentage of dither signal damping we are able to maintain a fixed sample-probe separation. Figure 3 shows the transmitted optical signal (upper trace) and change in the magnitude of the damping of the dither signal (lower trace) as the sample is brought closer to the tip as viewed from right to left. An interesting phenomena often observed in near-field microscopy is that their is an increase in intensity of the transmitted far-field light. In this particular case the increase in intensity is about 10 - 12 % above the base line. This increase in the observed intensity of the illumination source is caused by the sample perturbing the electric field of the near-field aperture. The fact that the specimen interacts with the both the illuminating light as well as the illumination source can cause image interpretation to be more difficult than conventional light microscopy where the sample only interacts with the illuminating light, but does perturb the illumination source.

5 Vibration Isolation

The shear force feedback mechanism maintains the tip sample separation of several nanometers. To maintain this constant separation and prevent the tip from being damaged by "crashing" into the sample, diligent precautions need to be taken. The two major precautions are vibration isolation and optimization of the z-feedback loop. To reduce vibration our instrument is mounted on a vibration isolation table. All necessary placement and manipulation of filters, apertures and shutters are performed prior to attaining feedback. In addition to reducing vibration in the instrument it is also necessary to reduce vibrations of the sample by securely mounting the coverslip. We have found that by using sticky tape to secure the edges of a coverslip to the scanning stage we can operate with a combined instrument-sample Ra noise floor of around 2.00 nm. The Ra value is the mean departure from a mean plane. This noise floor can be significantly improved upon if one uses immersion oil with an oil immersion lens, thus providing mechanical support and presumably vibration isolation directly below the area being imaged (Table I). Once feedback is attained, a repetitive topographic line scan is performed so that the z-piezo responsiveness can be optimized. In general samples with abrupt changes in topography need high gains and lower scan speeds than samples with a more gradual change in topography.

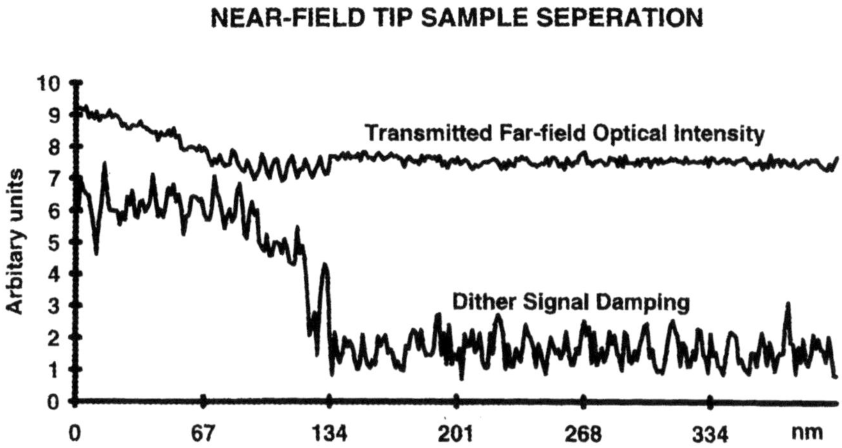

Fig 3. A plot of transmitted optical signal (upper trace) and magnitude of dither signal damping of a near-field probe as the sample is brought closer to the tip (right to left). As the separation between the sample and the near-field tip decreases the damping of the dither signal increases. Furthermore, the amount of light that is propagated to the far-field from the probe is often observed to increase, in this case by about 10 - 12 %. The reason for the increase in illumination is due to the close proximity of the sample perturbing the light at the aperture. Due to such effects care is required in NSOM image interpretation.

Table 1. Noise Floor Values of Instrument (nm)

Paramater	With Oil Immersion	Without Oil Immersion
Ra	1.46	2.5
RMS	2.00	3.3
Max Range	21.5	52.4

Table.1 Compares the combined instrument/sample noise floor with and without the sample being viscously coupled to an oil immersion lens . The values were determined by scanning freshly cleaved mica which is atomically flat. The Ra value represents the mean departure from a mean plane. The RMS value is the root mean square value for the entire scan area. The Max range represents the arithmetic difference between the highest peak and lowest valley for the entire area scanned.

Once the system is in feedback and scanning, the tip can be damaged by the mechanical vibrations of opening a camera shutter or by changing a filter cube. Damage to the tip is readily discernible by an increase in the amount of light being emitted from the tip by a least an order of magnitude and a corresponding loss of optical resolution. Both of these phenomenon can be attributed to an increase in the size of the aperture. In most cases topographic resolution also is reduced. However, occasionally a tip crash that results in the loss of optical resolution can sometimes result in an increased topographic resolution (data not shown). Our interpretation of these observations are that the tip breaks an angle or is jagged, consequently the tip that is profiling the surface is finer than before and would result in an increase in topographic resolution, while the larger optical aperture reduces the optical resolution.

6 Image Analysis

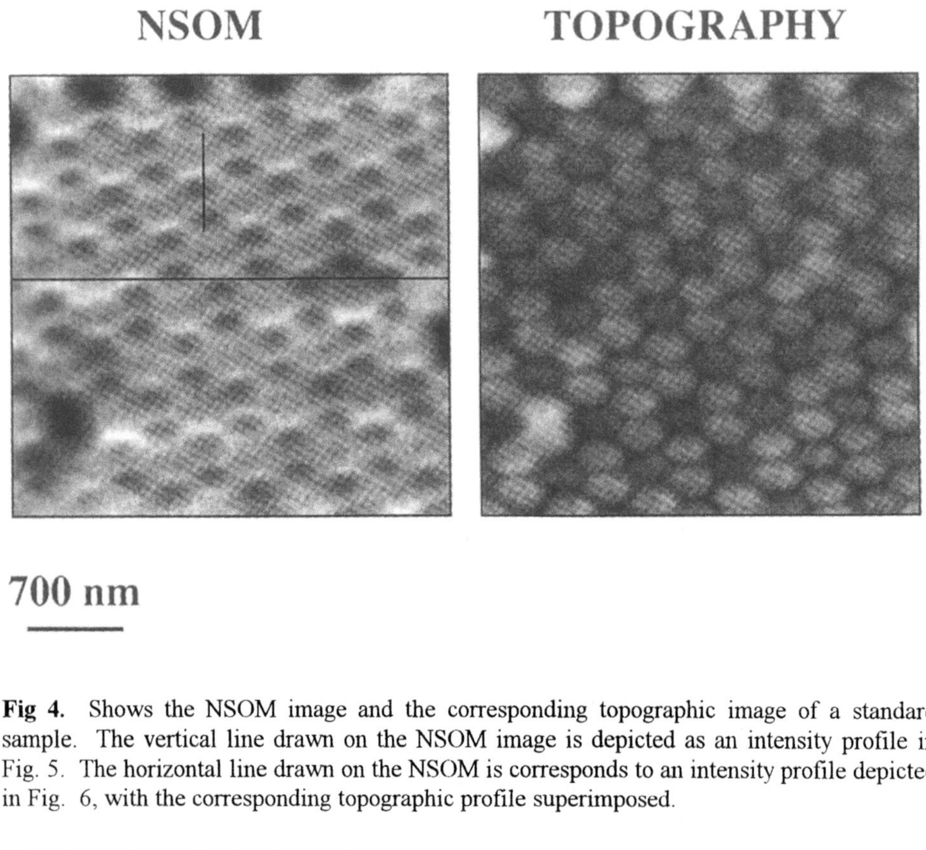

Fig 4. Shows the NSOM image and the corresponding topographic image of a standard sample. The vertical line drawn on the NSOM image is depicted as an intensity profile in Fig. 5. The horizontal line drawn on the NSOM is corresponds to an intensity profile depicted in Fig. 6, with the corresponding topographic profile superimposed.

The shear-force feedback mechanism generates a topographic image at the same time that the optical image is being generated. Figure 4 depicts a topographic and NSOM image of a standard test sample provided by TopoMetrix (Santa Clara, CA)

The sample consists of metal islands on a glass substrate arranged in an hexagonal pattern. The contrast in the topographic image is a function of z-height, the taller the topographic features the brighter they appear. Whereas in the corresponding NSOM image, contrast is a function of transmitted light intensity. A comparison of the images show that the raised metal islands seen in the topographic scan appear dark in the NSOM image. The sub-diffraction optical resolution is most discernible by examining the intensity profile in the NSOM image (Fig. 5).

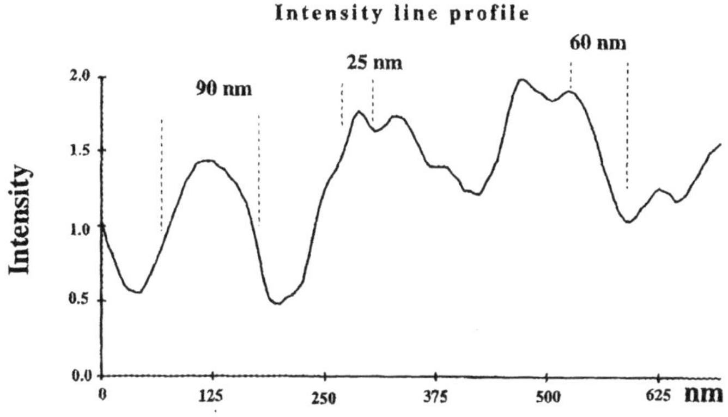

Fig 5. shows the intensity profile of the NSOM image depicted in Fig. 4. The line shows 3 resolution measurements. The highest resolution features are on the order of 25 nm, whereas the distance between minimum and maximum intensity is around 60 nm, and the FWHM measurement for features exhibiting high intensity is around 90 nm.

The highest resolution features that can be resolved on the line profile are 25 nm. While the spacing between maximum and minimum intensity (boundary between glass and metal island) is 60 nm, and the FWHM spacing for areas of high transmitance is 90 nm. Figure 6 shows a superposition of topographic and NSOM line profiles. A comparison of the line profiles show that where the topography is greatest , (i.e. metal islands) the transmitted intensity is lowest. The simplest interpretation of these results is that adsorption of light is the predominant factor affecting lateral optical resolution for this type of sample. An excellent paper on NSOM image interpretation is provided by Valaskovic et al. [5].

However, if one examines an image with abrupt large changes in topography the NSOM image is not so easy to interpret. Figure 7 is a 450 nm thick cross section of muscle prepared for conventional electron microscopy that has been etched with ethanoic NaOH for 5 minutes In both the topographic and NSOM images, the muscle fiber can be seen and the myofibrils discerned. An examination of superimposed topographic and NSOM line profiles (Fig. 8) show that the correlation between topography and transmitted intensity is not as straight forward as it was in Figure 6. While the intensity of the transmitted signal does generally increase when the sample becomes thinner, there are regions where the intensity does not readily correlate with the topography, as denoted by the asterisk.

7 NSOM of Living Systems

The near-field technique has been demonstrated to be most applicable for the analysis of surfaces with no abrupt topographic changes. The challenge in biology is how to apply this technique to samples that are relatively thick, wet, soft and translucent. We have taken the first step in this direction by using a near-field spectroscopic technique on living astrocytes and neurons [7].

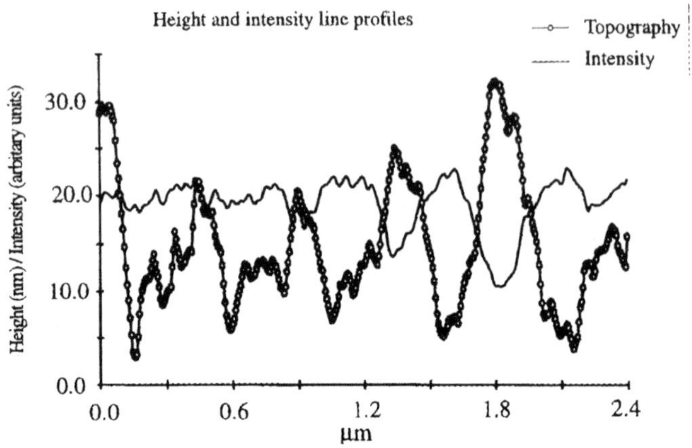

Fig 6. Shows a comparison of NSOM and topographic line profiles for the images in Fig. 4. The comparison shows that the highest features in the topographic image correspond to lower transmitted intensity in the NSOM image. The simplest interpretation of these results is that light adsorption is the predominant contrast mechanism affecting lateral optical resolution for this type of sample.

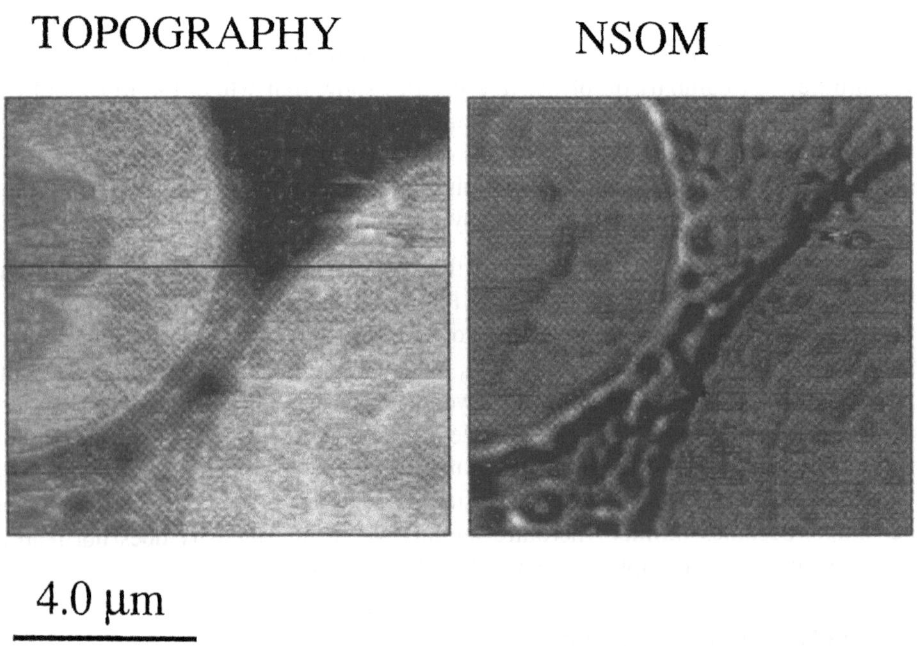

Fig 7. Shows a topographic and NSOM image of an etched thin section of muscle tissue. In both the topographic and NSOM image the muscle can be seen and the myofibrils discerned. The horizontal line drawn across the image denotes the region from which the height and intensity profiles are plotted in Fig. 8.

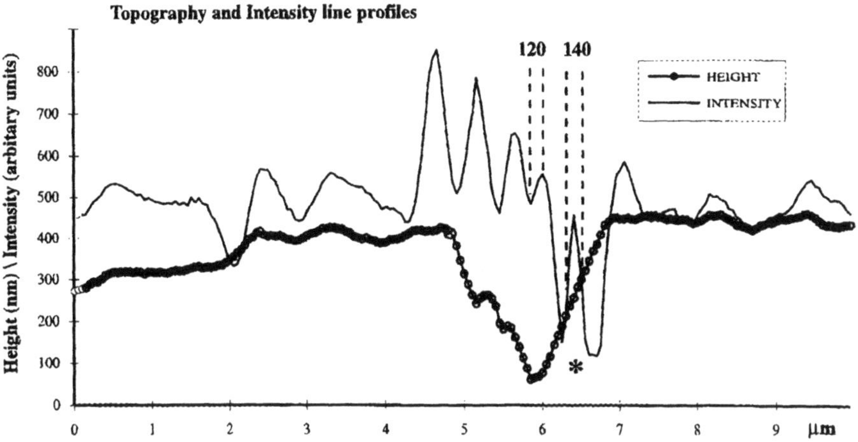

Fig 8. Shows a comparison of the topographic profile and intensity profile of the images depicted in Fig. 7. The comparison shows that in general the intensity increases as the topography decreases. However there are areas of increased transmission that can not be readily reconciled with topographic features, as denoted by the asterisk.

We have been able to demonstrate that the near-field probes can be placed in contact with a living cell without any apparent damage to the cell being observed. We used cultured cortical astrocytes loaded with the fluorescent calcium indicators fluo-3 as test specimens. We found that we could routinely place the probe on the cell without causing damage. We were able to monitor fluorescence intensity as a function of z-separation between the tip and the cell. As the probe approach the cell, the fluorescence signal increased to a maximum upon contact with the cell (Fig. 9). Advancing the tip further, and causing the cell to deform under the pressure of the tip, resulted in a decreased fluorescent intensity. However when the tip was retracted the intensity once again increased to its previous maximum value and then would decrease as the tip pulled away from the cell. By monitoring the fluorescent intensity a simple feedback method (photon-density feedback) can be employed to control the tip sample registration [6]. The left panel in Figure 9 shows an epi-fluorescence image of a neuronal cell loaded with the calcium sensitive indicator fluo-3, whereas the right panel shows the same area with near-field illumination. By using this technique to illuminate a small area of the cell immediately below the plasma membrane we have been able to perform point spectroscopy and measure rapid changes in fluorescence intensity after depolarization of the neuron using 50 mM KCl. The enhanced resolution in near-field technique is only achievable when the sample being illuminated is within the immediate vicinity of the near-field aperture. By contacting the cell with the probe we caused fluorescence to occur in both the near-field and the far-field. The far-field illumination resulted in an overall degradation of our optical resolution. Using optical principles borrowed from confocal microscopy, we incorporated a confocal pinhole in the optical detection pathway. When used in conjunction with a near-field probe a confocal pinhole significantly increases our near-field to far-field ratio. We anticipate that using Near-field Confocal Optical Spectroscopy (NCOS) it might be possible to study signal transduction events, such as vesicle fusion and the kinetics of individual ion channels.

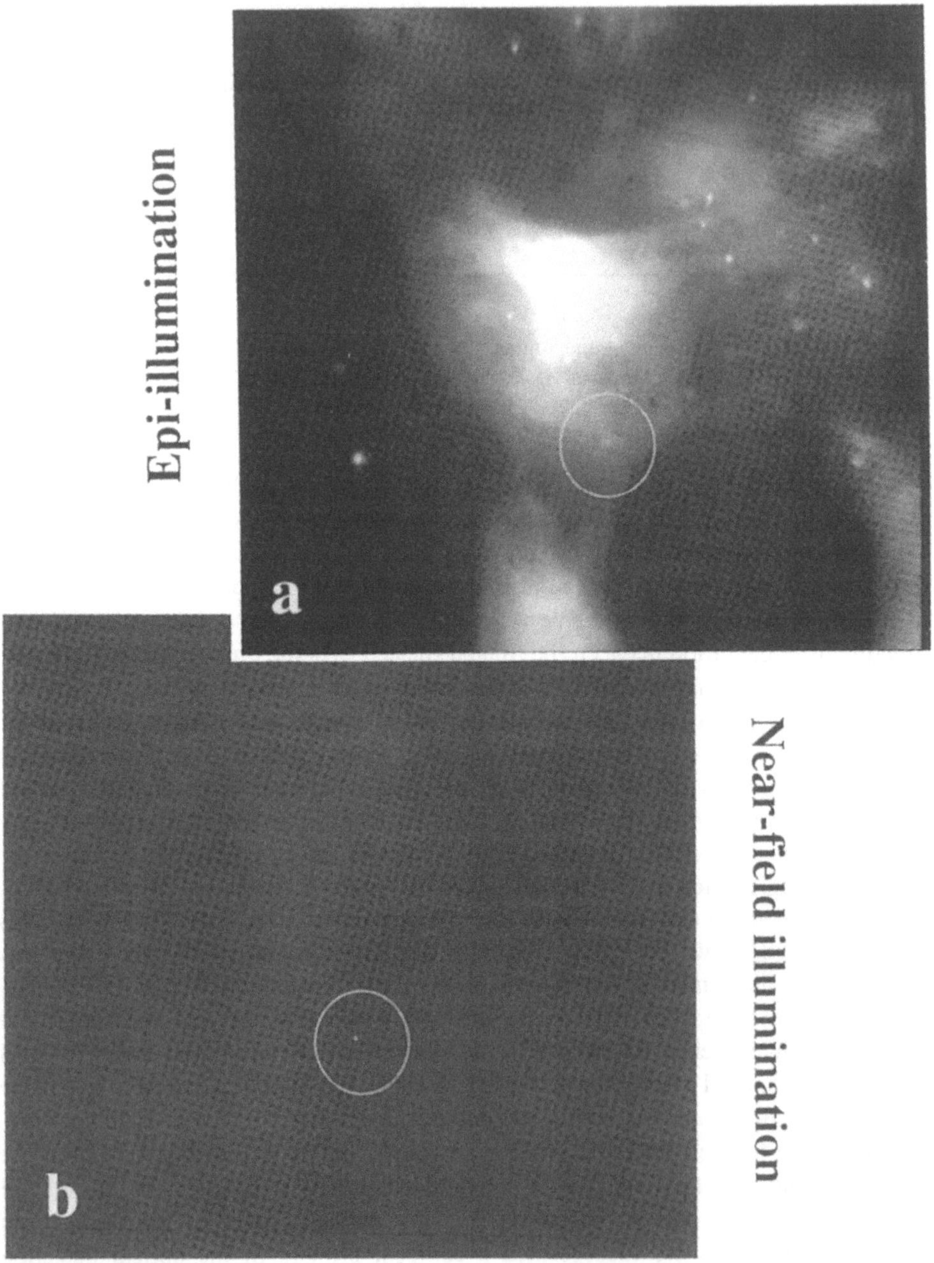

Fig 9. A) shows an epi-fluorescence image of an hippocampal neuron loaded with fluo-3, whereas **B)** shows the same region illuminated by a near-field probe with the epi-fluorescence turned off. The scale bar represents 5.0 μm. Only a small region of the cell beneath the probe is excited and fluoresces.

Acknowledgments

The authors are grateful to the following people: Steffan Kammer of TopoMetrix for providing the images and raw data for Figure 4. Clark Lindgren and Dennis Emery for the preparation of the thin sectioned material depicted in Fig. 7. Bruce Wagner for the SEM images of the pulled optical fibers depicted in Figures 2 and 3. This work was supported by grants from the ISU Laboratory of Cellular Signaling, NIH, the McKnight Foundation and TopoMetrix.

References

1. Synge, E. H. (1928) A suggested method for extending microscope resolution into the ultra microscopic region. Phil. mag. 6: 356-362.
2. Ash, E. A. and Nicholls, G. (1972) Super-resolution aperture scanning microscope. Nature 237: 510-512
3. Pohl, D. W., Denk, W. and Lanz, M. (1984) Optical stethoscopy. Image recording with resolution l/20. Phys. lett. 44: 651-653.
4. Sutter Instrument co. 40 Levroni court, Novato, CA 94949. USA. (personal communication).
5. Valaskovic, G. A., Holton, M. and Morrison, G. H. (1995) Image contrast of dielectric specimens in transmission mode near-field scanning optical microscopy: imaging properties and tip artefacts. J.Microsc. 179: 29-54.
6. Haydon, P. G., Marchese-Ragona, S. P., Basarsky, T., Szulczewski, M. and McCloskey (1996) Near-field Confocal Optical Spectroscopy (NCOS): sub-diffraction optical resolution for dynamic imaging of biological specimens. J. Microsc. (In press).

The Mediatophore, A Molecule Involved in Quantal Neurotransmitter Release

Yves Dunant, Alain Bloc, Philippe Corrèges, Jean Falk-Vairant[1] and Emmanuelle Roulet

Pharmacologie, Centre Médical Universitaire, CH-1211 Genève 4, Switzerland

Abstract

The release of hormones, neurotransmitters or other substances appears to be different in different secreting systems. Exocytosis and release without full fusion are well documented processes. We want to review here the roles played the plasmalemma and by the synaptic vesicles in rapid cholinergic synapses. In several experimental situations, vesicle fusion to the presynaptic membrane does not occur at the same time as transmitter release. The synaptic vesicles are nevertheless essential for the regulation of transmitter release, especially for the dynamic control of calcium homeostasis in nerve terminals. The production of acetylcholine quanta at the plasma membrane seems to require the mediatophore, a homo-oligomer of a 16 kDa proteolipid. The mediatophore is localised at the active zones. Reconstituted in proteoliposomes and in oocytes, or transfected in cell lines, it mimics several characteristic features of neurotransmission *in situ*. Transfection of mediatophore cDNA is sufficient to re-established quantal acetylcholine release in a cell line that was release-deficient.

1 Introduction

The requirements of the release process are quite different in different secreting systems, depending on the nature and properties of the compound to be released. Some substances are lipophilic others are not. In nerve cells, peptides are synthesised far from the site of secretion, rapid neurotransmitters are synthesised locally. Certain hormones have to be delivered in substantial amounts since they will be diluted in the whole organism; others act at a short distance on highly sensitive receptors. Rapid neurotransmitters are emitted in a jet-like manner; they operate at high concentrations with an extremely fast time course. It is easy to understand, therefore, that not only the process of release but also the underlying regulations might exhibit a variety of distinct mechanisms in different secreting systems.

Diffusion through the plasma membrane. Liposoluble substances such as NO or steroid hormones can freely diffuse across biological membranes. Thus, they do not need any specific releasing machinery. It is the rate of synthesis which in this case

[1] Deceased, November 1994

NATO ASI Series, Vol. H 100
Neurotransmitter Release and Uptake
Edited by Şakire Pöğün

regulate the rate of secretion. The time course of these regulations is thus expected to be relatively slow (seconds or minutes).

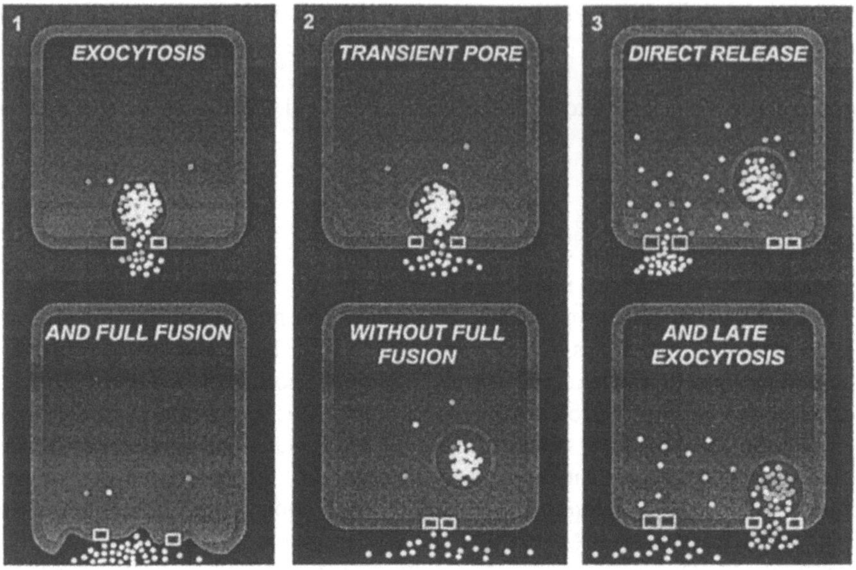

Fig. 1. Three different modes of vesicle assisted release. (1) Exocytosis with full fusion; (2) «kiss and run», or exocytosis without full fusion; (3) rapid release through the mediatophore oligomer, followed by delayed exocytosis. See explanations in the text.

Exocytosis with full fusion. This is the best described mechanism (see Burgoyne and Morgan, 1995; De Camilli, 1995; Südhof, 1995, among others), which accounts for a variety of secretions from practically all types of cells. As pictured in the first panel of Fig. 1, a granule or vesicle undergoes a full fusion with the plasma membrane and eventually releases its entire content into the extracellular space. Together with the hormone or mediator substance, other compounds are expelled, such as associated proteins or enzymes, ATP and various ions. Intragranullar calcium is also released in the process. As a matter of fact, accumulation of calcium into the organelle might represent a signal for granule maturation and subsequent fusion (Nicaise *et al.*, 1992). Exocytosis is preceded by establishment of a fusion pore that permits the most diffusible intragranular compounds to escape before expulsion of the core of the granule (Breckenridge and Almers, 1987). The whole process including vesicle recycling is also relatively slow (seconds or more). Full exocytosis can be regarded as a wasteful event for the cell since the whole content of the organelle is lost and the membrane of the granule has to be retrieved by endocytosis or another process before complete recovery.

"Kiss and run" : transient and partial release from a secretory granule. This type of secretion might be of major physiological significance. It was first proposed to account for observations made at the neuromuscular synapse (Ceccarelli and Hurlbut, 1980). Recently the process was more directly analysed by combining capacitance measurements and direct assessment of mediator release (Alvarez de Toledo *et al.*, 1993; see Neher, 1993). Partial release implicates the formation of a fusion pore between the interior of a granule and the extra cellular space. Diffusible intragranular substances are released through the pore but this is not followed by full exocytosis. The pore closes and the granule quits the plasma membrane without undergoing full fusion (panel 2 of Fig. 1). Such a process is much less wasteful since a single granule can be used for several release runs without loosing its whole structural equipment. It might explain why after a given stimulation of the adrenal medulla, the number of granules do not decrease but the density of their electron opaqueness diminishes, so that they look "empty" (Nahas *et al.*, 1967). It is not known whether the pore displays some specificity with respect to the substance to be released. The time course of such a process is expected be intermediate (milliseconds to seconds).

Release through a plasmalemmal device (mediatophore). Neurotransmitter emission at the neuromuscular junction, in the electric organ and in other rapidly operating synapses is a very fast process (less than 0.1 ms), which can be repeated at a high rate. The action potential, which is an electrotonic impulse, is converted in the axon terminal into a flash-like chemical impulse. Experiments on the *Torpedo* electric organ and other preparations have pointed toward a plasmalemmal protein (the mediatophore) able to release the neurotransmitter acetylcholine (ACh) in a quantal manner. However, the synaptic vesicles are also essential for the regulation of release in fast synapses. They contain a reserve stock of neurotransmitter, they insure protein traffic at the membrane and they are most probably involved in the dynamic control of intraterminal Ca^{2+}, which is essential the proper working of the synapse. This proposal (panel 3 of Fig. 1) was controversed. It is nevertheless supported by an increasing number of observations, some of them will be described thereafter.

2 The Vesicle Hypothesis for Quantal Acetylcholine Release

The vesicle hypothesis was formulated by De Robertis and Bennett (1955) and Del Castillo and Katz (1957) as an explanation for the quantal character of transmitter release, first observed by Fatt and Katz (1952). At the time of their finding, the spontaneous discharges giving rise to miniature end-plate potentials were attributed to activation of some specialised "transport molecule" located at the presynaptic membrane (Fatt and Katz, 1953), but the discovery soon after of the synaptic vesicles in the axoplasm of nerve terminals made it more likely the hypothesis that each Ach quantum would be pre-formed in one vesicle and release accomplished by vesicle fusion to the membrane. Thus the process of transmitter release would be the same as the "exocytosis", or "reverse pinocytosis", which was just described at that time for various gland cells (see reviews in Katz, 1969; MacIntosh and Collier, 1976; Israël *et al.*, 1979; Ceccarelli and Hurlbut, 1980; Tauc, 1982; Dunant, 1986, among others).

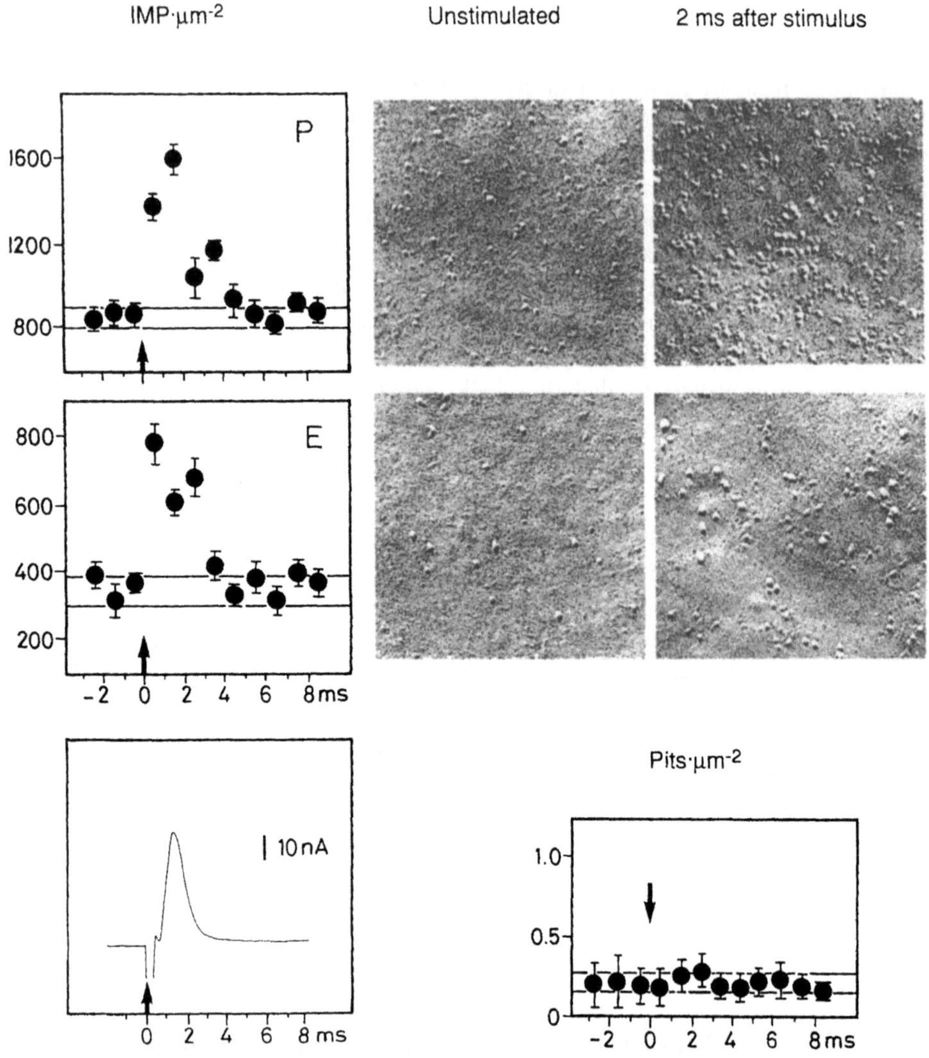

Fig 2. Ultrastructural changes accompanying the passage of a single nerve impulse at the *Torpedo* nerve-electroplate junction. Up to twenty tissue fragments were stimulated at 1 ms time intervals and cryofixed all at the same time. The specimens were subjected to cryofracture at the most superficial layers and replica were examined at the electron microscope. Transmission of a single nerve impulse (bottom trace) coincided with an abrupt rise in the number of intramembrane particles in both the P-face (upper row) and the E-face (lower row) of the presynaptic membrane. The density of exo-endocytosis pits did not rise during transmission of the single impulse (lower graph, right side). From Muller et al. (1987), with modifications.

Quantum size and vesicle content. The vesicular hypothesis received a strong support when it was demonstrated that synaptic vesicle of mammalian brain did contain ACh (De Robertis *et al.*, 1961; Whittaker *et al.*, 1964), and furthermore after isolation of vesicles from the *Torpedo* electric organ, a modified neuromuscular system with a purely cholinergic inervation (Israël *et al.*, 1968). However, attempts to correlate the quantum size with the ACh content of individual vesicles have been disappointing. The size of one quantum at the endplate has been evaluated to be 6000 - 10000 ACh molecules (Kuffler and Yoshikami, 1975). The amount of ACh present in a single vesicle has not been assessed yet by direct measurement at the neuromuscular junction. In the electric organ, the synaptic vesicles have been reported to contain 100000 - 200000 ACh molecules (Ohsawa *et al.*, 1879). In these synapses, vesicles are twice the size - and therefore height times the volume - of those of the neuromuscular junctions. It was therefore of interest to know whether the quantum size was also eight times larger in the electric organ. The response was negative. Transmitter release is indeed quantal in the *Torpedo* electric organ, but one quantum is composed of only 7000 - 10000 ACh molecules (Dunant and Muller, 1986). There is therefore enough ACh in a *Torpedo* vesicle to generate 10 or 20 quanta. Thus, despite of the large difference in vesicle size, the quantum size is practically the same at the endplate and in the electric organ.

The problem is even more critical if one considers that a quantum is most probably composed of subunits. Indeed, recordings made at the endplate with an appropriate signal to noise ratio revealed the occurrence of events smaller than the quantum. It was proposed therefore that the quantum was itself composed of discrete subunits (Kriebel and Gross, 1974; Kriebel *et al.*, 1976). Also in the *Torpedo* electric organ, the quantum has been shown to be composed of subunits. The release of a mean of ten subunits is well synchronised in about 85% of spontaneous miniature currents but desynchronisation in the remaining signals reveals the quantum substructure (Muller and Dunant 1987; Girod *et al.*, 1993). The «most elementary» component of transmitter release both in the endplate and the electric organ has therefore been estimated to contain about 1000 ACh molecules. This is one hundred times less than the content of a *Torpedo* vesicle. More recently, quantal release in the CNS have been at the centre of intensive research, reviewed by the groups of Sakman, Stevens, Stratford, Korn, Redman and others (see Chap. 16-20 of Stjärne *et al.*, 1994). Quantum sizes in CNS synapses exhibit a high degree of variability, but most reports point to quanta arising from the opening of 100 receptors or less, thus requiring just a few hundreds of transmitter molecules.

Vesicular ACh, a reserve pool of transmitter. The fate of vesicular ACh and extravesicular ACh was analysed as a function of time with several modes of nerve stimulation. From our very first experiments on the *Torpedo* electric organ, we found that the vesicular ACh pool remained surprisingly stable, even when stimulation was continued until exhaustion of transmission. In contrast, the cytosolic pool of ACh was used and renewed from the very beginning of activity, as seen in experiments where rapid-freezing was used at 10 ms time intervals to arrest processes in the tissues as rapidly as possible. The preferential utilisation of cytosolic ACh before the vesicular pool was demonstrated with the intact preparation as well as on isolated nerve terminals (Dunant *et al.*, 1972; Israël *et al.*, 1979; Dunant *et al.*,1982; Dunant and Israël, 1985).

Vesicle fusion and quantal ACh release. Another strong support to vesicular hypothesis came from experiments by Heuser *et al.* (1979), confirmed by the group of Ceccarelli (Torri-Tarelli *et al.*, 1985). They examined frog neuromuscular junctions that were subjected to rapid freezing at the time of impulse transmission. When transmitter release in the preparation was potentiated by 4-aminopyridine, both groups found numerous vesicle openings appearing in the membrane of motor nerve terminals during - and also after - impulse transmission. However, when the experiments were carried out in the absence of 4-aminopyridine, the picture was different. Transmission of a single impulse was signalled by clusters of intramembrane particles (IMPs) at the active zones. The clusters were interpreted as late images of a supposed exocytosis that would have been missed because they would have been to fast to be captured by the cryofixation (Heuser *et al.*, 1979).

Ultra-rapid freezing was also performed at the nerve-electroplate junction of *Torpedo*. Here, the characteristic change observed in the nerve terminal membrane at the precise time of transmitter release was a transient occurrence of large IMPs. The phenomenon lasted for only 2-3 ms during the passage of a single nerve impulse; it was not accompanied by vesicle openings in the presynaptic membrane (Fig. 2 and Muller *et al.*, 1987). The same IMP change lasted much longer in 4-aminopyridine treated *Torpedo* synapses, corresponding to a much longer time course of acetylcholine release under this condition (Garcia-Segura *et al.*, 1986). Also synaptosomes were rapidly frozen in a variety of conditions, all of them triggered the release of ACh; as in intact synapses, the common ultrastructural change was a transient formation of large IMPs in synaptosome membrane (Israël *et al.*, 1981). Large IMPs also occurred transiently in the membrane of proteoliposomes equipped with the constituents of the presynaptic plasma membrane (Israël *et al.*, 1984); or with the isolated mediatophore (Brochier *et al.*, 1992, see below). The latter results, performed in the absence of synaptic vesicles, were important in that they excluded the hypothesis that the new IMPs were vesicular particles left in the plasma membrane after an exocytosis that would have been missed by the freezing technique.

Dissociation between transmitter release and vesicle fusion. A time coincidence between transmitter release and vesicle fusion was only observed in exceptional situations. In recent experiments, the two processes could even be frankly dissociated. At the frog neuromuscular junction, synaptic vesicles could be stained with a fluorescent marker; then the rate of "destaining" was monitored together with the release of transmitter during the course of nerve stimulation. Under control conditions, the kinetics of destaining was roughly parallel to the rate of release for 1-2 minutes (Betz and Bewick, 1993). However, staurosporine, a protein kinase C inhibitor, fully abolished exocytosis but left synaptic transmission unaltered (Henkel and Betz, 1995).

A clear dissociation between ACh release and vesicle fusion was also demonstrated in the *Torpedo* electric organ submitted to a brief nerve stimulation (100 Hz during 12 s). Transmitter release took place during the first 10 s of the tetanus but images of vesicle fusion started to occur only after the end of activity, culminating at 1 min of the post-tetanic recovery period (Parducz *et al.*, 1994). A still more striking dissociation was found in Zn^{2+}-treated synapses. Incubation of small pieces of *Torpedo* electrogenic tissue in 250 μM $ZnCl_2$ irreversibly blocked the release of ACh but did not cause any significant alteration in the fine structure of resting nerve-electroplate junctions. Electrical stimulation at this stage resulted in Ca^{2+} accumulation and massive vesicle fusion's, in spite of the fact that no transmitter was released (Fig. 3). Many vesicles were lost; others were seen in different phases of fusion with the presynaptic membrane. This experimental schedule created an unusual situation where the exo-endocytotic activity of synaptic vesicles could be activated independently from the release of transmitter (Fig. 4; Parducz *et al.*, 1996).

Clostridial toxins and quantal release. Botulinum (BoTX) and tetanus (TeTX) toxins were recently found to be metalloproteases whose substrates are proteins establishing a complex between the synaptic vesicle and the plasma membrane. TeTX and BoTX B attack the vesicular VAMP-synaptobrevin; BoTX A cleaves SNAP-25 and BoTX C, syntaxin (Schiavo *et al.*, 1992; Blasi *et al.*, 1993). Thus, any exocytosis is expected to be abolished by the clostridial toxins. Indeed, in synapses intoxicated by botulinum neurotoxins, the release evoked by activation of the afferent nerve fibre is practically suppressed. A flaccid paralysis occurs and death will follow if respiratory muscles are affected.

What is less well known, but indeed established, is that some quantal transmitter release persists during the whole period of intoxication, which can last for several weeks. Spontaneous transmitter release recorded from intoxicated synapses is composed mainly of miniature potentials of small amplitude (Harris and Miledi, 1971; Kriebel *et al.*, 1976). Also larger events, sometimes of giant size, were recorded for a very long time at endplates where evoked release had been practically suppressed by botulinum toxin (Thesleff *et al.*, 1983). Moreover, evoked transmitter release could be partially restored by 4-aminopyridine after BoTX A intoxication. As for BoTX B or TeTX, they caused a desynchronisation of evoked release (Gansel *et al.*, 1987; Molgo *et al.*, 1990). The latter result suggests that VAMP-synaptobrevin might be necessary for the synchronization of quantal release.

Thus, by cleaving VAMP, SNAP 25, syntaxin and possibly other proteins, the clostridial toxins somewhat disorganize quantal release. Above all, they uncouple the release mechanism from its natural trigger, which is an abrupt elevation of calcium at the inner face of the plasma membrane. Release persists, however; at a low level. Therefore, the final pathway for quantal release still operates, but it is deprived of regulation. Therefore, blockade of vesicle fusion does not suppress all quantal release.

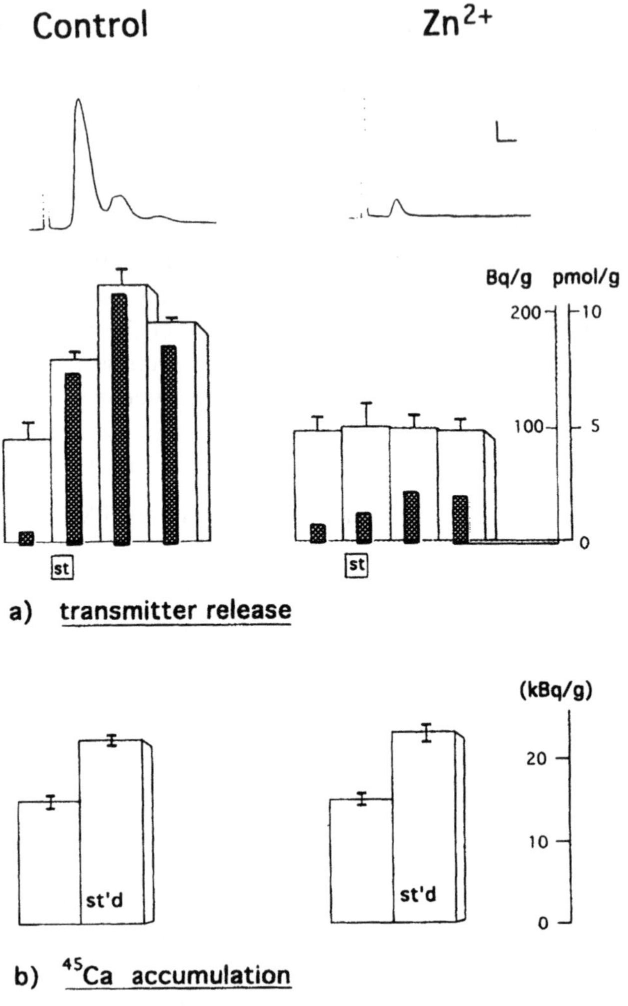

Fig 3. Inhibition of transmitter release in the *Torpedo* electric organ by zinc. *Upper record traces:* Effects of a 3 h incubation in the presence of 250 μM $ZnCl_2$ on the electrical responses generated by an isolated prism (column of electroplates). Calibration: 1 V and 2 ms. **a)** Inhibition of ACh release. Small fragments of electrogenic tissue were incubated overnight with ^{14}C-acetate and washed before the stimulation run. One min samples of the solution superfusing the prisms were taken before, during and after a 30 s stimulation at 10 Hz («St»). *Open columns*: amount of choline measured by luminescence in the samples (mean ± S.E.M. N=4). *Filled bars*: amount of ^{14}C-acetate in the same samples. Stimulation-evoked release was practically suppressed in Zn^{2+} -treated -synapses. **b)** Zn^{2+} did not prevent stimulation-dependent calcium accumulation in the synapses. The first column shows the cellular ^{45}Ca content of unstimulated prisms, the second column («St'd») that of prisms that were stimulated at 100 Hz for 12 s (mean ± S.E.M., N=12).

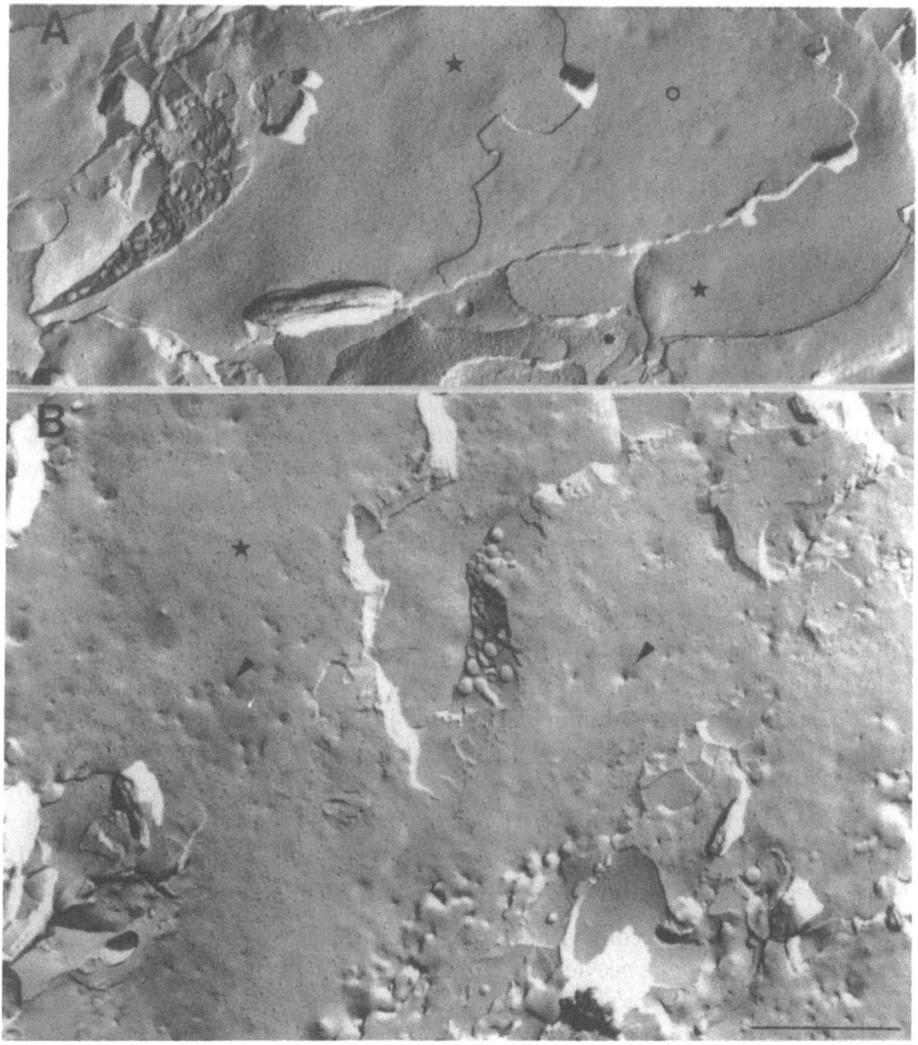

Fig 4. Freeze-fracture replicas of electrogenic tissue treated by Zn^{2+} (250 μM) for 2 h, and then fixed with paraformaldehyde and submitted to freeze-fracture following the «flatten-peeled» procedure (Corrèges *et al.*, 1996). Vast surfaces of presynaptic nerve terminal plasma membrane (P-face) were exposed (★). Fragment of Schwann cell membrane (O) and of receptor-rich postsynaptic membrane (✹) are also visible. **A**: Unstimulated Zn^{2+}-treated tissue. **B**: Zn^{2+}-treated specimen stimulated at 10 Hz for 2 min (as shown in Fig. 3, no transmitter was released under this condition). Note the extremely high density of pits (arrow heads) that perforate the P-leaflet of the presynaptic membrane. Bar = 1 μm.

3 The Mediatophore and ACh Release

The mediatophore is a proteolipid which was isolated from the plasma membranes of *Torpedo* cholinergic nerve endings (Israël *et al.*, 1984; 1986). The mediatophore was first identified by its capability to release ACh in a Ca^{2+}-dependent manner. For this assay, the candidate proteins were isolated from purified presynaptic membranes and incorporated into the membranes of liposomes filled with ACh. Transmitter release was then elicited by the successive addition of a Ca^{2+}-ionophore and of Ca^{2+} to the proteoliposome preparation.

Examined at the electron microscope, the isolated mediatophore appears as an oligomeric structure of 8 nm diameter with a central hole. The molecular weight of the native mediatophore is approximately 200 kDa. On SDS solubilisation and gel electrophoresis, the mediatophore is resolved as a homo-oligomer of a subunit of 16 kDa (Israël *et al.*, 1986; Birman *et al.*, 1986). Sequencing and molecular cloning of the 16 kDa subunit of the *Torpedo* mediatophore (Birman *et al.*, 1990) showed a high degree of homology with a 16 kDa subunit found in the membrane segment of the V-ATPase (Mandel *et al.*, 1988; Nelson, 1989), and also with ductin, a lipoprotein found in invertebrate gap-junctions (Finbow *et al.*, 1991).

The mediatophore has been localised histochemically by antibodies recognising the native (oligomeric) form of the proteolipid. A marked concentration of mediatophores is present in the presynaptic membranes of nerve terminals in the *Torpedo* electric organ. Gold immunolabelling forms small aggregates at sites facing the postsynaptic invaginations, a situation which probably corresponds to the active zones of the motor endplates (Brochier *et al.*, 1993). An extremely high concentration of mRNA coding for the mediatophore subunit has been found by *in situ* hybridisation in the brain of *Torpedo,* localised in the neurones of the electric lobe, that is, in the cholinergic cells innervating of the electric organs.

In addition to the proteoliposome assay described above, the mediatophore was reconstituted in *Xenopus* oocytes that were injected with the total set of poly $(A)^+$ mRNAs extracted from the neurones of the *Torpedo* electric lobe. When the mediatophore was expressed, ACh release was expressed as well (Fig. 5 and Cavalli *et al.*, 1991). To see whether the mediatophore was specifically required to support release, antisense oligonucleotides directed against segments of the mediatophore mRNA were injected together with the total poly $(A)^+$ mRNAs of the electric lobe. The antisense probes did not affect the rate of ACh synthesis, but they suppressed both the evoked ACh release and the expression of the mediatophore in oocyte membranes (Cavalli *et al.*, 1993). Oocytes were also primed with mRNA of the mediatophore 16 kDa subunit alone (Leroy *et al.*, 1994). Interestingly enough, on injection with the 16 kDa mRNA alone, more proteolipid was synthesised but the oocytes were less efficient in releasing ACh (as compared with oocytes injected with the total poly (A^+) mRNAs). It turned out that following priming with the pure mRNA most of the 16 kDa phospholipid accumulated in intracellular organelles; in contrast, with the total mRNAs substantial amounts of mediatophore were addressed

mediatophore, thereby activating and desensitising transmitter release. Zinc was more potent than calcium for both effects (Dunant *et al.*, 1996).

Intramembrane particles. Brochier *et al.* (1992) incorporated high amounts of pure mediatophore into ACh-containing liposomes. Transmitter release was elicited by successive addition of A23187 and calcium. As a result, large intramembrane particles were transiently seen in the proteoliposome membrane. Therefore the fleeting particle change caught by rapid freezing of the synapses during the act of release might well occur from an abrupt modification of the mediatophore molecules during the brief time of their activation.

Utilisation of the cytosolic pool of ACh. Oocytes primed with cholinergic neurone mRNAs accumulated substantial amounts of ACh. About 40% of this store was releasable in a single run, with either KCl or A23187 (Cavalli *et al.*, 1991). Did they utilise cytosolic ACh for immediate release? or the ACh stored in their own vesicular organelles? To answer this question, primed oocytes were disrupted in the presence of acetylcholinesterase and the amount of ACh content entrapped in vesicles was determined. Such bound ACh was at the limit of detection (about 1-2 % of the total content). Thus, almost all the oocyte ACh was either cytosolic or contained in some fragile compartment, clearly different from neural synaptic vesicles (Cavalli *et al.*, 1993). This was in line with previous results on the electric organ *in situ*, where the free pool of ACh was shown to be the first to be used and renewed on stimulation.

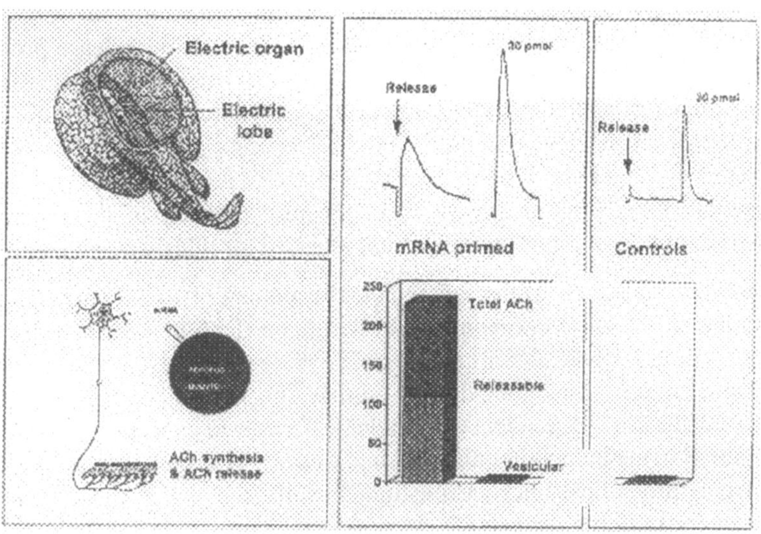

Fig 5. Expression of acetylcholine (ACh) release in *Xenopus laevis* injected with cholinergic cell mRNAs. Oocytes were primed with poly(a$^+$) mRNAs extracted from the *Torpedo* electric lobe. Four days later they were shown to contain a substantial amount of transmitter and to release it on KCl depolarization (arrow). Control oocytes did not contain ACh and did not release the transmitter. Apparently, the ACh contained in primed oocytes was entirely cytosolic since disruption of the oocytes in the presence of cholinesterase caused hydrolysis of the whole transmitter content.

to the plasma membrane (Leroy and Meunier, 1995). Another protein present in cholinergic neurons is probably be necessary for addressing the 16 kDa to the plasmalemma where the homo-oligomer mediatophore has to be assembled.

4 Functional Characteristics of the Mediatophore, as Compared to the Natural Working of the Synapses

Calcium activation. As a trigger for transmitter release in natural synapses, Ca^{2+} acts at a high concentration for a brief moment at delimited spots (microdomains) localised at the inner face of the presynaptic membrane (see G. Augustine, this symposium; Llinas *et al.*, 1992). The activation ACh release by the mediatophore was analysed as a function of the Ca^{2+} concentration in different preparations, which were : i) *Torpedo* synapses *in situ*, ii) synaptosomes prepared from the electric organ, iii) proteoliposomes reconstituted with the purified mediatophore, iv) oocytes primed with mediatophore mRNA. The curves describing the amount of transmitter release as a function of Ca^{2+} concentration could be superimposed with the different systems, indicating that the mediatophore has the precise Ca^{2+} sensitivity expected for the mechanism which is at work in the native synapses (Cavalli *et al.*, 1991).

Calcium desensitisation. What is less known, but fully documented, is a second action of Ca^{2+}, namely, the desensitisation of the release mechanism. Desensitisation displays a relatively slow time course. It has been described *in situ* at the squid giant synapse and in other junctions (Katz and Miledi, 1969; Adams et al, 1985). The process takes some time to develop (about half of a minute), but requires only micromolar Ca^{2+} concentrations. Calcium-desensitisation of the release process was well analysed with synaptosomes isolated from the *Torpedo* electric organ. In proteoliposomes reconstituted with the pure mediatophore, desensitisation was also demonstrated; its dependency on time and Ca^{2+} concentration was the same as in synaptosome experiments (Israël *et al.*, 1987). Therefore, the Ca^{2+}-induced desensitisation of transmitter release is an intrinsic property of the mediatophore molecule; the support of other proteins is not required for this phenomenon.

Pharmacology. Basically, drugs can change the amount of ACh emitted by a given nerve impulse in two different manners: by acting on the Ca^{2+} influx, or by altering the release process itself. A variety of drugs of the second category act on the mediatophore directly, since they display similar effects i) on synaptic transmission *in situ*, ii) on the Ca^{2+}-dependent ACh release from synaptosomes, iii) on oocytes primed with the mediatophore mRNA, and, most important, iv) on the release elicited from proteoliposomes reconstituted with the pure mediatophore. Of particular interest are the compounds derived from cetiedil. Most of them are potent blockers of ACh release, while a particular derivative potentiates the release. Either with the pure mediatophore or with intact nerve terminals, the inhibitors depressed ACh release, while the derivative enhanced it (Morot-Gaudry-Talarmain *et al.*, 1989). Zinc was recently found to affect synaptic transmission by exerting a dual action on the mediatophore. Like calcium, zinc both activated and desensitised the

5 Mediatophore and Quantal ACh Release

Quantal release evoked by stimulation of cultured cell lines. The question to be solved now is whether the mediatophore is necessary for the pulsatile and quantal character of transmitter release. This could not be addressed in previous experiments where the release was measured with biochemical techniques whose time resolution was at the best in the range of seconds. By using a modification of the *Xenopus* myoball approach (see M-m. Poo, this symposium; Evers *et al.*, 1989), Jean Falk-Vairant recently developed a procedure for investigating evoked ACh release from cells previously loaded with this neurotransmitter (Falk-Vairant *et al.*, 1994, 1996a, 1996b). Different types of cells can indeed take up ACh from the external medium under appropriate conditions (Israël *et al.*, 1994). After dissociation, ACh-filled cells were transferred into a culture of *Xenopus* embryonic myocytes. One spherical myocyte (myoball) was patched in a whole-cell configuration and served as a sensitive and rapid ACh detector. Individual ACh-filled cells were then moved into contact with the myoball and held by a fine-tipped glass pipette through which they could be electrically stimulated. Evoked ACh release was detected as nicotinic inward currents recorded from the voltage-clamped myoball. Under these conditions, certain cell lines did release the neurotransmitter in a quantal and pulse-like manner, while other lines were unable to do so. Surprisingly, some non-neuronal cell lines such as glioma C6-Bu-1 cells or LTK fibroblasts were quite efficient for quantal release. Following a treatment with dibutyryl-cAMP for 3-6 days, LTK fibroblasts changed their shape and acetylcholine release was greatly enhanced. The currents recorded in myocytes upon electrical stimulation of differentiated fibroblats exhibited a protracted time course, suggesting that the release sites can be maintained in an open state for a very long time (Falk-Vairant *et al.*, 1996c).

Restoration of quantal ACh release upon transfection with mediatophore cDNA. Using the above technique, we found that cells of the mouse neuroblastoma N18TG-2 line were unable to release ACh. Actually, parallel work indicated that these cells could establish synaptic contacts in coculture with myocytes, but these contacts were not functional : no quanta could be recorded from the myocyte, even when ACh synthesis had been restored by transfection with choline acetylase cDNA (Zhong *et al.*, 1995). We found in addition that N18TG-2 cells contained only very low levels of mediatophore in their membranes. Were them release-deficient because they lacked the mediatophore? To test this, transfection of the N18TG-2 cells was performed by using a plasmid coding for the *Torpedo* mediatophore 16kDa subunit. In the transfected cells, the *Torpedo* protein was expressed in the plasma membrane and quantal ACh release was re-established. On electrical stimulation, they generated in the adjacent myoball evoked currents that were curare-sensitive, calcium-dependent and displayed discrete amplitude levels, as in naturally-occurring synapses (Fig. 6 and Falk-Vairant *et al.*, 1996a; 1996b).

Individual transfected cells were able to produce a total of 5 to 77 evoked responses. The amplitude of evoked currents ranged from 0 (failures) to 1500 pA. The quantal nature of release was analysed in a series of 55 consecutive currents applied to a cell of the T7 clone (Fig. 6). From the Poisson low (Katz, 1969) a mean quantal content of m=2.4 was calculated on the basis of the number of failures (5/5). The predicted number of responses composed of 0, 1, 2, 3, or more quanta was compared with the recorded distribution. An acceptable correspondence was found for a quantal size of 200 pA. From an experiment to another, quantal sizes of 150-200 pA were calculated. These values compare well with those found at newly formed *Xenopus* neuromuscular junction (94 pA) (Evers *et al.*, 1989) and also for the mean size of miniature currents generated by *Torpedo* synaptosomes (65 pA) (Girod *et al.*, 1992), which were the starting material for the present study. One can estimate that opening of one receptor produces a current of 1-2 pA under experimental conditions. Now, two ACh molecules are needed to open one receptor and > 50 % of transmitter molecules may be wasted in the process. Hence, the quantal size measured here should correspond to 400-1000 ACh molecules. This is less than the quantum found in the adult neuromuscular or nerve- electroplate junctions (6-10 000) (Kuffler and Yoshikami, 1975; Dunant and Muller, 1986) but close to the value described at the embryonic junctions (Evers *et al.*, 1989).

6 Relationships of Mediatophore with Synaptic Vesicle Proteins

How did mediatophore transfection restore release? In the discussion of this finding, it is recalled that the original (release-deficient) N18TG-2 cells had very low levels of the 16 kDa protein in their plasma membrane. We believe that transfection with a plasmid encoding the 16 kDa *Torpedo* protein could have over-expressed the subunit and overcome the defect. The cells were now provided with enough *Torpedo* 16kDa subunits to assemble mediatophore oligomers.

Another possibility would be that mediatophore transfection has switched on the expression of other missing proteins necessary for exocytosis. The release deficient N18TG-2 cell line is actually deficient for several proteins like choline acetyltransferase (Zhong *et al.*, 1995), synaptophysin and the vesicular ACh transporter (unpublished data). Since we know that choline acetyltransferase and the transporter are co-regulated (Berrard *et al.*, 1995), the easiest way to test the hypothesis was to check the choline acetyltransferase activity before and after mediatophore transfection. The activity was close to the background level in parental N18TG-2 cells (3.67 ± 0.95 pmol ACh per min per 10^6 cells; s.e.m., N=3). It was not increased in the T3 mediatophore-transfected clone (1.25 ± 0.31 pmol, N=3), and in the T11 clone (1.50 pmol, N=1). It would have been anyway surprising that a single transfection could switch on the expression of a large set of proteins.

Another proposal might be that the mediatophore is involved in the creation of a «fusion pore» (see Fig. 1). The mediatophore subunit is homologous to the 16 kDa subunit of V-ATPase membrane sector (Nelson, 1989). The 16 kDa protein (and other components of the membrane sector of V-ATPase) was recently shown to form a complex with synaptobrevin and synaptophysin (Galli *et al.*, 1996). Also physophilin, the membrane target of synaptophysin, was found to be homologous to the 39 kDa subunit of the V-ATPase (Siebert *et al.*, 1994). Hence, a mediatophore-physophilin-synaptophysin complex might be a candidate to form a "fusion pore" between synaptic vesicle and plasma membrane. Mediatophore, which opens upon calcium action at a low affinity (see above), would give the "fusion pore" its characteristic calcium dependency.

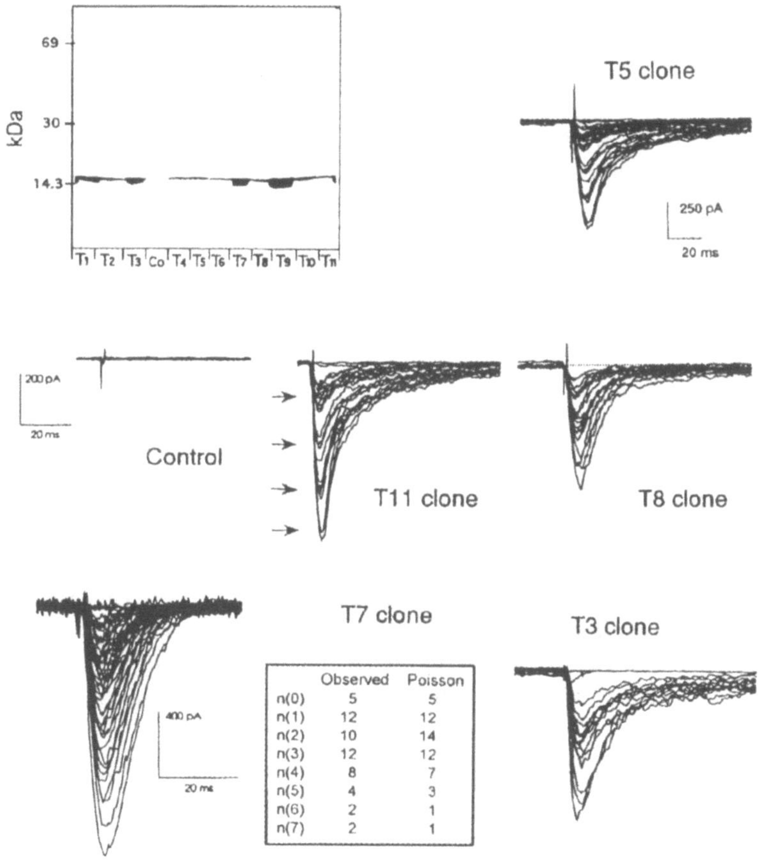

	Observed	Poisson
n(0)	5	5
n(1)	12	12
n(2)	10	14
n(3)	12	12
n(4)	8	7
n(5)	4	3
n(6)	2	1
n(7)	2	1

Fig 6. Restoration of quantal ACh release by transfection of mediatophore cDNA. Western blot: expression of the *Torpedo* mediatophore subunit in the membranes of transfected neuroblastoma cells. The control clone («C») was transfected with the expression vector only, without the insert coding for the 16 kDa subunit. «T1» to «T11» refer to monoclonal transfected clones. All expressed the mediatophore subunit.
Electrophysiological traces: Control and mediatophore-transfected cells were tested for quantal ACh release recorded with a patch-clamped Xenopus myoball. No release could be elicited from control N18TG-2 cells. In contrast, cells from the transfected clones generated clear inward currents in response to a constant electrical stimulation. The peaks of currents displayed discrete fluctuations at preferred levels (arrows) separated by approximately 200 pA intervals. A Poisson analysis was made from the currents evoked from a cell of the T7 clone. The observed responses containing 0, 1, 2, 3, 4 or more quanta were compared to the values predicted by Poisson distribution (Katz, 1969). The good fit suggests that the recorded release resulted from the summation of an integral number of elementary events.

Our own opinion, however, is that the mediatophore can release ACh directly from the cytoplasmic pool. This would be in line with the previous results obtained with the nerve terminals of the electric organ, as well as with the mediatophore reconstituted in proteoliposomes or oocytes. This will also account for experiments where transmitter release was found to occur in the absence of vesicle fusion. Whatever will be the final issue, the mediatophore will have to be considered as one of the essential proteins for quantal transmitter release.

Acknowledgements

A large part of this work was carried out in collaboration with the Laboratoire de neurobiologie cellulaire et moléculaire, C.N.R.S., Gif sur Yvette, France (M. Israël and coll.). We are grateful to N. Collet, F. Loctin, F. Pillonel and P. Sors for technical assistance and help with the manuscript. This work was supported by the Fonds national suisse pour la Recherche scientifique (grant Nr 31-37703.93).

Bibliography

Adams, D.J., Takeda, K. and Umbach, J.A. (1985) Inhibitors of calcium buffering depress evoked transmitter release at the squid giant synapse. *J. Physiol. Lond.* **369**, 145-159.

Alvarez de Toledo, G., Fernandez Chacon, R. and Fernandez, J.M. (1993) Release of secretory products during transient vesicle fusion. *Nature* **363**, 554-558.

Berrard, S., Varoqui, H., Cervini, R., Israël, M., Mallet, J. and Diebler, M.-F. (1995) Coregulation of two embedded gene products, choline acetyltransferase and the vesicular acetylcholine transporter. *J. Neurochem.* **65**, 939-942.

Betz, W.J. and Bewick, G.S. (1993) Optical monitoring of transmitter release and synaptic vesicle recycling at the frog neuromuscular junction. *J. Physiol. Lond.* **460**, 287-309.

Birman, S., Israël, M., Lesbats, B. and Morel, N. (1986) Solubilization and partial purification of a presynaptic membrane protein ensuring calcium-dependent acetylcholine release from proteoliposomes. *J. Neurochem.* **47**, 433-444.

Birman, S., Meunier, F.M., Lesbats, B., LeCaer, J.P., Rossier, J. and Israël, M. (1990) A 15 kD proteolipid found in mediatophore preparations from *Torpedo* presents high sequence homology with the bovine chromaffin granule protonophore. *FEBS Lett.* **261**, 303-306.

Blasi, J., Chapman, E.R., Link, E., Binz, T., Yamasaki, S., De Camilli, P., Südhof, T.C., Nieman, H. and Jahn, R. (1993) Botulinum neurotoxin A selectively cleaves the synaptic protein SNAP-25. *Nature* **365**, 160-163.

Breckenridge, L.J. and Almers, W. (1987) Currents through the fusion pore that forms during exocytosis of a secretory vesicle. *Nature* **328**, 814-817.

Brochier, G., Gulik-Krzywicki, T., Lesbats, B., Dedieu, J. and Israël, M. (1992) Calcium-induced acetylcholine release and intramembrane particle occurrence in proteoliposomes equipped with mediatophore. *Biol.Cell* **74**, 225-230.

Brochier, G., Israël, M. and Lesbats, B. (1993) Immunolabelling of the presynaptic membrane of *Torpedo* electric organ nerve terminals with an antiserum towards the acetylcholine releasing protein mediatophore. *Biol.Cell* **78**, 145-154.

Burgoyne, R.D. and Morgan, A. (1995) Ca^{2+} and secretory-vesicle dynamics. *Trends. Neurosci.* **18**, 191-196.

Cavalli, A., Dunant, Y., Leroy, C., Meunier, F.M., Morel, N. and Israël, M. (1993) Antisense probes against mediatophore block transmitter release in oocytes primed with neuronal mRNAs. *Eur.J.Neurosci.* **5**, 1539-1544.

Cavalli, A., Eder-Colli, L., Dunant, Y., Loctin, F. and Morel, N. (1991) Release of acetylcholine *Xenopus* oocytes injected with nRNAs from cholinergic neurons. *EMBO J.* **10**, 1671-1675.

Ceccarelli, B. and Hurlbut, W.P. (1980) Vesicle hypothesis of the release of quanta of acetylcholine. *Physiol. Rev.* **60**, 396-441.

Corrèges, P., Sors, P. and Dunant, Y. (1996) Flatten-peeled : a new approach to freeze fracture morphology. *Microscopy Research and Technique* (in press).

De Camilli, P. (1995) The eighth Datta Lecture. Molecular mechanisms in synaptic vesicle recycling. *FEBS. Lett.* **369**, 3-12.

De Robertis, E. and Bennett, H.S. (1955) Some features of the submicroscopic morphology of synapses in frog and earthworm. *J.Biophys.Biochem.Cytol.* **1**, 47-58.

De Robertis, E., Pellegrino de Iraldi, A., Rodriguez De Lores Arnaiz, G. and Gomez, C.J. (1961) The isolation of nerve endings and synaptic vesicles. *J.Biophys.Biochem.Cytol* **3**, 611-614.

Del Castillo, J. and Katz, B. (1957) La base "quantale" de la transmission neuromusculaire. In: CNRS, (Ed.) *Microphysiologie comparée des éléments excitables*, pp. 245-258.

Dunant, Y. (1986) On the mechanism of acetylcholine release. *Prog. Neurobiol.* **26**, 55-92.

Dunant, Y., Gautron, J., Israël, M., Lesbats, B. and Manaranche, R. (1972) Les compartiments d'acétylcholine de l'organe électrique de la Torpille et leurs modifications par la stimulation. *J. Neurochem.* **19**, 1987-2002.

Dunant, Y. and Israël, M. (1985) The release of acetylcholine. *Scientific American* **252**, 58-66.

Dunant, Y., Jones, G.J. and Loctin, F. (1982) Acetylcholine measured at short time intervals during transmission of nerve impulses in the electric organ of *Torpedo*. *J. Physiol. Lond.* **325**, 441-460.

Dunant, Y., Loctin, F., Vallée, J.-P., Parducz, A., Lesbats, B. and Israël, M. (1996) Activation and desensitization of acetylcholine release by zinc in *Torpedo* nerve terminals. *Pflügers Arch.* (submitted).

Dunant, Y. and Muller, D. (1986) Quantal release of acetylcholine evoked by focal depolarisation at the *Torpedo* nerve-electroplaque junction. *J. Physiol. Lond.* **379**, 461-478.

Evers, J., Laser, M., Sun, Y., Xie, Z. and Poo, M.M. (1989) Studies of nerve-muscle interactions in *Xenopus* cell culture: Analysis of early synaptic currents. *J. Neurosci.* **9**, 1523-1539.

Falk-Vairant, J., Corrèges, P., Eder-Colli, L., Salem, N., Meunier, F., Lesbats, B., Loctin, F., Synguelakis, M., Israël, M. and Dunant, Y. (1996) Evoked acetylcholine release expressed by transfection of mediatophore cDNA. *J. Neurochem.* **66**, 1322-1325.

Falk-Vairant, J., Corrèges, P., Eder-Colli, L., Salem, N., Roulet, E., Bloc, A., Meunier, F., Lesbats, B., Loctin, F., Synguelakis, M., Israël, M. and Dunant, Y. (1996) Quantal acetylcholine release induced by mediatophore transfection. *Proc.Natl.Acad.Sci.USA* **93** (in press).

Falk-Vairant, J., Dunant, Y. and Israël, M. (1994) Quantal acetylcholine release in reconstituted systems. *J. Neurochem.* **63**, S90

Falk-Vairant, J., Israël, M., Bruner, J., Stinnakre, J., Meunier, F.M., Gaultier, P., Meunier, F., Lesbats, B., Synguelakis, M., Corrèges, P. and Dunant, Y. (1996) Evoked transmitter release from fibroblasts loaded with acetylcholine, enhancement by cAMP. *Neuroscience* (in press).

Fatt, P. and Katz, B. (1952) Spontaneous subthreshold activity at motor nerve endings. *J. Physiol. Lond.* **117**, 109-128.

Fatt, P. and Katz, B. (1953) Chemo-receptor activity at the motor end-plate. *Acta Physiol. Scand.* **29**, 117-127.

Finbow, M.E., Pitts, J.D., Goldstein, D.J., Schlegel, R. and Findlay, B.C. (1991) The E5 oncoprotein target : A 16-kDa channel- forming protein with diverse functions. *Molec.Carcinog.* **4**, 441-444.

Galli, T., McPherson, P.S. and De Camilli, P. (1996) The Vo sector of the V-ATPase, synaptobrevin, and synaptophysin are associated on synaptic vesicles in a Triton X-100-resistant, freeze-thawing sensitive, complex. *J. Biol. Chem.* **271**, 2193-2198.

Gansel, M., Penner, R. and Dreyer, F. (1987) Distinct sites of action of clostridial neurotoxins revealed by double-poisoning of mouse motor nerve terminals. *Pflügers Arch.* **409**, 533-539.

Garcia-Segura, L.M., Muller, D. and Dunant, Y. (1986) Increase in the number of presynaptic large intramembrane particles during synaptic transmission at the *Torpedo* nerve-electroplaque junction. *Neuroscience* **19**, 63-79.

Girod, R., Corrèges, P., Jacquet, J. and Dunant, Y. (1993) Space and time characteristics of transmitter release at the nerve-electroplaque junction of Torpedo. *J. Physiol. Lond.* **471**, 129-157.

Girod, R., Eder-Colli, L., Medilanski, J., Dunant, Y., Tabti, N. and Poo, M.M. (1992) Pulsatile release of acetylcholine by isolated nerve terminals (synaptosomes) isolated from the *Torpedo* electric organ. *J. Physiol. Lond.* **450**, 325-340.

Harris, A.J. and Miledi, R.B. (1971) The effect of type D botulinum toxin on frog neuromuscular junctions. *J. Physiol. Lond.* **217**, 497-515.

Henkel, A.W. and Betz, W.J. (1995) Staurosporine blocks evoked release of FM1-43 but not acetylcholine from frog motor terminals. *J.Neurosc.* **15**, 8246-8258.

Heuser, J.E., Reese, T.S., Dennis, M.J., Jan, Y., Jan, L. and Evans, L. (1979) Synaptic vesicle exocytosis captured by quick freezing and correlated with quantal transmitter release. *J. Cell. Biol.* **81**, 275-300.

Israël, M., Dunant, Y. and Manaranche, R. (1979) The present status of the vesicular hypothesis. *Prog. Neurobiol.* **13**, 237-275.

Israël, M., Gautron, J. and Lesbats, B. (1968) Isolement des vésicules synaptiques de l'organe électrique de la Torpille et localisation de l'acétylcholine à leur niveau. *C.R.Acad.Sci.(Paris)* **266**, 273-275.

Israël, M., Lesbats, B., Morel, N., Manaranche, R., Gulik-Krzywicki, T. and Dedieu, J. (1984) Reconstitution of a functional synaptosomal membrane possessing the protein constituents involved in acetylcholine translocation. *Proc.Natl.Acad.Sci. USA.* **81**, 277-281.

Israël, M., Lesbats, B., Synguelakis, M. and Joliot, A. (1994) Acetylcholine accumulation and release by hybrid NG108-15, glioma and neuroblastoma cells - Role of a 16 kDa membrane protein in release. *Neurochem.Int.* **25**, 103-109.

Israël, M., Manaranche, R., Morel, N., Dedieu, J., Gulik-Krzywicki, T. and Lesbats, B. (1981) Redistribution of intramembrane particles related to acetylcholine release by cholinergic synaptosomes. *J.Ultrastruct.Res.* **75**, 162-178.

Israël, M., Meunier, F.M., Morel, N. and Lesbats, B. (1987) Calcium-induced desensitization of acetylcholine release from synaptosomes or proteoliposomes equipped with mediatophore, a presynaptic membrane protein. *J. Neurochem.* **49**, 975-982.

Israël, M., Morel, N., Lesbats, B., Birman, S. and Manaranche, R. (1986) Purification of a presynaptic membrane protein that mediates a calcium-dependent translocation of acetylcholine. *Proc.Natl.Acad.Sci.USA* **83**, 9226-9230.

Katz, B. (1969) The release of neural transmitter substances. *University Press:Liverpool* 60P.

Katz, B. and Miledi, R.B. (1969) Tetrodotoxin-resistant electric activity in presynaptic terminals. *J. Physiol. Lond.* **203**, 459-487.

Kriebel, M.E. and Gross, C.E. (1974) Multimodal distribution of frog miniature endplate potentials in adult, denervated and tadpole leg muscle. *J. Gen. Physiol.* **64**, 85-103.

Kriebel, M.E., Llados, F. and Matteson, D.R. (1976) Spontaneous subminiature end-plate potentials in mouse diaphragm muscle: evidence for synchronous release. *J. Physiol. Lond.* **262**, 553-581.

Kuffler, S.W. and Yoshikami, D. (1975) The number of transmitter molecules in a quantum : an estimate from iontophoretic application of acetylcholine at the neuromuscular synapse. *J. Physiol. Lond.* **251**, 465-482.

Leroy, C. and Meunier, F.M. (1995) Differential targeting to the plasma membrane of the Torpedo 15-kDa proteolipid expressed in oocytes. *J. Neurochem.* **65**, 1789-1797.

Leroy, C., Meunier, F.M., Lesbats, B. and Israël, M. (1994) In vitro expression of the 15 kDa subunit of the mediatophore and functional reconstitution of acetylcholine release. *Gen. Pharmacol.* **25**, 245-255.

Llinas, R., Sugimori, M. and Silver, R.B. (1992) Microdomains of high calcium concentration in a presynaptic terminal. *Science USA* **256**, 677-679.

MacIntosh, F.C. and Collier, B. (1976) Neurochemistry of cholinergic terminals. In: Zaimis, E. (Ed.) *Handbook of Experimental Pharmacology*, pp. 99-228. Berlin: Springer Verlag.

Mandel, M., Moriyama, Y., Hulmes, J.D., Pan, Y.C.E., Nelson, H. and Nelson, N. (1988) cDNA sequence encoding the 16-kDa proteolipid of chromaffin granule implies gene duplication in the evolution of H^+-ATPases. *Proc.Natl.Acad.Sci. USA* **85**, 5521-5524.

Molgo, J., Comella, J.X., Angaut Petit, D., Pécot-Dechavassine, M., Tabti, N., Faille, L., Mallart, A. and Thesleff, S. (1990) Presynaptic actions of botulinal neurotoxins at vertebrate neuromuscular junctions. *J. Physiol. Paris.* **84**, 152-166.

Morot-Gaudry-Talarmain, Y., Diebler, M.-F., Robba, M., Lancelot, J.-C., Lesbats, B. and Israël, M. (1989) Effect of cetiedil analogs on acetylcholine and choline fluxes in synaptosomes and vesicles. *Eur. J. Pharmacol.* **166**, 427-433.

Muller, D. and Dunant, Y. (1987) Spontaneous quantal and subquantal transmitter release at the *Torpedo* nerve-electroplaque junction. *Neuroscience* **20**, 911-921.

Muller, D., Garcia-Segura, L.M., Parducz, A. and Dunant, Y. (1987) Brief occurrence of a population of large intramembrane particles in the presynaptic membrane during transmission of a nerve impulse. *Proc.Natl.Acad.Sci.(USA)* **84**, 590-594.

Nahas, G.G., Zagury, D., Milhaud, A., Manger, W.M. and Pappas, G.D. (1967) Acidemia and catecholamine output of the isolated canine adrenal gland. *Am.J.Physiol.* **213**, 1186-1192.

Neher, E. (1993) Secretion without full fusion. *Nature* **363**, 497-498.

Nelson, N. (1989) Structure, molecular genetics and evolution of vacuolar H^+-ATPases. *J.Bioenerg.Biomembr.* **21**, 553-571.

Nicaise, G., Maggio, K., Thirion, S., Horoyan, M. and Keicher, E. (1992) The calcium loading of secretory granules. A possible key event in stimulus secretion coupling. *Biol.Cell* **75**, 89-99.

Ohsawa, K., Dowe, G.H.C., Morris, S.J. and Whittaker, V.P. (1979) The lipid and protein content of cholinergic synaptic vesicles from the electric organ of *Torpedo marmorata* purified to constant composition : implication for vesicle structure. *Brain Res.* **161**, 447-457.

Parducz, A., Corrèges, P., Sors, P. and Dunant, Y. (1996) Impairment of the exo-endocytosis cycle by zinc in Torpedo nerve terminals. Submitted.

Parducz, A., Loctin, F., Babel-Guérin, E. and Dunant, Y. (1994) Exo-endocytotic images following tetanic stimulation at a cholinergic synapse. A role in calcium extrusion? *Neuroscience* **62**, 93-103.

Siebert, A., Lottspeich, F., Nelson, N. and Betz, H. (1994) Purification of the synaptic vesicle-binding protein physophilin identification as the 39-kDa subunit of the vesicular H^+ ATPase. *J. Biol. Chem.* **269**, 28329-28334.

Stjärne, L., Greengard, P., Grillner, S.E., Hokfeld, T.G.M. and Eds (1994) Molecular and cellular mechanisms of neurotransmitter release. *Adv. in Sec. Mess. Phosphoprotein Res.* **29**, 1-569.

Südhof, T.C. (1995) The synaptic vesicle cycle: a cascade of protein-protein interactions. *Nature* **375**, 645-653.

Tauc, L. (1982) Non-vesicular release of neurotransmitter. *Physiol. Rev.* **62**, 857-893.

Torri-Tarelli, F., Grohovaz, F., Fesce, R. and Ceccarelli, B. (1985) Temporal coincidence between synaptic vesicle fusion and quantal secretion of acetylcholine. *J.Cell Biol.* **101**, 1386-1399.

Whittaker, V.P., Michaelson, I.A. and Kirkland, R.J. (1964) The separation of synaptic vesicles from nerve endings particles ("synaptosomes"). *Biochem. J.* **90**, 293-303.

Zhong, Z.G., Misawa, H., Furuya, S., Kimura, Y., Noda, M., Yokoyama, S. and Higashida, H. (1995) Overexpression of choline acetyltransferase reconstitutes discrete acetylcholine release in some but not all synapse formation-defective neuroblastoma cells. *J. Physiol. (Paris)* **89**, 137-145.

Stimulation of Exocytosis in Cultured Cerebellar Granule Cells by Electrical Field Stimulation

Cousin M.A., Hurst, H., Held B.[1] and Nicholls D.G.

Neurosciences Institute, Department of Pharmacology, University of Dundee, Dundee, DD1 9SY.
[1]Department of Pharmacology, University of Leeds, Leeds.

1 Introduction

External excitatory stimuli are required in order to evoke neurotransmitter release from cultured neurones. Traditionally, chemical stimuli have been employed to mimic synaptic input, with elevated KCl being the most widely used method to depolarize the plasma membrane. However, KCl stimulation has disadvantages in that the plasma membrane of the preparation is permanently clamped in a depolarized state allowing no repolarization to take place. In addition by definition, no modulation of K^+ channels are able to be studied by this technique.

Field stimulation is routine for various brain slice preparations, including hippocampal (Dooley *et al.*, 1987, 1988) and striatal (Herdon and Nahorski, 1989) sections, recently this more physiologically relevant stimulus has been applied to initiate Na^+ channel activation and thus action potentials in cultured neurones including dorsal root ganglion, sympathetic and hippocampal neurones (Holz *et al.*, 1988; Wakade and Wakade, 1988; Wakade *et al.*, 1990; Schwartz and Malik, 1991; Ryan and Smith, 1995) with monitoring of Ca^{2+}-dependent exocytosis.

1.1 Electrical field stimulation

The electrical field stimulation technique involves application of a potential difference of known frequency, pulse width, current and time between two parallel platinum wires. Since the neuronal cell interior is at isopotential, the drop in voltage across the cell is concentrated entirely at the plasma membrane, allowing activation of voltage-sensitive Na^+ channels. In accordance with this, a large increase in intracellular free Na^+ in the somata of individual granule cells has been demonstrated in response to field stimulation (Cousin *et al.*, 1995a).

As electrical field stimulation only produces a transient depolarization of the plasma membrane, multiple stimuli can be applied with no loss of cellular activity and modulation of voltage-sensitive K^+ and Ca^{2+} channels can be observed. It has recently been demonstrated for example that voltage-sensitive Ca^{2+} channels localized at the cell somata of granule cells are unable to be activated by field stimuli that are able to evoke large increases in neurite intracellular free Ca^{2+} ($[Ca^{2+}]_c$) and exocytosis (Cousin *et al.*, 1995a). Also, application of the K^+ channel antagonist 4-aminopyridine , but not the specific K^+_A channel antagonist dendrotoxin, increased

NATO ASI Series, Vol. H 100
Neurotransmitter Release and Uptake
Edited by Şakire Pöğün

$[Ca^{2+}]_c$ in populations and individual somata and neurites of granule cells (Beani *et al.*, 1994; Cousin *et al.*, 1995a, 1995b) demonstrating that K^+ channels were involved in repolarization of the plasma membrane.

2 Techniques Employed

In order to investigate exocytosis evoked by electrical field stimulation from granule cells two different approaches were employed. (1) An assay for the release of soluble contents from small synaptic vesicles using the non-metabolizable isotope [^{3}H]-D-aspartate, and (2) a visualization of recycling vesicles in living tissue in real time, using the fluorescent probe FM1-43 with an inverted fluorescence microscope. By employing both of these techniques in parallel, exocytosis can be followed at the resolution of a single synaptic varicosity.

2.1 Relevance of [^{3}H]-D-aspartate to Label Transmitter Pools

Although the isotope [^{3}H]-D-aspartate has been employed for many years its ability to label the transmitter pool of endogenous glutamate has still been questioned (Neville *et al.*, 1987; McMahon and Nicholls, 1990). This has been mainly due to that fact that, although [^{3}H]-D-aspartate is widely recognized to be accumulated via the plasma membrane amino acid transporter, researchers have never been able to demonstrate aspartate accumulation inside small synaptic vesicles (Naito and Ueda, 1985; Maycox *et al.*, 1988; Carlson *et al.*, 1989; Shioi and Ueda, 1990). Also, apart from the fact that aspartate can be a relatively selective antagonist for the NMDA receptor, it has little or no action at AMPA receptors which mediate most fast excitatory amino acid synaptic transmission. In addition, the ability of [^{3}H]-D-aspartate to be released in a Ca^{2+}-dependent manner from synaptosomal preparations is a matter of some debate (Poli *et al.*, 1985; Levi and Gallo, 1986; Simonato *et al.*, 1993). However some Ca^{2+}-dependent release of endogenous aspartate has been demonstrated in synaptosomal preparations, particularly using lower levels of depolarization (Maura *et al.*, 1989, 1991; Terrian *et al.*, 1990; Raiteri *et al.*, 1990; Poli *et al.*, 1991; Pende *et al.*, 1993; Zhuo *et al.*,1995). This Ca^{2+}-dependent component can be inhibited in cerebrocortical synaptosomes by the clostridial toxin tetanus toxin (TeTx) (McMahon *et al.*, 1992), which cleaves the synaptic vesicle protein synaptobrevin II implicated in the exocytotic process (Schiavo *et al.*, 1992), indicating that endogenous aspartate can be released in a vesicular manner.

Slice preparations from the hippocampus have always suggested a transmitter role for endogenous aspartate with numerous studies demonstrating a clear KCl- (Burke and Nadler, 1988, 1989; Burke *et al.*, 1993; Galarreta *et al.*, 1993; Fleck *et al.*, 1993; Kimura *et al.*, 1995) or electrical field-stimulated (Szerb, 1988) Ca^{2+}-dependent release. Slice preparations from other brain regions also demonstrate a robust Ca^{2+}-dependent release of endogenous aspartate, stimulated either electrically or with KCl (Flint *et al.*, 1981; Collins *et al.*, 1982; Maura *et al.*, 1991; Collins, 1993; Palmer and Reiter, 1994; Kimura *et al.*, 1995). Ca^{2+}-dependent release of [^{3}H]-D-aspartate from slice preparations has been reported in cerebellar (Bosley *et al.*,

1983), hippocampal (Simonato *et al.*, 1993; Gaur *et al.*, 1994; Keith *et al.*, 1994) and cortical slices (Palmer and Reiter, 1994, but see Beani *et al.*, 1992).

The KCl-stimulated Ca^{2+}-dependent release of [^{3}H]-D-aspartate and endogenous aspartate from the cerebellar granule cell preparation has been well documented in many studies (Zhu and Chuang, 1987; Kingsbury *et al.*, 1988; Rogers *et al.*, 1991; Westergaard *et al.*, 1991; Bianchi *et al.*, 1992; Belhage *et al.*, 1992). However simple Ca^{2+}-dependency for the release of the analogue from this preparation is not sufficient to prove exocytosis. Therefore it was important to confirm that the analogue did label the vesicular amino acid pool in the first instance.

2.2 Use of FM1-43 to Investigate Vesicle Dynamics

The fluorescent styryl dye FM1-43 has been utilized to observe synaptic vesicle recycling in intact cells at the resolution of a single varicosity in both motor and nerve terminals (Betz *et al.*, 1992; Betz and Bewick, 1992; Ryan *et al.*, 1993; Reuter, 1995; Ryan and Smith, 1995; Smith and Betz 1996). FM1-43 is a divalent cation with a hydrophilic head containing two quaternary nitrogen groups, and a dibutyl lipophilic tail. Therefore by virtue of its structure, FM1-43 is able to insert into the outer leaflet of membranes but not diffuse across them, enabling the dye to label the lumen of vesicles retrieved via endocytosis. The labeling has been shown to be specific for synaptic vesicles in a recent electron microscopy study (Henkel *et al.*, 1996). Also, since FM1-43 only forms a weak interaction with lipid bilayers, the dye can be removed from the exposed lumen upon exocytosis by continuous superfusion, rendering it unavailable for reuptake. Therefore in theory, both exo- and endocytosis can be observed in intact preparations, and indeed in a recent paper both processes were monitored simultaneously using capacitance methods and FM1-43 staining (Smith and Betz 1996).

3 Field-Evoked Exocytosis in Cerebellar Granule Cells

3.1 Exocytotic Nature of [^{3}H]-D-aspartate Release

When cerebellar granule cells are loaded for 60 min with 0.5μCi [^{3}H]-D-aspartate, and then stimulated with 50mM KCl, the release of the analogue mirrors that observed with endogenous and exogenous glutamate (Gallo and Levi, 1982; Dolphin and Prestwich, 1985; Van-Vliet *et al.*, 1989), displaying a Ca^{2+}-dependent release in mature but not immature cells (Fig 1A). This inability to release the analogue in a Ca^{2+}-dependent manner can be attributed to the lack of synapses of immature cells (Van-Vliet *et al.*, 1989). Also, again in parallel with endogenous glutamate exocytosis (Pocock *et al.*, 1993), stimulation of Ca^{2+}-dependent [^{3}H]-D-aspartate release was only observed with KCl concentrations above 30mM (Fig. 1B). KCl-evoked Ca^{2+}-dependent release was also completely inhibited by TeTx when preincubated for 24 hours at 5μg/ml and 1μM bafilomycin A1 (Table 1). Bafilomycin A1 inhibits the V-type ATPase on the synaptic vesicle, causing depletion of accumulated amino acids which leak out down their concentration gradient into the cytoplasm (Bowman *et al.*, 1988; Moriyama and Futai, 1990; Floor *et al.*, 1990; Wang and Floor, 1994).

The KCl-evoked Ca^{2+}-dependent release of [^{3}H]-D-aspartate has been demonstrated to be completely inhibited by the P/Q-type Ca^{2+} channel antagonists Agatoxin-IVA and Conotoxin-MVIIC in hippocampal slices (Burke *et al.*, 1993; Gaur *et al.*, 1994; Kimura *et al.*, 1995), while release can be blocked in granule cells by addition of the non-selective Ca^{2+} and Na^+ channel antagonist flunarizine (Cousin *et al.*, 1993). One ambiguity in these studies is that endogenous glutamate released by exocytosis might exchange with cytoplasmic [^{3}H]-D-aspartate via the plasma membrane cotransporter to give the appearance of a vesicular efflux of the analogue. Therefore it was important to determine that the Ca^{2+}-dependent release of [3H]-D-aspartate was not an artifact of the stimulation protocol. This was tested in two ways, firstly, the release was not affected by preincubation with the ionotropic NMDA and AMPA/KA receptor antagonists MK-801 (10μM) and CNQX (2μM) indicating that no postsynaptic receptors were participating in the response. Secondly, [^{3}H]-D-aspartate release could not be stimulated by the addition of 5μM exogenous glutamate in the presence of MK-801 and CNQX (a concentration ten fold larger than that released by KCl stimulation; (Pocock *et al.*, 1993) eliminating the possibility that exocytosed endogenous glutamate was exchanging with a cytosolic pool of the analogue across the plasma membrane amino acid transporter.

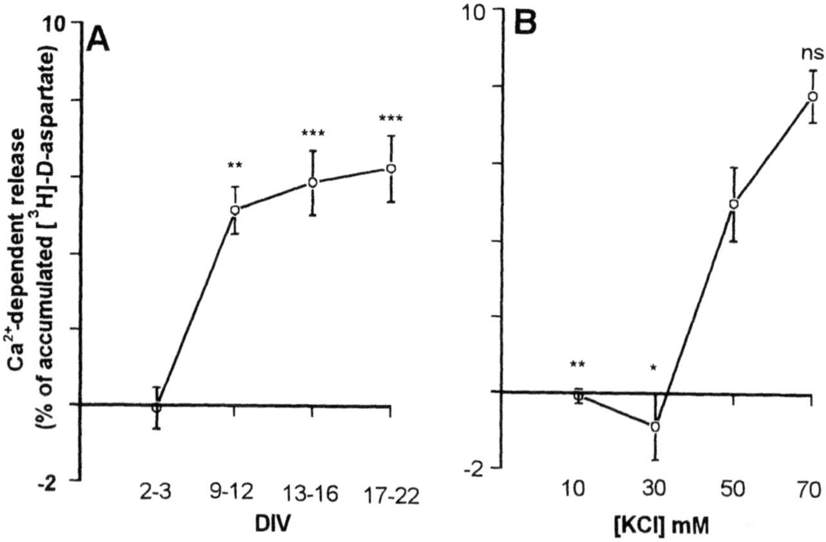

Fig 1. Culture age- and voltage-dependence of [^{3}H]-D-aspartate release. (**A**) Granule cells of increasing culture age (DIV) were loaded with 0.5μCi [^{3}H]-D-aspartate for 60 min and stimulated with 50mM KCl in either 1.3mM or low Ca^{2+} medium as indicated. (**B**) Cells were loaded with [^{3}H]-D-aspartate for 60 min and release was stimulated with increasing concentrations of KCl for 2 min as indicated. Results are presented as Ca^{2+}-dependent release of [^{3}H]-D-aspartate as a percentage of the total accumulated isotope. Levels of significance (Student's t test) $p<0.001$ = ***, $p<0.01$ = **, $p<0.05$ = *. n>9±SEM.

When these results are taken together it is clear that the Ca^{2+}-dependent release of [^{3}H]-D-aspartate in cerebellar granule cells is exocytotic.

3.1.1 Electrical Field Stimulation of [^{3}H]-D-aspartate Release

In addition to KCl, [^{3}H]-D-aspartate can also be released from granule cells by stimulation with an applied electrical field (Bianchi *et al.*, 1992). When granule cells are stimulated with a uniform electrical field of 40Hz for 10s (150mA current, 1ms pulse width), a Ca^{2+}-dependent release of [3H]-D-aspartate is observed, however the magnitude of the release is four fold smaller to that compared with KCl. When the number of stimuli are increased to four 10s 40Hz trains with a 10s delay between each, an efflux comparable to KCl-evoked release is observed with a similar Ca^{2+}-dependency. In accordance with a technique that relies upon action potential propagation, the Ca^{2+}-dependent release of [^{3}H]-D-aspartate was completely inhibited by 200nM of the voltage-sensitive Na^+ channel antagonist tetrodotoxin (TTx) (Table 1). In addition, the K^+ channel antagonist 4-aminopyridine (4-AP) at 1mM, was able to significantly increase the field-stimulated release (Table 1), an enhancement which could be completely inhibited by TTx (Table 1) indicating that the phenomenon was due to action potential modulation. This is in contrast with KCl-evoked exocytosis in that neither TTx nor 4-AP had any significant effect on the Ca^{2+}-dependent release of [^{3}H]-D-aspartate (Table 1), consistent with a transmitter release independent of Na^+ and K^+ channel activation. As observed with KCl stimulation, the field-evoked Ca^{2+}-dependent [^{3}H]-D-aspartate efflux is also exocytotic, as preincubation with 5μg/ml TeTx or 1μM bafilomycin A1 completely inhibited the release (Table 1).

Table 1. Characterization of KCl- and field-evoked Ca^{2+}-dependent [^{3}H]-D-aspartate release.

	50mM KCl	**40Hz 10s**
Control	100 ± 7.6	100 ± 3.8
Tetanus toxin	0.0 ± 6.0**	-7.25 ± 2.25***
Bafilomycin A1	3.6 ± 6.1*	5.5 ± 21.2*
TTx	100.4 ± 3.5	6.9 ± 11.0***
4-AP	86.9 ± 5.4	144.9 ± 10.8**
TTx + 4-AP	-	2.4 ± 7.1***

Granule cells loaded with [^{3}H]-D-aspartate were preincubated with the K^+ and Na^+ channel antagonists 4-aminopyridine (4-AP, 1mM) and tetrodotoxin (TTx, 200nM) for 2 min prior to stimulation as indicated with 4 trains of 40Hz for 10s (1ms, 150mA) whose polarity was reversed after every train. Cells were also preincubated with tetanus toxin (5μg/ml) and bafilomycin A1 (1μM) for 24 hours and 75 min respectively before KCl stimulation. Results are expressed as a percentage of release above basal normalized to control levels. Levels of significance with respect to control (Student's t test) $p<0.001$ = ***, $p<0.01$ = **, $p<0.05$ = *. n=6±SEM.

3.2 Frequency-Dependency of Exocytosis

In other neuronal systems amino acid release on average seems to require higher stimulation frequencies to evoke exocytosis. In slice preparations, where field stimulation has been utilized for a number of years, higher frequencies have usually been demonstrated to be required to induce amino acid release (Collins *et al.*, 1982; Szerb, 1988; Kihara *et al.*,1989; Kangrga and Randic, 1990; Israel *et al.*, 1993) but see Corradetti *et al.*, 1984, when compared with other transmitters such as dopamine (Dooley *et al.*, 1987; Herdon and Nahorski, 1989; Kennedy *et al.*, 1992), acetylcholine (Dooley *et al.*, 1987; Israel *et al.*, 1993; Wessler *et al.*, 1995), noradrenaline (Dooley *et al.*, 1987, 1988; Wessler *et al.*, 1990; Fredholm and Hu, 1993), or 5-hydroxytryptamine (Dooley *et al.*, 1987; Wichmann *et al.*, 1989). Transmitter release from neuronal cultures has also been demonstrated with low frequency electrical field stimulation, with substance P (Holz *et al.* 1988), ACh (Prado *et al.*, 1993) and [^{3}H]-noradrenaline (Wakade and Wakade, 1988; Wakade *et al.*, 1990; Schwartz and Malik, 1991) released from cultured dorsal root ganglion and sympathetic neurons respectively. Not all transmitter release is facilitated by an increase in frequency however, as [^{3}H]-noradrenaline and ACh efflux are depressed at higher stimulation frequencies (Wakade and Wakade, 1988; Agoston and Lisziewicz, 1989; Wessler *et al.*, 1990). In addition, in neurones which store and release two transmitters, a frequency-dependent release can be observed in

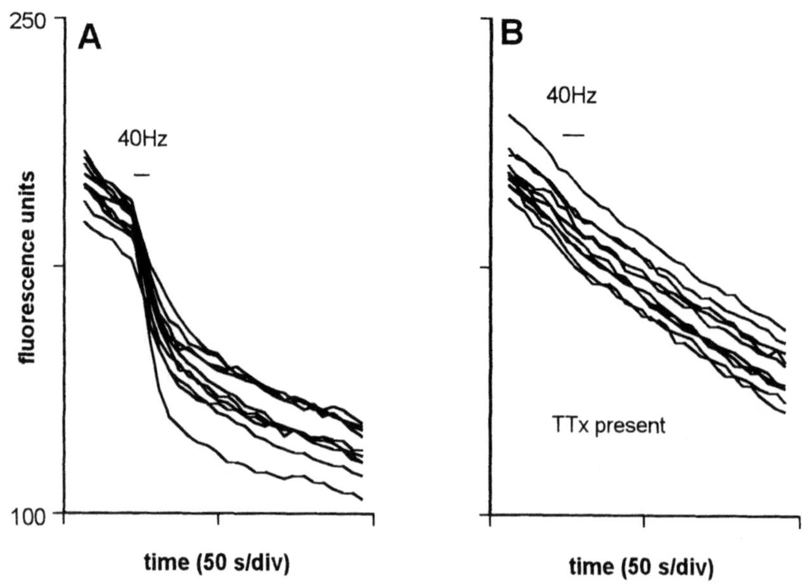

Fig 2. Field-evoked FM1-43 destaining. Granule cell neurites were loaded with FM1-43 by stimulation with 70mM KCl, then stimulated with an electrical field of 40Hz for 10s (100mA, 1ms pulse width). In **B**, cells were preincubated with 200nM TTx for 2 min before stimulation.

preparations such as myenteric neurons, spinal cord and Torpedo electric organ slices which release their different classical and peptide transmitters in response to low and high frequencies of stimulation (Agoston and Lisziewicz, 1989; Israel *et al.*, 1993; Franck *et al.*, 1993).

Capacitance analysis in secretory cells have demonstrated that vesicles are able to transiently fuse and retrieve to release their contents without being incorporated into the plasma membrane (Alvarez de Toledo *et al.*, 1993; Monck and Fernandez, 1994), the transient fusion event being termed "kiss and run". However, little is known about transient vesicle fusion in neuronal cells to the point where it is not clearly known whether it occurs at all. FM1-43 may be able to give an insight into this process as the washout kinetics of the dye ($t_{1/2}$ approx 10s) should be too slow to be removed via a transient fusion event but not full fusion with the plasma membrane.

3.2.3 Field-Stimulated FM1-43 Staining in Granule Cells

When granule cells are loaded with FM1-43 in the presence of Ca^{2+} under stimulation with 70mM KCl, the neurite regions accumulate the marker via endocytosis, which follows exocytosis, and take on a punctate appearance indicating the presence of clusters of small synaptic vesicles, which correspond with specific varicosities on the neurites of the cells (Cousin *et al.*, 1995a). When these KCl-loaded neurites are stimulated with an electrical field of 40Hz for 10s, a destaining is observed corresponding to exocytosis of synaptic vesicles (Fig 2A). It has been demonstrated previously that both electrical field- and KCl-stimulation both release FM1-43 from the same vesicle pool (Cousin *et al.*, 1995a) in granule cells although the release evoked by KCl stimulation is larger. As with evoked [^{3}H]-D-aspartate release, the field-evoked destaining of FM1-43-loaded neurites was completely inhibited by 200nM TTx (Fig 2B). Field stimulation can also be employed to load the cells, resulting in a staining pattern similar to that observed with KCl. These loaded neurites can be destained by either KCl or field stimulation. The field-stimulated loading of FM1-43 can also be inhibited by TTx and low extracellular Ca^{2+} as would be expected for a process dependent on Na^{+} and Ca^{2+} channel activation. When FM1-43-loaded granule cells were exposed to electrical fields of increasing frequency, while keeping action potential number constant, destaining was also increased in a non-linear manner (Table 2). This finding supports the possibility that at low frequencies, vesicles may be undergoing transient fusion although this cannot be confirmed by FM1-43 data alone.

2.1 Frequency-Dependency of [3H]-D-aspartate Exocytosis in Granule Cells

As destaining of FM1-43-loaded neurites cannot resolve whether soluble vesicle contents are released, the effect of stimulation frequency on [^{3}H]-D-aspartate exocytosis was examined. In order to keep the number of pulses constant, the cells were exposed to four trains of stimuli composed of 400 action potentials with 10s intervals between each train. Because of this, the stimulation protocol was completed more quickly with high frequency stimulation, so all experimental conditions were sampled after a period corresponding to the longest stimulation to control for transporter mediated efflux. In an attempt to reduce the basal outflow via the amino acid transporter, the antagonist dihydrokainate was applied to the cells, however as found by Bianchi *et al.* (1992) and Huston *et al.* (1995), the compound actually increased basal outflow and was therefore not employed. When cells were stimulated at increasing frequencies, a non-linear stimulation above baseline similar to the FM1-43 destaining was observed (Table 2), indicating that for the same number of action potentials, evoked release of [^{3}H]-D-aspartate can be increased at higher frequencies.

3.2.4 Frequency-Dependency of Exocytosis

The facilitation of exocytosis observed in the granule cells confirms the findings of Bianchi *et al.* (1992) who also found field-evoked release to be frequency-dependent. However, additionally in this study the same number of pulses were applied to the cells over the same time period. The mechanism of this facilitation is not known but would probably involve a build up of Ca^{2+} in the presynaptic area due to the higher frequency of Ca^{2+} channel activation not allowing time for diffusion to clear the area before the next stimulus arrives, in a manner similar to paired pulse facilitation. The rationale behind this theory is that the Ca^{2+} sensor for exocytosis cooperatively binds four Ca^{2+} ions, therefore if high concentrations of Ca^{2+} enter the terminal, ions at the highest affinity sites may not come off before the next stimulus arrives, thereby increasing the efficiency of release.

Table 2. Frequency-dependency of field-evoked exocytosis.

Frequency	[^{3}H]-D-aspartate	FM1-43
5Hz	21.4 ± 1.4	5.5 ± 7.4
10Hz	60.1 ± 3.0	42.5 ± 5.5
20Hz	72.8 ± 7.1	69.0 ± 5.2
40Hz	100 ± 8.2	100 ± 3.6
80Hz	135 ± 2.1	-

Granule cells loaded with [^{3}H]-D-aspartate were stimulated with electrical fields of increasing frequency four times with a 10s interval. Each period contained 400 stimuli and cells release was sampled after 6 min for all cells. Alternatively, granule cells were loaded with FM1-43 in the presence of 70mM KCl and then stimulated with electrical fields of increasing frequency. Results are normalized to release or destaining above basal levels evoked by 40Hz stimulation. n=3±SEM.

Imaging with the ratiometric Ca^{2+} indicator fura-2 confirmed that maximal neurite $[Ca^{2+}]_c$ levels were proportional to the frequency of stimulation (Fig 3). Additionally, the [^{3}H]-D-aspartate data argue against the possibility that transmitter is released by transient vesicle fusion at low stimulation frequencies.

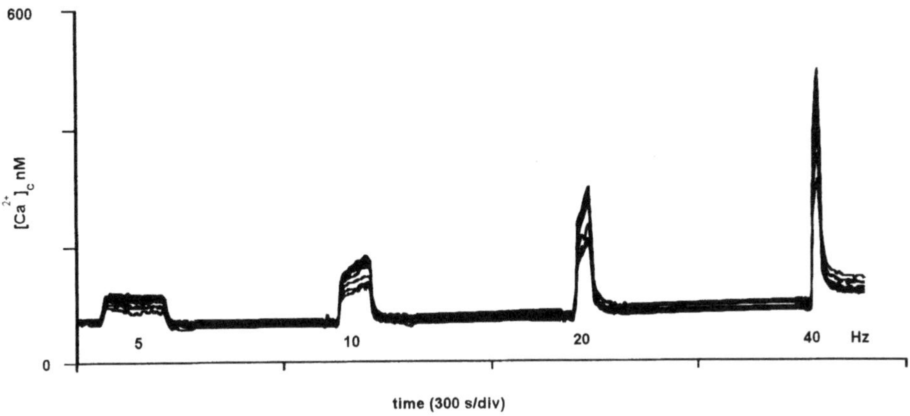

Fig 3 - Frequency-dependent increases in neurite $[Ca^{2+}]_c$ Fura-2-loaded granule cells were stimulated with 400 action potentials of increasing frequency as indicated.

The observation that facilitation occurs at higher stimulation frequencies agrees with the proposed type of stimuli that granule cells *in situ* would encounter from mossy fibre inputs, reported to be up to 150Hz (De Schutter and Bower, 1994; Gabbiani *et al.*, 1994). This is probably different to the frequency-dependency demonstrated in neurons that co-store two transmitters in discrete vesicle populations, as peptides stored in large dense cored vesicles require a large bulk elevation in $[Ca^{2+}]_c$ produced by high frequency stimulation for release, whereas classical transmitters can use Ca^{2+} influx localized at active zones (Verhage *et al.*, 1991).

Near complete destaining of FM1-43-labeled neurites have been reported to occur in response to lower stimulation frequencies in mixed hippocampal cultures (Reuter, 1995; Ryan and Smith, 1995). This is at odds with observations in granule cells and may reflect differences in the efficiency of synaptic transmission between the culture preparations. Alternatively, the lack of complete destaining may be due to a higher level of background staining in the granule cell preparation due to constitutive membrane trafficking.

3.3 Repetitive Field Stimulation of [^{3}H]-D-aspartate Release

The magnitude of [^{3}H]-D-aspartate release can be elevated as described before by extending the number of 40Hz 10s stimuli, and as Fig 4 shows, release of [^{3}H]-D-aspartate linearly increases with stimuli number, while retaining a comparable Ca^{2+}-dependency for four stimulations (74% of increase above basal) compared to 71% for one 40Hz stimulation. This would tend to suggest that granule cells are able to respond to repeated trains of high frequency stimulation without losing the ability to exocytose transmitter due to depletion. The Ca^{2+}-dependency of the response and the return to near baseline values after field application is a good indication that cell damage is not occurring during these multiple stimuli.

3.3.1 Repetitive Field Stimulation of FM1-43 Labeled Varicosities

When FM1-43 labelled granule cell varicosities were repetatively stimulated stimulated with the same protocol to evoke the release of [^{3}H]-D-aspartate (4 trains of 40Hz for 10s with 10s intervals), the amount of destaining followed an exponential course (Fig 5). This is in contrast to the [^{3}H]-D-aspartate release data where efflux increases linearly with stimuli number. However, as only one population of vesicles is loaded with KCl stimulation, the pool of labeled vesicles will be diluted with unlabeled vesicles (due to the fact that the dye is perfused away from the membrane to prevent reuptake upon subsequent stimuli). Therefore although fewer of the vesicles which are exocytosed by the latter trains of stimuli are labeled with FM1-43, they are full of [^{3}H]-D-aspartate, suggesting that these vesicles are either recruited from a pool other than that which recycle normally, or that the same pool is able to reprime itself very quickly for subsequent release.

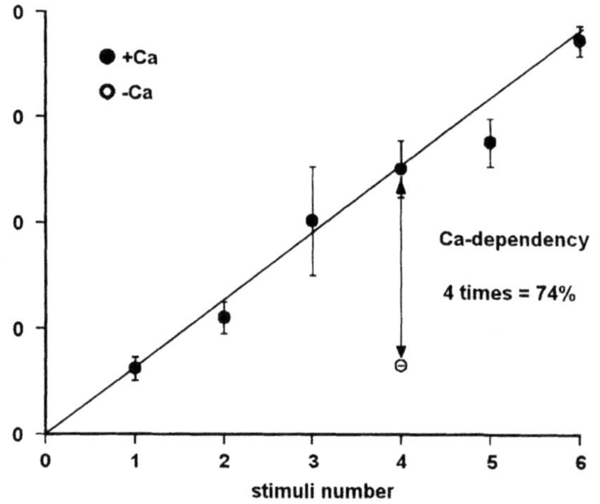

Fig 4. Increasing [^{3}H]-D-aspartate release with repetitive field stimulation. Granule cells loaded with [^{3}H]-D-aspartate were stimulated with an electrical field 40Hz frequency, (1ms pulse width, 100mA current) with the indicated number of 10s trains. Polarity was reversed between each stimulation. Closed circles denote 1.3mM Ca^{2+} medium, open low Ca^{2+} medium. Results are normalized to release evoked by one 40Hz stimulation above basal levels. n=3±SEM.

3.3.2 Time Course of [^{3}H]-D-aspartate Labeling of Exocytotic Pool

If the vesicle pool participating in the latter rounds of exocytosis recycle very quickly from the plasma membrane back to the active zone, the vesicles would have to take up transmitter very rapidly to participate in another round of exocytosis. Studies in slices suggest that changes in the metabolic pool of aspartate compared to glutamate are rapidly reflected with a corresponding increase in aspartate release with either electrical or KCl stimulation (Szerb, 1988; Burke and Nadler, 1989), however in synaptosomes, increased loading of the cytosolic aspartate pool had no effect on either Ca^{+}-dependent aspartate or glutamate release (McMahon and Nicholls, 1990).

The question was addressed in granule cells by loading with [^{3}H]-D-aspartate for increasing periods of time to investigate how quickly the isotope could enter the vesicle from the cytoplasm. Even after only 1 minute of exposure to the isotope, a Ca^+-dependent release was observed of the same percentage of total accumulation as that seen with 60 min exposure (Fig 6). This release was also able to be inhibited with bafilomycin A1, indicating the transmitter was accumulated inside the exocytotic vesicle. Therefore [^{3}H]-D-aspartate can be transported across the vesicle membrane with great rapidity, fast enough to replenish a rapidly recycling pool of synaptic vesicles.

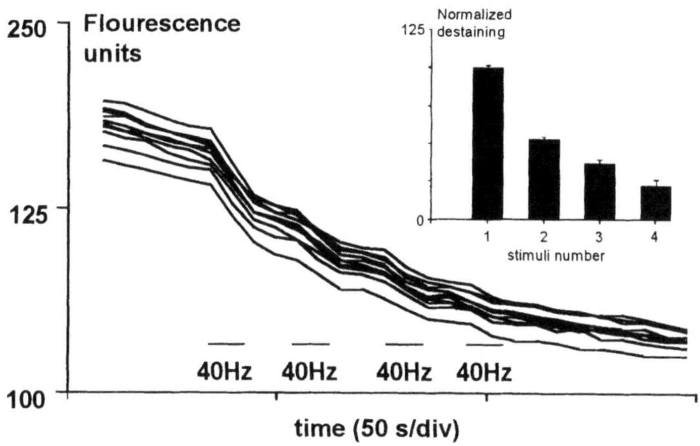

Fig 5. Destaining of granule cell neurites with repetitive field-stimulation. Granule cells were loaded with FM1-43 in the presence of 70mM KCl and stimulated with 4 trains of electrical stimuli (40Hz, 10s, 1ms, 100mA) with a 10s interval. Bars indicate stimulation periods. Polarity was reversed after each train. **Inset** - Bar graph of relative destaining after each train of action potentials normalized to the first destaining. n=30±SEM

3.3.4 Sustained Exocytosis in Response to Repetitive High Frequency Stimulation

It has been demonstrated that granule cells possess the capability to release a large amount of transmitter exocytotically (approximately 35%) when stimulated with the Na^+/K^+ATPase antagonist ouabain. Ouabain was demonstrated to increase the excitability of the membrane allowing a Na^+ and K^+ channel-generated prolonged exocytosis to proceed before ATP depletion occurred (Cousin *et al.*, 1995c); therefore granule cells can repeatedly exocytose transmitter.This observation infers that one of two things must be happening in response to repeated high frequency stimulation - either the same vesicles are rapidly recycling and refilling with transmitter, but because the dye is washed away no reuptake is possible or, unlabeled vesicles from a reserve pool are recruited by the high frequency stimulation and take part in the latter rounds of exocytosis. In support of the latter hypothesis, evidence has been reported that at high stimulation frequencies a reserve pool can be utilized which is usually bound to the synaptic vesicle protien synapsin using anti-synapsin antibodies in lamprey knockout mice and synapsin I and II knockout mice (Li *et al.*, 1995; Pieribone *et al.*, 1995; Rosahl *et al.*, 1995).

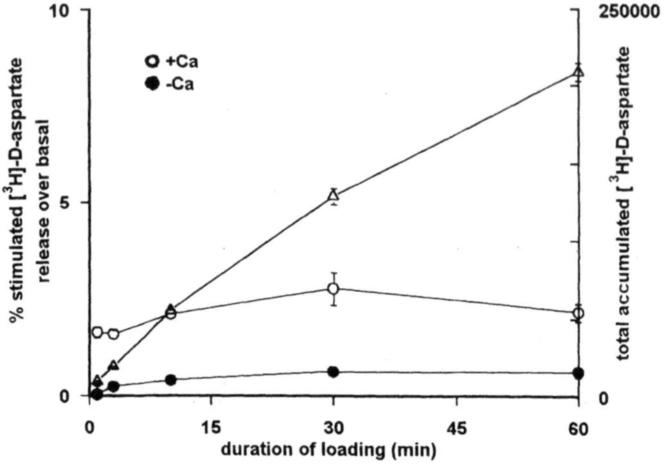

Fig 6. [^{3}H]-D-aspartate labeling of exocytotic pool. Granule cells were loaded with [^{3}H]-D-aspartate for increasing periods of time as indicated. Cells were stimulated with an electrical field of 40Hz for 10s four times reversing polarity. after each stimulation in either 1.3mM Ca^{2+} (open circles) or low Ca^{2+} (closed circles) medium. Release is presented as stimulated over basal efflux. Accumulated [^{3}H]-D-aspartate is presented as total counts (open triangles).

The kinetics of small synaptic vesicle recycling and repriming have been estimated to require approximately a $t_{1/2}$ of 20s and a delay of 15s respectively in hippocampal neurons (Ryan and Smith, 1995) however no indication is given as to whether the reprimed vesicles are refilled with transmitter. As [^{3}H]-D-aspartate release has been demonstrated after only 1 minute of loading in this preparation it is possible that vesicles could recycle and refill in time for latter round of exocytosis.

4 Conclusions

1) The non-metabolizable analogue [^{3}H]-D-aspartate has been demonstrated to label the vesicular amino acid pool in cultured cerebellar granule cells with exocytosis of the transmitter being able to be evoked by either KCl^-, or electrical field-stimulation.
2) Field stimulated exocytosis is dependent on action potential generation as it can be modulated by Na^+ and K^+ channel antagonists.
3) Field-evoked exocytosis in granule cells is frequency-dependent with no contribution from transient vesicle fusion. This would agree with the proposed frequencies that granule cells *in situ* would encounter from mossy fibre inputs, reported to be up to 150Hz.
4) Exocytosis of [^{3}H]-D-aspartate, but not FM1-43 destaining, is linear with increasing stimuli number indicating that [^{3}H]-D-aspartate-loaded, non-FM1-43-

labeled vesicles are contributing to the release. The vesicle pool utilized in the latter stimuli may be either a reserve pool or a rapidly recycling pool at the plasma membrane. The ability of $[^3H]$-D-aspartate to rapidly enter synaptic vesicles does not discount the latter hypothesis.

References

Agoston DV, Lisziewicz J (1989) Calcium uptake and protein phosphorylation in myenteric neurons, like the release of vasoactive intestinal polypeptide and acetylcholine, are frequency dependent. J Neurochem 52: 1637-1640.

Alvarez de Toledo G, Fernandez Chacon R, Fernandez JM (1993) Release of secretory products during transient vesicle fusion [see comments]. Nature 363:554-558.

Beani L, Bianchi C, Antonelli T, Calo G, Morari M, Ferioli V, Gaist G (1992) Comparison of $[^3H]$choline and D-$[^3H]$aspartate efflux from guinea pig and human neocortex. J Neurochem 58:1454-1459.

Beani L, Tomasini C, Govoni BM, Bianchi C (1994) Fluorimetric determination of electrically evoked increase in intracellular calcium in cultured cerebellar granule cells. J Neurosci Methods 51:1-7.

Belhage B, Rehder V, Hansen GH, Kater SB, Schousboe A (1992) ^{3}H-D-aspartate release from cerebellar granule neurons is differentially regulated by glutamate and K-stimulation. J Neurosci Res 33:436-444.

Betz WJ, Mao F, Bewick GS (1992) Activity-dependent fluorescent staining and destaining of living vertebrate motor nerve terminals. J Neurosci 12:363-375.

Betz WJ, Bewick GS (1992) Optical analysis of synaptic vesicle recycling at the frog neuromuscular junction. Science 255:200-202.

Bianchi C, Beani L, Antonelli T, Vedovato M, Calo G, Tomasini C (1992) A simple method for electrical field stimulation of cultured granule cells. J Neurosci Methods 45:175-182.

Bosley TM, Woodhams PL, Gordon RD, Balazs R (1983) Effects of anoxia on the stimulated release of amino acid neurotransmitters in the cerebellum in vitro. J Neurochem 40:189-201.

Bowman EJ, Siebers A, Altendorf K (1988) Bafilomycins: a class of inhibitors of membrane ATPases from microorganisms, animals cells and plants cells. Biochemistry 85:7972-7976.

Burke SP, Adams ME, Taylor CP (1993) Inhibition of endogenous glutamate release from hippocampal tissue by Ca2+ channel toxins. Eur J Pharmacol 238:383-386.

Burke SP, Nadler JV (1988) Regulation of glutamate and aspartate release from slices of the hippocampal CA1 area: effects of adenosine and baclofen. J Neurochem 51:1541-1551.

Burke SP, Nadler JV (1989) Effects of glucose deficiency on glutamate/aspartate release and excitatory synaptic responses in the hippocampal CA1 area in vitro. Brain Res 500:333-342.

Carlson MD, Kish PE, Ueda T (1989) Characterization of the solubilized and reconstituted ATP-dependent vesicular glutamate uptake system. J Biol Chem 264:7369-7376.

Collins GG, Anson J, Kelly EP (1982) Baclofen: effects on evoked field potentials and amino acid neurotransmitter release in the rat olfactory cortex slice. Brain Res 238:371-383.

Collins GG (1993) Actions of agonists of metabotropic glutamate receptors on synaptic transmission and transmitter release in the olfactory cortex. Br J Pharmacol 108:422-430.

Corradetti R, Lo Conte G, Moroni F, Passani MB, Pepeu G (1984) Adenosine decreases aspartate and glutamate release from rat hippocampal slices. Eur J Pharmacol 104:19-26.

Cousin MA, Nicholls DG, Pocock JM (1993) Flunarizine inhibits both calcium-dependent and -independent release of glutamate from synaptosomes and cultured neurones. Brain Res 606:227-236.

Cousin MA, Held B, Nicholls DG (1995a) Selective neurite calcium responses in cerebellar granule cells during field stimulation. Eur J Neurosci 17:2379-2388.

Cousin MA, Pocock JM, Nicholls DG (1995b) Intracellular free calcium responses in electrically-stimulated cerebellar granule cells. Biochem Soc Trans 23:648-652.

Cousin MA, Nicholls DG, Pocock JM (1995c) Modulation of ion gradients and glutamate release in cultured cerebellar granule cells by ouabain. J Neurochem 64:2097-2104.

De Schutter E, Bower JM (1994) An active membrane model of the cerebellar Purkinje cell II. Simulation of synaptic responses. J Neurophysiol 71:401-419.

Dolphin AC, Prestwich SA (1985) Pertussis toxin reverses adenosine inhibition of neuronal glutamate release. Nature 316:148-150.

Dooley DJ, Lupp A, Hertting G (1987) Inhibition of central neurotransmitter release by omega-conotoxin GVIA, a peptide modulator of the N-type voltage-sensitive calcium channel. Naunyn Schmiedebergs Arch Pharmacol 336:467-470.

Dooley DJ, Lupp A, Hertting G, Osswald H (1988) Omega-conotoxin GVIA and pharmacological modulation of hippocampal noradrenaline release. Eur J Pharmacol 148:261-267.

Fleck MW, Henze DA, Barrionuevo G, Palmer AM (1993) Aspartate and glutamate mediate excitatory synaptic transmission in area CA1 of the hippocampus. J Neurosci 13:3944-3955.

Flint RS, Rea MA, McBride WJ (1981) In vitro release of endogenous amino acids from granule cell-, stellate cell-, and climbing fiber-deficient cerebella. J Neurochem 37:1425-1430.

Floor E, Leventhal PS, Schaeffer SF (1990) Partial purification and characterization of the vacuolar H(+)-ATPase of mammalian synaptic vesicles. J Neurochem 55:1663-1670.

Franck J, Brodin E, Fried G (1993) Differential release of endogenous 5-hydroxytryptamine, substance P, and neurokinin A from rat ventral spinal cord in response to electrical stimulation. J Neurochem 61:704-711.

Fredholm BB, Hu PS (1993) Effect of an intracellular calcium chelator on the regulation of electrically evoked [^{3}H]-noradrenaline release from rat hippocampal slices. Br J Pharmacol 108:126-131.

Gabbiani F, Midtgaard J, Knöpfel T (1994) Synaptic integration in a model of cerebellar granule cells. J Neurophysiol 72:999-1009.

Galarreta M, Solis JM, Menendez N, Conejero C, Martin del Rio R (1993) Nicotinamide adenine dinucleotides mimic adenosine inhibition on synaptic transmission by decreasing glutamate release in rat hippocampal slices. Neurosci Lett 159:55-58.

Gallo V, Levi G (1982) Selective release of glutamate from cerebellar granule cells differentiating in culture. Proc Natl Acad Sci (USA) 79:7919-7923.

Gaur S, Newcomb R, Rivnay B, Bell JR, Yamashiro D, Ramachandran J, Miljanich GP (1994) Calcium channel antagonist peptides define several components of transmitter release in the hippocampus. Neuropharmacol 33:1211-1219.

Henkel AW, Lübke J, Betz WJ (1996) FM1-43 dye ultrastructural localization in and release from frog motor nerve terminals. Proc Natl Acad Sci USA 93:1918-1923.

Herdon H, Nahorski SR (1989) Investigations of the roles of dihydropyridine and omega-conotoxin-sensitive calcium channels in mediating depolarisation-evoked endogenous dopamine release from striatal slices. Naunyn Schmiedebergs Arch Pharmacol 340:36-40.

Holz GG, Dunlap K, Kream K (1988) Characterization of the electrically evoked release of substance P from dorsal root ganglion neurons: methods and dihydropyridine sensitivity. J Neurosci 8:463-471.

Huston E, Cullen JP, Burley JR, Dolphin AC (1995) The involvement of multiple calcium channel sub-types in glutamate release from cerebellar granule cells and its modulation by $GABA_B$ receptor activation. Neuroscience 68: 465-478.

Israel M, Lesbats B, Bruner J (1993) Glutamate and acetylcholine release from cholinergic nerve terminals, a calcium control of the specificity of the release mechanism. Neurochem Int 22:53-58.

Kangrga I, Randic M (1990) Tachykinins and calcitonin gene-related peptide enhance release of endogenous glutamate and aspartate from the rat spinal dorsal horn slice. J Neurosci 10:2026-2038.

Keith RA, Mangano TJ, DeFeo PA, Ernst GE, Warawa EJ (1994) Differential inhibition of neuronal calcium entry and [^{3}H]-D-aspartate release by the quaternary derivatives of verapamil and emopamil. Br J Pharmacol 113:379-384.

Kennedy RT, Jones SR, Wightman RM (1992) Dynamic observation of dopamine autoreceptor effects in rat striatal slices. J Neurochem 59:449-455.

Kihara M, Misu Y, Kubo T (1989) Release by electrical field stimulation of glutamate, gamma-aminobutyric acid, and other amino acids from slices of the rat medulla oblongata. J Neurochem 52:261-267.

Kimura M, Yamanishi Y, Hanada T, Kagaya T, Kuwada M, Watanabe T, Katayama K, Nishizawa Y (1995) Involvement of P-type calcium channels in high potassium-elicited release of neurotransmitters from rat brain slices. Neurosci 66:609-615.

Kingsbury A, Gallo V, Balázs R (1988) Stimulus-coupled release of amino acids from cerebellar granule cells inculture. Brain Res 448:46-52.

Levi G, Gallo V (1986) Release studies related to the neurotransmitter role of glutamate in the cerebellum: an overview. Neurochem Res 11:1627-42.

Li L, Chin LS, Shupliakov O, Brodin L, Sihra TS, Hvalby O, Jensen V, Zheng D, McNamara JO, Greengard P, Andersen P (1995) Impairment of synaptic vesicle clustering and of synaptic transmission, and increased seizure propensity, in synapsin I-deficient mice. Proc Natl Acad Sci USA 92:9235-9239.

Maura G, Carbone R, Raiteri M (1989) Aspartate-releasing nerve terminals in rat striatum possess D-2 dopamine receptors mediating inhibition of release. J Pharmacol Exp Ther 251:1142-1146.

Maura G, Barzizza A, Folghera S, Raiteri M (1991) Release of endogenous aspartate from rat cerebellum slices and synaptosomes - inhibition mediated by a 5-HT2 receptor and by a 5-HT1 receptor of a possibly novel subtype. Naunyn-Schmied Arch Pharmacol 343:229-236.

Maycox PR, Deckwerth T, Hell JW, Jahn R (1988) Glutamate uptake by brain synaptic vesicles. J Biol Chem 263:15423-15428.

McMahon HT, Foran P, Dolly JO, Verhage M, Wiegant VM, Nicholls DG (1992) Tetanus toxin and botulinum toxins type A and B inhibit glutamate, gamma-aminobutyric acid, aspartate, and met-enkephalin release from synaptosomes. Clues to the locus of action. J Biol Chem 267:21338-21343.

McMahon HT, Nicholls DG (1990) Glutamine and aspartate loading of synaptosomes: A reevaluation of effects on calcium-dependent excitatory amino acid release. J Neurochem 54:373-380.

Monck JR, Fernandez JM (1994) The exocytotic fusion pore and neurotransmitter release. Neuron 12:707-716.

Moriyama Y, Futai M (1990) H(+)-ATPase, a primary pump for accumulation of neurotransmitters, is a major constituent of brain synaptic vesicles. Biochem Biophys Res Commun 173:443-448.

Naito S, Ueda T (1985) Characterization of glutamate uptake into synaptic vesicles. J Neurochem 44:99-109.

Neville LF, Arvin B, Roberts PJ (1987) Is D-aspartate a useful tool for assessing excitatory amino acid neurotransmitter function? Biochem Soc Trans 16:315-316.

Palmer AM, Reiter CT (1994) Comparison of the superfused efflux of preaccumulated D-[^{3}H]aspartate and endogenous L-aspartate and L-glutamate from rat cerebrocortical minislices. Neurochem Int 25:441-450.

Pende M, Lanza M, Bonanno G, Raiteri M (1993) Release of endogenous glutamic and aspartic acids from cerebrocortex synaptosomes and its modulation through activation of a gamma-aminobutyric acid$_B$ (GABA$_B$) receptor subtype. Brain Res 604:325-330.

Pieribone VA, Shupliakov O, Brodin L, Hilfiker-Rothenfluh S, Czernik AJ, Greengard P (1995) Distinct pools of synaptic vesicles in neurotransmitter release. Nature 375:493-497.

Pocock JM, Cousin MA, Nicholls DG (1993) The calcium channel coupled to glutamate exocytosis from cerebellar granule cells is inhibited by the spider toxin Aga-GI. Neuropharmacol 32:1185-1194.

Poli A, Contestabile A, Migani P, Rossi L, Rondelli C, Virgili M, Bissoli R, Barnabei O (1985) Kainic acid differentially affects the synaptosomal release of endogenous and exogenous amino acidic neurotransmitters. J Neurochem 45:1677-1686.

Poli A, Lucchi R, Vibio M, Barnabei O (1991) Adenosine and glutamate modulate each other's release from rat hippocampal synaptosomes. J Neurochem 57:298-306.

Prado MAM, Gomez MV, Collier B. (1993) Mobilization of a vesamicol-insensitive pool of acetylcholine from a sympathetic neuron by ouabain. J Neurochem 61: 45-56.

Raiteri M, Marchi M, Costi A, Volpe G (1990) Endogenous aspartate release in the rat hippocampus is inhibited by M2 'cardiac' muscarinic receptors. Eur J Pharmacol 177:181-187.

Reuter H (1995) Measurements of exocytosis from single presynaptic nerve terminals reveal heterogeneous inhibition by Ca^{2+}-channel blockers. Neuron 14:773-779.

Rogers KL, Philibert RA, Duttton GR (1991) K-stimulated amino acid release from cultured cerebellar neurons: comparison of static and dynamic stimulation paradigms. Neurochem Res 16:899-904.

Rosahl TW, Spillane D, Missler M, Herz J, Selig DK, Wolff JR, Hammer RE, Malenka RC, Südhof TC (1995) Essential functions of synapsins I and II in synaptic vesicle regulation. Nature 375:488-493.

Ryan TA, Reuter H, Wendland B, Schweizer FE, Tsien RW, Smith SJ (1993) The kinetics of synaptic vesicle recycling measured at single presynaptic boutons. Neuron 11:713-724.

Ryan TA, Smith SJ (1995) Vesicle pool mobilization during action potential firing at hippocampal synapses. Neuron 14:983-989.

Schiavo G, Benfenati F, Poulain B, Rossetto O, Polverino-de-Laureto P, Dasgupta BR, Montecucco C (1992) Tetanus and botulinum-B neurotoxins block neurotransmitter release by proteolytic cleavage of synaptobrevin. Nature 359:832-835.

Schwartz DD, Malik KU (1991) Mechanism of neurotransmitter release elicited by the preferential alpha 1-adrenergic receptor agonist phenylephrine in superfused superior cervical ganglion cells in culture. J Neurochem 57:831-838.

Shioi J, Ueda T (1990) Artificially imposed electrical potentials drive L-glutamate uptake into synaptic vesicles of bovine cerebral cortex. Biochem J 267:63-68.

Simonato M, Bregola G, Muzzolini A, Bianchi C, Beani L (1993) Characterization of the K-evoked ^{3}H-D-aspartate outflow in the rat hippocampus in vitro. Neurochem Int 23:555-560.

Smith CB, Betz WJ (1996) Simultanious independent measurement of endocytosis and exocytosis. Nature 380: 531-534.

Szerb JC (1988) Changes in the relative amounts of aspartate and glutamate released and retained in hippocampal slices during stimulation. J Neurochem 50:219-224.

Terrian DM, Gannon RL, Rea MA (1990) Glutamate is the endogenous amino acid selectively released by rat hippocampal mossy fiber synaptosomes concomitantly with prodynorphin- derived peptides. Neurochem Res 15:1-5.

Van-Vliet BJ, Sebben M, Dumuis A, Gabrion J, Bockaert J, Pin JP (1989) Endogenous amino acid release from cultured cerebellar neuronal cells: effect of tetanus toxin on glutamate release. J Neurochem 52:1229-1239.

Verhage M, McMahon HT, Ghijsen WEJM, Boomsma F, Wiegant V, Nicholls DG (1991) Differential release of amino acids, neuropeptides and catecholamines from nerve terminals. Neuron 6:517-524.

Wakade AR, Wakade TD (1988) Comparison of transmitter release properties of embryonic sympathetic neurons growing in vivo and in vitro. Neurosci 27:1007-1019.

Wakade TD, Bhave SV, Bhave A, Przywara DA, Wakade AR (1990) Ca^{2+} mobilized by caffeine from the inositol 1,4,5-trisphosphate-insensitive pool of Ca^{2+} in somatic regions of sympathetic neurons does not evoke [^{3}H]norepinephrine release. J Neurochem 55:1806-1809.

Wang Y, Floor E (1994) Dynamic storage of glutamate in rat brain synaptic vesicles. Neurosci Lett 180:175-178.

Wessler I, Dooley DJ, Werhand J, Schlemmer F (1990) Differential effects of calcium channel antagonists (omega-conotoxin GVIA, nifedipine, verapamil) on the electrically-evoked release of [^{3}H]acetylcholine from the myenteric plexus, phrenic nerve and neocortex of rats. Naunyn Schmiedebergs Arch Pharmacol 341:288-294.

Wessler I, Dooley DJ, Lohr B (1995) P-type Ca2+ channels trigger stimulus-evoked [^{3}H]acetylcholine release from mammalian motor endplates. Eur J Pharmacol 278:83-86.

Westergaard N, Fosmark H, Schousboe A (1991) Metabolism and release of glutamate in cerebellar granule cells cocultured with astrocytes from cerebellum or cerebral cortex. J Neurochem 56:59-66.

Wichmann T, Limberger N, Starke K (1989) Release and modulation of release of serotonin in rabbit superior colliculus. Neurosci 32:141-151.

Zhou M, Peterson CM, Lu Y-B, Nadler JV (1995) Release of glutamate and aspartate from CA1 synaptosomes:selective modulation of aspartate release by ionotropic glutamate receptor ligands. J Neurochem 64: 1556-1566.

Zhu XZ, Chuang DM (1987) Modulation of calcium uptake and D-aspartate release by $GABA_B$ receptors in cultured cerebellar granule cells. Eur J Pharmacol 141:401-408.

Modulation of Synaptic NMDA Current by Hydrogen Ion and Histamine in Hippocampal Slices

Hale Saybaşılı

Boğaziçi University, Biomedical Engineering Institute, 80815, Bebek, İstanbul, Turkey

Abstract

The postsynaptic receptor proteins that are excited by excitatory amino acid, L-glutamic acid can be subdivided into four main classes depending on their sensitivity to selective agonists: N-methyl-D- Aspartate (NMDA); Kainate; (S)-2-amino-3- (3-hydroxy-5-methylisoxazol-4-yl)propionic acid (S-AMPA) and metabotropic receptors. One of these receptors, the NMDA receptor has captured the attention of researchers because of its involvement in physiological processes, such as Long Term Potentiation (LTP) as well as neurological disorders and excitotoxicity. The NMDA receptor ionophore complex is coupled to an ion channel that is permeable to sodium, potassium, calcium ions and blocked by magnesium ions in a voltage dependent way. This receptor complex has distinct modulatory sites. Control of the ion channel through these multiple modulatory sites is important in the behavior of this multifunctional receptor.In this study, proton and histamine modulation of synaptic NMDA current recorded from CA1 pyramidal cell layer of rat hippocampal slices is presented. These modulatory mechanisms may have important implications during alkaline shifts that occur during normal synaptic release of glutamate and under ischemia/anoxia condition when the pH of the extracellular medium is lowered.

1 Introduction

"Synapses" as defined by Sherrington in 1897 are the sites of contact between nerve cells. The presynaptic neuron is the place where the neurotransmitter vesicles are stored. Neurotransmitters are released to the synaptic cleft with an increase in calcium ion concentration in the presynaptic terminal. Upon their release to the synaptic cleft, the neurotransmitter molecules interact with the postsynaptic membrane from the sites known as receptors. Depending on the nature of the permeability change to ions, the generated response can be inhibitory or excitatory. The powerful excitatory actions of the dicarboxylic amino acids, L-glutamic and L-aspartic acid were first shown by Hayashi in 1954. L-glutamic acid activates at least four subtypes of pharmacologically distinguishable postsynaptic receptors:N-methyl-D-aspartate (NMDA); kainate; (S)-2-amino-3- (3-hydroxy-5-methylisoxazol-4-yl)propionic acid ((S)-AMPA) (1)and metabotropic receptors. Kainate and AMPA receptors are collectively referred as non-NMDA receptors. NMDA and non-NMDA

NATO ASI Series, Vol. H 100
Neurotransmitter Release and Uptake
Edited by Şakire Pöğün

receptors have different properties. Activation of the NMDA receptor is voltage dependent and causes an increase in the permeability to sodium, potassium and calcium ions (2). NMDA receptor is blocked by magnesium ions under resting state conditions (3). This voltage dependent component shows a slow activation and the response is abolished by a number of phosphono amino acids i.e., (R)-2-amino-5-phosphonopentanoic acid (AP-5) or 2-amino-7-phosphonoheptanoic acid (AP-7). Non-NMDA receptors do not show voltage dependence and are permeable to sodium and potassium ions. Activation of non-NMDA receptors mediates a fast component and upon their activation they rapidly desensitize (4). These receptors are blocked by selective non-NMDA antagonists; 6-cyano-7-nitroquinoxaline-2,3-dione (CNQX) (5) or with a recently identified one, 2,3-dihydro-6-nitro-7-sulfamyl-benzo (F)quinoxaline (NBQX) (6). Autoradiographic studies showed that non-NMDA receptors are widely distributed in the central nervous system and except for the cerebellum this distribution corresponds closely to that of the NMDA receptors (7). NMDA and non-NMDA receptors are colocalized at CA1 region of the hippocampus (8). The density of NMDA receptors in Schaffer collateral pathway and in perforant path synapses in hippocampus seems to be the highest in the entire brain.

1.1 The NMDA Receptor-Channel Complex and Its Modulatory Sites

The NMDA receptor channel is a ligand gated ion channel that has several identified and suggested modulatory sites. Those are: Agonist binding site, Glycine binding site, Magnesium ion binding site, Dissociative anesthetic binding site, Zinc ion binding site, Redox modulatory site, Hydrogen ion binding site and Polyamine binding site. The channel is blocked by magnesium ion in the negative voltage range and blockage is relieved by membrane depolarization, increasing the permeability of the channel to calcium ion besides sodium and potassium ions (2).

L-glutamic acid or NMDA binds to the agonist binding site of the receptor. The receptor can be blocked by D-AP5or D-AP7, these antagonists have good selectivity and low micromolar affinity for the transmitter recognition site.

Johnson and Ascher in 1987 showed that the classical inhibitory neurotransmitter, glycine potentiated the NMDA receptor mediated responses in a strychnine insensitive way in cultured brain neurons (9). They suggested that glycine acts allosterically to increase the frequency of NMDA induced channel opening. After the discovery of the glycine potentiating effect on NMDA receptors, it was immediately noted that there was a nearly identical neuroanatomical distribution between [^{3}H]-glycine binding sites and those of NMDA sensitive [^{3}H]- Glutamate binding sites. When glycine was removed from the receptor environment virtually no NMDA response could be elicited indicating that glycine acts as a coagonist at the NMDA receptor complex, thus agonist and glycine bindings are simultaneously required for channel activation. However electrophysiological studies in vivo and in slices of adult tissue failed to show any consistent potentiating effect of glycine on NMDA responses, suggesting that the glycine site was maximally activated by the extracellular concentration of glycine in slices. 7-Chlorokynurenic acid antagonizes the NMDA response by acting at the glycine site. It was found that the acquisition of spatial memory is impaired in rats that are subjected to the chronical application of 7-Chlorokynurenic acid (10).

Ault et al. reported in 1980 that magnesium ion selectively antagonized NMDA induced spinal motorneuron depolarization where it had no effect on non-NMDA activity (11). Later, Mayer and Westbrook suggested that the NMDA channel block by magnesium ions and the voltage sensitivity of the channel might be related (3). In magnesium free solutions, NMDA agonists open the channel with a linear current-voltage relation, however addition of magnesium ion produces a dramatic reduction of the NMDA induced currents recorded at negative membrane potentials. The effect is detectable with micromolar concentrations of magnesium ion and nearly a total block is obtained with physiological concentrations (i.e.1-2mM). The blockage is voltage dependent and abolished by depolarization.

The dissociative anesthetics such as phencyclidine (PCP), ketamine and (±) 5methyl-10,11-dihydro5H-dibenzo[a, d]cyclohepten-5,10-iminemaleate (MK-801) bind to a receptor site inside the ion channel (12). Once bound to PCP receptor, the ligand can be trapped in the channel. The neurotoxic side effects of these anesthetics limit their application, since these drugs induce acute pathomorphological changes in specific population of brain neurons (13). Also MK-801 exerts its dissociative effect on seizure activity of hippocampal kindled cats at a dose level that produces severe behavioral toxicity (14). Dissociative anesthetics are also referred as open channel blockers, because these non-competitive antagonists can gain access to the channel only in the open state. The time required for the dissociation of the open channel blockers from their binding site limit their clinical application (15).

NMDA receptor complex is also modulated by endogenous zinc ions. It has been indicated that zinc ions selectively attenuate the NMDA receptor mediated neurotoxicity without effecting the non-NMDA receptor caused neurotoxicity (16). Zinc ions are suggested to be released in to the extracellular space during excitation of hippocampal slices. The corelease of zinc ions together with the endogenous glutamate might be related with the control of the resultant NMDA response. High potassium induced depolarization in hippocampal slices cause the release sufficient zinc ions to achieve a uniform extracellular concentration of 300 μM. Maximal extrusion of zinc ions were coincided with the period of hyper excitability of the tissue (17). Ionized zinc concentration is shown to be important in the blockage of the NMDA channel, e.g., at pH 9.0, 20 μM zinc ion depressed NMDA current by 6 per cent and at pH 7.3, 20 μM zinc ion depressed NMDA current by 55 per cent (18). Zinc ions may constitute an important negative feedback mechanism to control excessive synaptic excitation during prolonged exposure to glutamate.

The redox modulatory site, which is also an integral part of the NMDA receptor consists one or more sulfhydryl groups. Under reducing conditions, where disulfide bond is unformed the opening frequency of the NMDA receptor channel increased thus leading to the NMDA receptor mediated neurotoxicity (15,19).

NMDA receptor modulation by proton concentration was analyzed in cultured hippocampal neurons (18,20). Increased external hydrogen ion concentration suppressed, whereas decreased external hydrogen ion concentration enhanced the NMDA receptor mediated current in a voltage insensitive way. Decrease in pH from 7.3 to 6.0 reduced the NMDA response by 33 per cent and an increase in pH from 7.3 to 8.0 potentate the NMDA response by 14 per cent. The non-NMDA receptor mediated response was only slightly changed with this altered pH. Ischemia related decrease in extracellular pH is accompanied by an increased amount of extracellular glutamic acid concentration. Thus the hydrogen ion interaction with NMDA receptor may influence the time course of the post-ischemic neuronal damage. Ischemia and its relation with the damage of the neurons in the CA1 region of the hippocampus have been shown in a human case, a patient who exhibited a marked memory impairment (21). After transient forebrain ischemia the extracellular concentrations of excitatory aminoacids were elevated in cultured hippocampal cells. Hippocampal astrocytes are also suggested to be involved in the induced release (22).

Ransom and Stec in 1988 have reported that in the extensively washed rat cortical neurons the polyamine spermine and spermidine increased the binding of the open channel blocker MK-801 to the NMDA receptor, suggesting that polyamines enhance NMDA receptor activity. Spermine was unable to activate NMDA receptors in the absence of glutamate and glycine, suggesting an interaction with an allosteric site that is independent from those for the coagonists (23). The mechanism of action of polyamines appears to be complex at NMDA receptors; at low micromolar concentrations (1-10 μM) spermine enhanced the NMDA current by increasing channel opening frequency and at higher concentrations (>10 μM) a voltage dependent decrease in channel current amplitude and average open time have suggested a modulation through an independent site on the NMDA receptor complex (24).

1.2 Histaminergic System in the Brain

The 2- (4-imidazolyl)ethylamine or histamine (HA) is an endogenous amine and its release from histamine fibers in the brain affects attention and arousal state (Figure 1). Histamine is a neuromodulator and the histaminergic pathway projects throughout the brain including all fields of the hippocampus. Histamine poorly penetrates the blood-brain barrier, so it is synthesized from L-histidine with the decarboxylation by L-histidine decarboxylase in the tuberomammillary nucleus of the hypothalamus. L-Histidine decarboxylase activity is distributed in a heterogeneous fashion between cerebral regions, with the highest levels being found in hypothalamus. Garbarg et al. showed that diencephalic lesions involving the medial forebrain bundle induce a progressive decrease in both the histamine concentration and L-histidine decarboxylase activity of the telencephalon (25). Histamine is synthesized and released from a discrete set of neurons ascending through the lateral hypothalamic area and widely projecting the telencephalon. Histamine containing neurons uniquely possess a specific histidine decarboxylating enzyme and those neurons can synthesize and store labeled histamine. Histamine acts on target cells in mammalian brain, indicating the widespread distribution of this monoamine in the brain and the sedative properties of most antihistaminics in therapeutic doses shows the importance of this monoamine system.

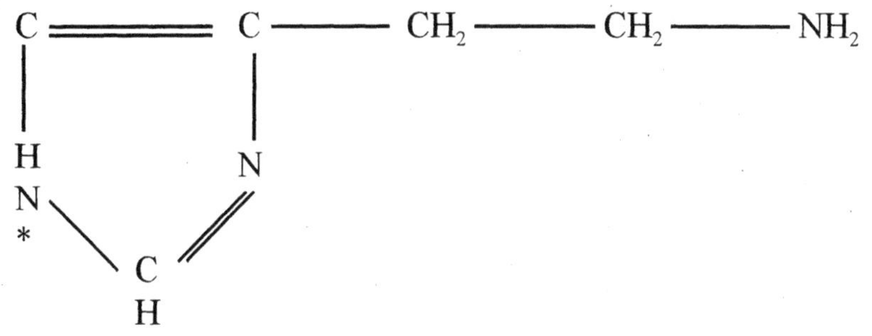

Fig 1. Structure of Histamine. Histamine has an amine group (*) on the imidazole ring and has a pK_R value near 6.0.

Till now, three histamine receptors have been identified which mediate the modulatory histamine action in the brain: H1, H2 and H3 receptors. All three histamine receptors mediate their action through the activation of second messenger systems (26). Classical histamine receptors that are present in brain tissue are linked to different intracellular messenger system, through which they may influence several physiological functions. Histamine is released to the synaptic cleft upon depolarization and its release is completely inhibited in the absence of calcium ions. H1 histamine receptor actions are almost exclusively excitatory. H1 receptors are coupled to phosphoinositide turnover and participate in H1 receptor mediated cyclic guanosine monophosphate formation, they can also potentiate the activators of adenylate cyclase. Activation of this receptor type increases the firing rate, generates depolarization and facilitates signal transmission. In the brain stem, this type of activity is generated due to the reduced potassium current. Inhibitory action of H1 receptor agonists can also be observed, this effect is found to be sensitive to intracellular calcium concentration. Excitation is generated by reduced and inhibition is generated by increased intracellular calcium through calcium dependent potassium channels. H2 histamine receptor actions may be inhibitory or excitatory. H2 receptor activation can open or close calcium dependent potassium channels causing depression or potentiation of excitation. H2 receptor activation in hippocampus causes the potentiation of excitation. This effect is mediated through the suppression of calcium activated potassium conductance ($g_{Ca^{2+}(K^+)}$) via the activation of adenosine 3'-5' monophosphate (cAMP); as a result of this activity action potential is prolonged and accommodation is blocked. H2 histamine receptor agonist impromidine decreases the afterhyperpolarization in hippocampal pyramidal neurons (27). Histamine synthesis and release are regulated by H3 receptors. H1 and H2 histamine receptors are located on the postsynaptic site whereas H3 histamine receptors are located presynaptically and they display distinct pharmacological specificity. Presence of 1 μM exogenous histamine in the medium inhibited the evoked release of histamine. The inhibitory action of 1 μM exogenous histamine on the release process was progressively reversed in the presence of impromidine at increasing concentrations, indicating that impromidine is an antagonist of the autoreceptor H3 (impromidine is H2 receptor agonist as well). It was also observed that the histamine autoinhibitory process operates at concentrations of the amine at least hundred times lower than those required to stimulate the postsynaptic receptors in the brain slice preparation, probably indicating that autoreceptors of histamine are more sensitive to the neurotransmitter than corresponding postsynaptic receptor (28).

Histamine application has selectively enhanced the NMDA receptor mediated currents in isolated hippocampal pyramidal neurons. Histamine (100 μM) enhanced the NMDA autaptic current, the same effect was mimicked with selective H3 agonist, R-α-methyl histamine (100 μM), however it was impossible to block this effect with H3 antagonist thioperamide (29). Enhancement of NMDA current by histamine has been blocked by spermine in isolated hippocampal neurons, suggesting an interaction of histamine with the polyamine binding site of the NMDA receptor complex (30).

2 Experimental Results

Presented experimental results were obtained from 350 μm thick hippocampal slices of young (15-28 days old) Wistar male rats. Upon low frequency stimulation of Schaffer collateral commisural pathway (0.1 Hz) with twisted bipolar tungsten electrodes and by using patch clamp tight-seal whole cell recording technique the excitatory postsynaptic currents (EPSCs) of CA1 pyramidal cells were recorded. Holding potential was -50 mV and in the presented figures the EPSCs were averaged ten times. The NMDA receptor mediated synaptic currents were recorded in the presence of 12 μM 6,7-dinitroquinoxaline-2,3-dione (DNQX). The composition of the artificial cerebrospinal fluid (ACSF) was (in mM): 125 NaCl, 3 KCl, 1.2 KH_2PO_4, 2.0 $CaCl_2$, 1.6 $MgCl_2$, 10 glucose, 26 $NaHCO_3$. Pipette intracellular solution composition was as follows (in mM):140 CsF,10 HEPES, 5 EGTA, 0.5 Ca_{2+}, 0 ATP (pH 7.4). Resistances of patch electrodes were between 3-5 MΩ. Chloride mediated post synaptic currents were suppressed by replacement of intracellular chloride with fluoride; fluoride is an inhibitor of phosphoinositol mediated responses (31).

Alteration of the pH of the perfusion medium was confirmed following equilibration of the solution with carbogen and phenol red was present in the medium as a second indicator of pH. Reported pH values reflect the values for the gassed reservoir of ACSF. Histamine hydrochloride (10 μM) was added to ACSF and the pH of ACSF was adjusted before applying to the perfusion system. Signals were recorded with an Axoclamp 2A amplifier and for the data collection and analysis pCLAMP version 5.5 (Axon Instruments) was used.

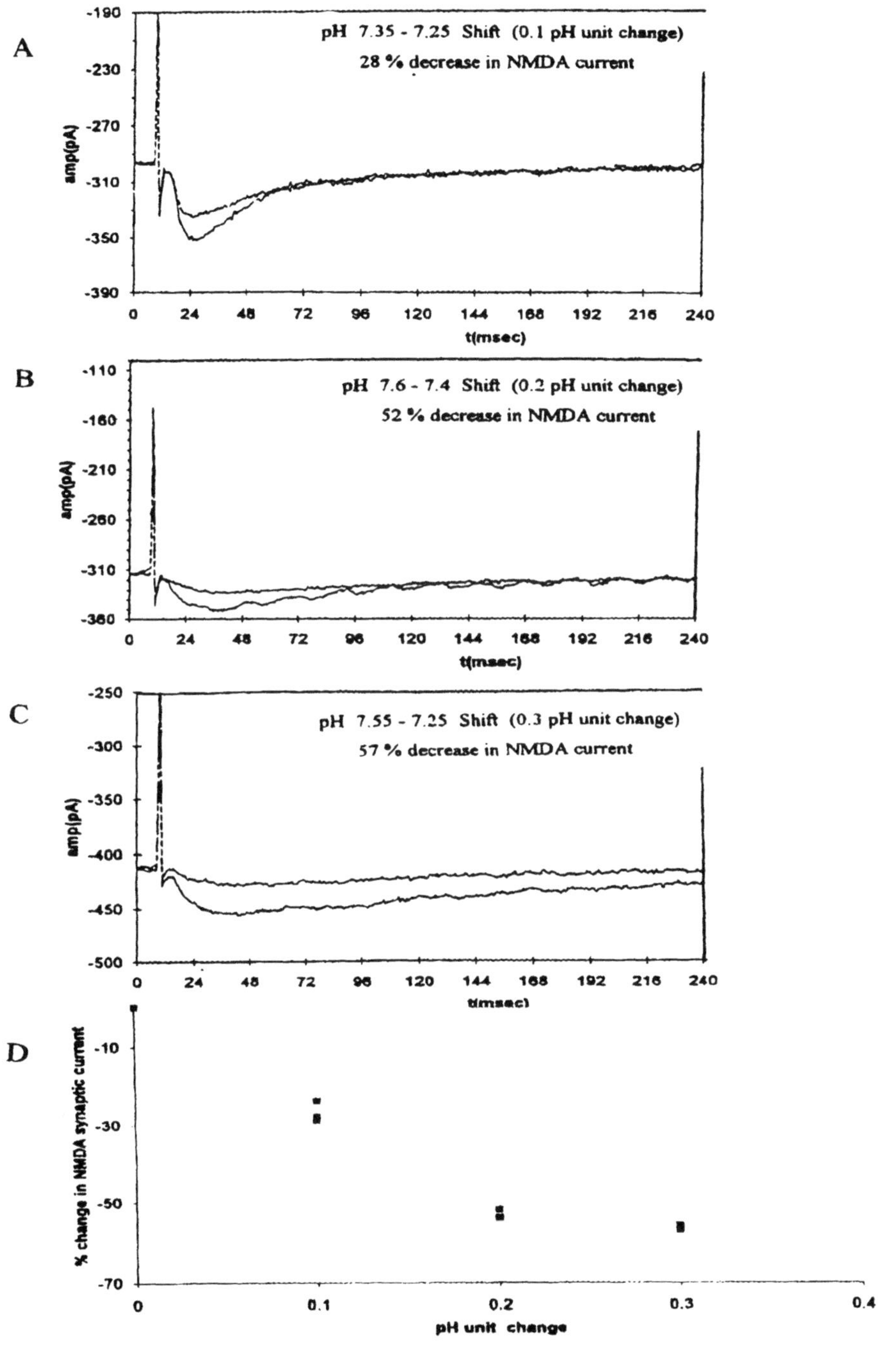

Fig 2. Effect of Extracellular pH Change on Synaptic NMDA Current in the Presence of DNQX (12 μM) **A.** pH 7.35-7.25 shift decreased the NMDA current by 28% (n=3). **B.** pH 7.6-7.4 shift decreased the NMDA current by 52% (n=2).**C.** pH 7.55-7.25 shift decreased the NMDA current by 57% (n=2). Synaptic currents were blocked by AP5 (50 μM) at the end of the experiments. **D.** The relation between extracellular pH and NMDA current component.

Changing the pH of the ACSF in the acidic direction (0.1,0.2,0.3 pH units) depressed the Schaffer collateral mediated NMDA receptor current of the CA1 pyramidal cell (Figure 2). The depression was reversible and non-NMDA current component was not influenced with these small pH shifts (figures not shown here) that is, the pH effect is selective for the NMDA current component of EPSC. Since NMDA and non-NMDA receptors are colocalized at the CA1 region of the hippocampus (8), the pH effect seems to be a postsynaptically mediated effect.

Histamine effect on the NMDA component of the Schaffer collateral mediated EPSC shown to be a pH dependent effect (32). At pH 7.4 histamine has no effect on the pharmacologically isolated NMDA current (Figure 3A). At pH 7.0-7.2 histamine has enhanced the synaptic NMDA current (Figure 3B); at elevated pH values histamine has depressed the synaptic NMDA current (Figure 3C). The isolated non-NMDA current component of the EPSC was not affected with histamine application at pH 7.2 (Figure 3D). The relation between pH and the action of histamine is shown in Figure 3E.

In the above presented two effects, that is, pH and histamine effects on the synaptic NMDA current, the involvement of second messenger systems was excluded, since in the pipette intracellular solution ATP was not included and furthermore fluoride was present.

3 NMDA Receptor and Its Relation with Physiological and Pathological Conditions

It has widely been accepted that NMDA receptor activation is a requirement for Long Term Potentiation (LTP) induction (33). Activity dependent synaptic plasticity was initially described in the hippocampus by Bliss and Lømo in 1973 (34). LTP is believed to be the underlying mechanism of learning and memory. Long Term Potentiation is the sustained increase (i.e.hours or more) in the amplitude of the synaptic response evoked in a cell by a test stimulus delivered to an afferent hippocampal pathway following a brief tetanic stimulation (e.g.100 Hz for 1 sec) of that pathway. Using calcium imaging techniques it has been shown that tetanic stimulation elevates calcium within dendrites and dendritic spines (33). Increased intracellular calcium, evoked by excitatory amino acids could be caused by either transmembrane influx of calcium or release of calcium from intracellular stores. The involvement of metabotropic glutamate receptors in the LTP generation has been reported. This fact was shown by the blockage of the tetanus induced LTP upon the application of selective metabotropic glutamate receptor antagonist, (RS)-α-methyl-4 -carboxyphenylglycine (35).

Cultured hippocampal neurons exhibit acute toxic swelling upon 30 minutes exposure to excitatory amino acids. Replacement of extracellular sodium with an impermeant cation attenuated the neuroexcitatory action of excitatory amino acids and blocked the acute toxic swelling. Similar protection was also achieved by replacing extracellular chloride with an impermeant anion that is expected to secondarily decrease the net sodium and water influx (36). Upon the extracellular removal of both sodium and calcium ions cortical neurons exposed to glutamate showed neither acute swelling nor delayed degeneration. It has been suggested that sodium dependent component is responsible from immediate cell swelling and calcium dependent component is responsible from delayed cell degeneration.

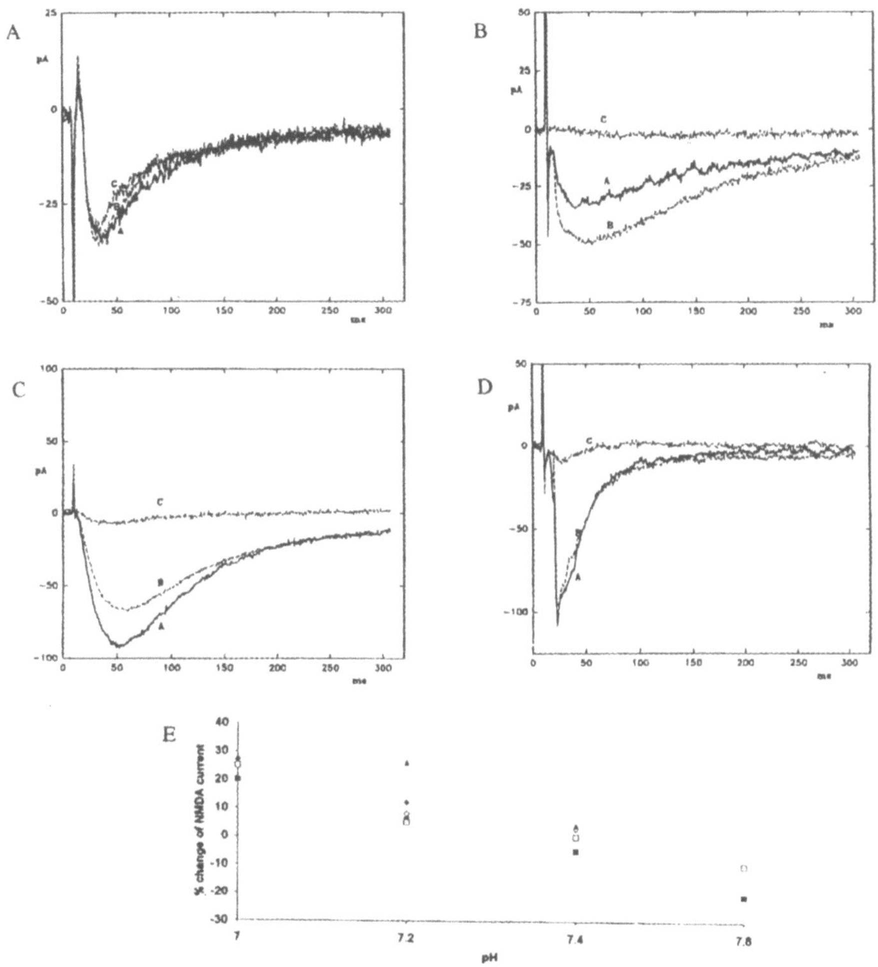

Fig 3. Action of Histamine on the Excitatory Postsynaptic Current Components. **A.** At standard bath pH of 7.4 histamine had no consistent effect on the NMDA current (A: control; B; HA application; C: wash). **B.** When the pH value of the perfusion system was lowered to pH 7.0-7.2 the NMDA current was enhanced with histamine application (A: control; B; HA application; C: blockage by AP5 (50 μM)). **C.** At a raised pH value of 7.6 histamine treatment depressed the NMDA current component of EPSC (A: control; B: HA application; C: blockage by kynurenate). Synaptic NMDA currents are recorded in the presence of DNQX (12 μM). **D.** Action of histamine on the non-NMDA current at an extracellular pH value of 7.2 in the presence of 50 μM AP5 (A: control; B: HA application; C: blockage by DNQX (12 μM)). **E.** Graph of the relation between pH and the action of Histamine. (32)

It was also reported that blockage of NMDA receptors with the focal injections of 2-amino-7-phosphonoheptanoic acid (AP7) immediately before ischemia protects all cell types in the hippocampus against ischemic damage in vivo. This indicates that NMDA receptors are functional in brain ischemia (37).

It is extremely important to keep the intracellular calcium concentration between 50-150 nM range against an extracellular concentration of millimolar range under resting conditions. The resting free intracellular calcium appears to be controlled primarily by sequestration into organelles such as smooth endoplasmic reticulum and mitochondria. ATP depletion initiates an immediate linear increase in intracellular calcium concentration and prolonged elevation of intracellular calcium concentration is deleterious for neuronal function. When energy production is impaired in the brain as a result of anoxia, hypoglycemia or ischemia there is a massive release of glutamate into the extracellular milieu. Under these conditions both non-NMDA and NMDA receptors are stimulated which lead to the activation of conductance mechanisms for sodium, potassium and calcium ions. In a glutamatergic terminal the electrogenic reuptake of glutamate together with sodium can cause a depolarization of the plasma membrane in corticolcerebral preparation. This depolarization can be sufficient to activate voltage dependent calcium channels and induce exocytotic release of transmitter from synaptosomes. During long periods of synaptic activity when glutamate accumulates in the synaptic cleft, the reuptake of the transmitter into the terminal causes a short term facilitation of further transmitter release (38). Cultured hippocampal astrocytes also release glutamate in to the extracellular medium under anoxia/ischemia conditions (22).

In this study, anoxia/ischemia condition was mimicked with acidic pH shifts of the ACSF; under this condition the NMDA current component of EPSC was selectively depressed. During ischemia, conditions for maximal NMDA channel activation are present. Therefore, selective modulation of synaptic NMDA current by proton concentration, which is shown in this study, may represent an intrinsic protection mechanism for preventing the damage that can occur in hippocampal CA1 pyramidal cell.

Histamine application to the perfusion system having pH 7.2 caused an enhancement of the synaptic NMDA current. Under slightly alkaline conditions, application of histamine to the slice resulted in a decrease in the amplitude of the NMDA mediated synaptic current. Under physiological pH 7.3-7.4, histamine has no effect on the NMDA mediated component of the EPSC (32). The histamine effect was selective for the NMDA current component of EPSC, indicating that this is a postsynaptic effect. The enhancing effect of histamine on NMDA receptor current under slightly acidic conditions can be explained in the following way; there is a large negative surface potential on the extracellular and intracellular side of the NMDA receptor. Histidine which is the precursor of histamine possesses weakly basic imidazolum function having a pK_R value of 6.0. At pH 6.0 over 50 per cent of histidine molecules contain a protonated R group and at pH 7.0 approximately 10 per cent have a positive charge. Protonated form of histamine under slightly acidic conditions might be a potent regulator of NMDA current. At slightly alkaline pH values histamine possess no charge on it so the probability of the molecule for binding to NMDA receptor is low, thus causing a reduction in the NMDA current component of the EPSC. The effect of histamine was not mediated with the involvement of second messenger systems, since ATP was absent and fluoride was present in the intracellular solution. This indicates that, the involvement of classical histamine receptors, H1,H2 and H3 is not possible. Furthermore, the selective agonists of H1 (2-Br-phenylhistamine, 1 μM); H2 (impromidine, 1 μM) and H3 (R-α-methylhistamine, 1 μM) were not able to mimic the histamine effect on the synaptic NMDA current at pH 7.2 (Figures are not shown here). The action of histamine was opposite to that of hydrogen ions in the presented experiments. A recently published report by Traynelis et al. indicated that polyamine may interact with pH sensing site of the NMDA receptor (39). It is possible that histamine can mask the effect of protons on the NMDA receptor or occupy polyamine binding site of the NMDA receptor.

Extracellular pH may be transiently lowered during ischemia/anoxia; under this condition histamine can maintain the transmission or aggravate the excitotoxicity. On the other hand when the alkaline transients occur during normal synaptic release of glutamate, the depressant action of histamine may act to protect the system. It is obvious that, endogenous monoamine histamine is an important neuromodulator of central nervous system especially when its classical receptor effects coincide with its effect on the NMDA receptor ionophore complex.

References

1. Foster, A. C., Fagg, G. E. (1984). "Acidic Amino Acid Binding Sites in Mammalian Neuronal Membranes: Their Characteristics and Relationship to Synaptic Receptors" Brain Research Reviews, Vol. 7, pp. 103-164.
2. MacDermott, A. B., Mayer, M. L., Westbrook, G. L. Smith, S. J., Barker, J. L. (1986). "NMDA Receptor Activation Increases Cytoplasmic Calcium Concentration in Cultured Spinal Cord Neurons, Nature" Vol. 321, pp. 519-522.
3. Mayer, M. L., Westbrook, G. L. (1984). "Channel Block by Magnesium Ions may Underlie Voltage Sensivity of N-Methyl-D-Aspartic Acid Action on Mouse Spinal Neurons in Culture" Journal of Physiology (London), Vol. 349, 14P.
4. Hestrin, S., Nicoll, R. A., Perkel, D. Y. Sah, P. (1990). "Analysis of Excitatory Synaptic Action in Pyramidal Cells using Whole Cell Recording from Rat Hippocampal Slices" Journal of Physiology (London), Vol. 422, pp. 203-225.
5. Honore, T. Davies, S. N., Drejer, J., Fletcher, E. J., Jacobsen, P., Lodge, D., Nielsen, F. E. (1988). "Quinoxalinediones: Potent Competative Non-NMDA Glutamate Receptor Antagonists" Science, Vol. 241, pp. 701-703.
6. Diemer, N. H., Jørgensen, M. B., Johansen, F. F., Sheardown, M., Honore, T. (1992). "Protection Against Ischemic Hippocampal CA1 Damage in the Rat with a New Non-NMDA Antagonist, NBQX" Acta Neurologica Scandinavica, Vol. 86, pp. 45-49.

7. Monaghan, D. T., Holets, V. R., Toy, D. W. Cotman, C. W. (1983). "Anatomical Distributions of Four Pharmacologically Distinct [^{3}H]-L-Glutamate Binding Sites" Nature, Vol. 306, pp. 176-179.
8. Bekkers, J. M., Stevens, C. F. (1989). "NMDA and non-NMDA Receptors are colocalized at Individual Excitatory Synapses in Cultured Rat Hippocampus" Nature, Vol. 341, pp. 230-233.
9. Johnson, J. W., Ascher, P. (1987). "Glycine Potentiates the NMDA Response in Cultured Mouse Brain Neurons," Nature, Vol. 325, pp. 529-531.
10. Watanabe, M., Himi, T., Saito, H., Abe, K. (1992). "Involvement of Glycine Site Associated with the NMDA Receptorin hippocampal LTP and Acquisition of Spatial Memory in Rat" Brain Research, Vol. 582, pp. 58-64.
11. Ault, B., Evans, R. H., Francis, A. A., Oakes, D. J., Watkins, J. C. (1980). "Selective Depression of Excitatory Amino Acid Induced Depolarizations by Magnesium Ions in Isolated Spinal Cord Preparations" Journal of Physiology (London), Vol. 307, pp. 413-428.
12. MacDonalt, J. F., Bartlett, M. C., Mody, I., Pahapill, P., Reynolds, J. N., Salter, M. W. Schneiderman, J. H., Pennefather, P. S. (1991). "Action of Ketamine, Phencyclidine and MK-801on NMDA Receptor Currents in Cultured Mouse Hippocampal Neurons" Journal of Physiology (London), Vol. 432, pp. 483-508.
13. Olnay, J. W., Labruyere, J., Price, M. T. (1989). "Pathological Changes Induced in Cerebrocortical Neurons by Phencyclidineand Related Drugs" Science, Vol. 244, pp. 1360-1362.
14. Wada, Y., Hasegava, H., Nakanura, M., Yamaguchi, N. (1992). "The NMDA Receptor Antagonist MK-801 has a Dissociative Effect on Seizure Activity of Hippocampal Kindled Cats" Pharmacology Biochemistry and Behaviour, Vol. 43, pp. 1269-1272.
15. Lipton, S. (1993). "Prospects for Clinically Tolarated NMDA Antagonist:Open Channel Blockers and Alternative Redox States of Nitric Oxide" Trends in Neuroscience, Vol. 16, No. 12, pp. 527-532.
16. Peters, S., Koh, J., Choi, D. W. (1987). "Zinc Selectively Blocks the Action of N-Methyl-D-Aspartate on Cortical Neurons" Science, Vol. 236, pp. 589-592.
17. Assaf, S. Y., Chung, S. H. (1984). "Release of Endogeneous Zinc Ion from Brain Tissue During Activity" Nature, Vol. 308, pp. 734-736.
18. Vyklicky, L., Vlachova, V., Krusek, J. (1990). "The Effect of External pH Changes on Responses to Excitatory Amino Acids in Mouse Hippocampal Neurons" Journal of Physiology (London), Vol. 430, pp. 497-517.
19. Köhr, G., Eckard, S., Lüddens, H., Monyer, H., Seeburg, P. H. (1994). "NMDA Receptor Channels:Subunit Specific Potentiation by Reducing Agents" Neuron, Vol. 12, pp. 1031-1040.
20. Tang, C., Dichter, M., Morad, M. (1990). "Modulation of the NMDA Channel by Extracellular Hydrogen Ion" Proceedings of National Academy of Sciences (USA), Vol. 87, pp. 6445-6449.
21. Squire, L. R., Zola-Morgan, S. (1988). "Memory:Brain Systems and Behaviour" Trends in Neurosciences, Vol. 11, No. 4, pp. 170-175.

22. Ogata, T., Nakamura, Y., Shibata, T., Kataoka, K. (1992). "Release of Excitatory Amino Acids from Cultured Hippocampal Astrocytes Induced by a Hypoxic-Hypoglycemic Stimulation" Journal of Physiology (London), Vol. 58, pp. 1957-1959.
23. Ransom, R. W., Stec, N. L. (1992). "Cooperative Modulation of [^{3}H]-MK-801 Binding to the N-Methyl-D-Aspartate Receptor Ion Channel Complex by L-Glutamate, Glycine and Polyamines" Journal of Neurochemistry, Vol. 51, pp. 830-836.
24. Rock, D. M., MacDonald, R. L. (1992). "The Polyamine Spermine has Multiple Actions on N-Methyl-D-Aspartate Receptor Single Channel Currents in Cultured Cortical Neurons" Molecular Pharmacology, Vol. 41, pp. 83-88.
25. Garbarg, M., Barbin, G., Feger, J., Schwartz, J. C. (1974). "Histaminergic Pathway in Rat Brain Evidenced by Lesions of the Medial Forebrain Bundle" Science, Vol. 186, pp. 833-835.
26. Schwartz, J. C., Arrang, J. -M., Garbarg, M., Pollard, H. Ruat, M. (1991). "Histaminergic Transmission in the Mammalian Brain" Physiological Reviews, Vol. 71, No. 1, pp. 1-51.
27. Haas, H. L., Konnerth, A. (1983). "Histamine and Noradrenaline Decrease Calcium Activated Potassium Conductance in Hippocampal Pyramidal Cells" Nature, Vol. 302, pp. 432-434.
28. Arrang, J. M., Garbarg, M., Schwartz, J. C. (1983). "Autoinhibition of Brain Histamine Release Mediated by a Novel Class (H3) of Histamine Receptor" Nature, Vol. 302, pp. 832-837.
29. Bekkers, J. M. (1993). "Enhancement by Histamine of NMDA Mediated Synaptic Transmission in the Hippocampus" Science, Vol. 261, pp. 104-106.
30. Vorobjev, S. V., Sharonova, I. N., Walsh, I. B., Haas, H. L. (1993). "Histamine Potentiates N-Methyl-D-Aspartate Responses in Acutely Isolated Hippocampal Neurons" Neuron, Vol. 11, pp. 827-844.
31. Kay, A. R., Miles, R., Wong, R. K. S. (1986). "Intracellular Fluoride Alters the Kinetic Properties of Calcium Currents Facilitating the Investigation of Synaptic Events in Hippocampal Neurons" Journal of Neuroscience, Vol. 6, No. 10, pp. 2915-2920.
32. Saybaşılı, H., Stevens, D. R., Haas, H. L. (1995). "pH Dependent Modulation of N-Methyl-D-Aspartate Receptor Mediated Synaptic Currents by Histamine in Rat Hippocampus in Vitro" Neuroscience Letters, Vol. 199, pp. 225-227.
33. Bliss, T. V. P., Collingridge, G. L. (1993). "A Synaptic Model of Memory:Long-Term Potentiation in the Hippocampus" Nature, Vol. 361, pp. 31-39.
34. Bliss, T. V. P., Lømo, T. (1973). "Long Lasting Potentiation of Synaptic Transmission in the Dentate Area of the Anesthetized Rabbit Following Stimulation of the Perforant Path" Journal of Physiology (London), Vol. 232, pp. 331-356.
35. Bashir, Z. I., Bortolotto, Z. A., Davies, C. H. Beretta, N. Irwing, A. J., Seal, A. J., Henley, J. M., Jane, D. E., Watkins, J. C., Collingridge, G. L. (1993). "Induction of LTP In the Hippocampus Needs Synaptic Activation of Glutamate Metabotropic Receptors" Nature, Vol. 363, pp. 347-350.

36. Rotman, S. M. (1985). "The Neurotoxicity of Excitatory Amino Acids is Produced byPassive Chloride Influx" Journal of Neuroscience, Vol. 5, pp. 1483-1489.
37. Simon, R. P., Swan, J. H., Griffiths, T., Meldrum, B. S. (1984). "Blockage of N-Methyl-D-Aspartate Receptors May Protect Against Ischemic Damage in the Brain" Science, Vol. 226, pp. 850-852.
38. Nicholls, D. G. (1989). "Regulation of Calcium in Isolated Nerve Terminals (Synaptosomes):Relationship to Neurotransmitter Release,"in Khachaturian, Z. S., Cotman, C. W., Pettegrew, J. W. (Eds.), Calcium, Membranes, Aging and Alzheimer's Disease, Annals of the New York Academy of Sciences, Vol. 568, New York.
39. Traynelis, S. F., Hartley, M. Heineman, S. F. (1995). "Control of Proton Sensitivity of the NMDA Receptor by RNA Splicing and Polyamines" Science, Vol. 268, pp. 873-875.

Maturation of Pre- and Post-Synaptic Compartments in Hippocampal Neurons Developing in Culture

Silvia Coco, Claudia Verderio, Paolo Forbicini, Alberto Bacci and Michela Matteoli

CNR Molecular and Cellular Pharmacology and B. Ceccarelli Centers, University of Milano, Italy.

The cellular machinery responsible for cell-to-cell communication reaches its maximum efficiency and spatial complexity in neurons. In these cells, mechanisms for both transmission and reception of chemical signals coexist. In most cases, these two functions are carried out by distinct neuronal districts, the axon, which conducts action potentials and releases neurotransmitters, and the dendrites, specialized to receive and integrate the chemical signals. The differentiation of axonal and dendritic processes has therefore important implications for the ability of neurons to find, recognize and form functional contacts with a target cell. Formation of synaptic contacts coincides with a regional redistribution of organelles and cytoskeletal proteins (Fletcher et al., 1991; Sanes, 1989; Burgoyne, 1991).This reorganization leads to the formation of highly specialized surface and cytoplasmic domains, which define the pre- and post -synaptic compartments. At the mature synapse, synaptic vesicles are selectively clustered at specialized regions of the presynaptic plasmalemma, the so called active zones. At these sites synaptic vesicles undergo cycles of exo-endocytosis and release neurotransmitter in the extracellular space. Receptors for neurotransmitters are mainly localized on the postsynaptic plasmamembrane where they mediate the postsynaptic response to the released neurotransmitter. Synaptic transmission is usually terminated by neurotransmitters' uptake system which, in the case of neurons, is localized in the presynaptic terminals, where it acts by removing the neuromediators from the synaptic cleft.

The lack of suitable experimental models has strongly impaired the study of the mechanisms by which neurons control the localization of synaptic components. Cultured hippocampal neurons present unique advantages in the study of neuronal development and synaptogenesis. When maintained in primary culture, they acquire characteristic axons and dendrites by a stereotyped sequence of developmental events (Dotti et al., 1988) and eventually they form physiologically active synapses (Bartlett and Banker, 1984). The maturation of synaptic contacts can been studied in hippocampal neurons by using proteins of synaptic vesicles as markers for presynaptic maturation (Fletcher et al., 1991; Matteoli et al., 1991 and 1992). At the postsynaptic level, several subtypes of ionotropic glutamate receptors can be used as markers for postsynaptic organization and maturation (Craig et al.,1993; Verderio et al., 1994). For these reasons, cultured hippocampal neurons represent a particularly suitable model to study maturation of pre- and post-synaptic compartments during synaptogenesis.

NATO ASI Series, Vol. H 100
Neurotransmitter Release and Uptake
Edited by Şakire Pöğün

1 Presynaptic Specialization

The presynaptic structure is typically a portion of an axon characterized by a cluster of tightly packed synaptic vesicles closely apposed to the portion of the axonal plasmalemma which participates in the synaptic contact. Synaptic vesicles are connected to each other by a cytoskeletal matrix, which must be highly dynamic because synaptic vesicles appear to freely intermix within the cluster (Betz et al., 1992). In addition, a pool of vesicles, those which are juxtaposed just beneath the plasma membrane, appear to be directly docked to the membrane and may be already partially engaged in a fusion reaction which is blocked "half-way" by a Ca^{2+} sensitive clamp (Sollner et al.,1993). These vesicles are the first to undergo exocytosis. The presence of docked synaptic vesicles only at regions of the presynaptic plasmalemma which are part of presynaptic specializations (active zones) demonstrates that this region of the neuronal surface must be highly specialized. The corresponding portion of the plasmalemma contains voltage sensitive Ca^{2+} channels, mediating the specific rise of Ca^{2+}, and molecules involved in the selective docking of synaptic vesicles. Whereas the identity of Ca^{2+} channels concentrated at active zones of CNS neurons remains unclear, different molecules have been recently identified which mediate synaptic vesicle docking and fusion (for reviews see Rothman, 1994; Sudhof, 1995; Chapman et al., 1995; Schweizer et al., 1995). Two soluble proteins, N-ethylmaleimide-sensitive factor (NSF) and soluble NSF attachment proteins (SNAPs), which are essential for membrane fusion in cell free assay systems (Sollner et al, 1993), have been shown to play a major role also at the synapse. At the nerve terminal, receptors for these proteins have been identified. These receptors, the so called SNAREs, are synaptobrevin, a protein of synaptic vesicles, and two proteins of the plasma membrane, syntaxin and SNAP25. The SNAREs have been recently shown to be the molecular targets of clostridial neurotoxins which act as potent inhibitors of neurotransmitter release (Schiavo et al., 1992a and b; Link et al., 1992; reviewed in Huttner 1993; Niemann et al.,1994). The formation of the fusion complex is controlled, at least in part, by negative regulation which involves syntaxin and synaptobrevin. Syntaxin interacts with the soluble factor n-sec1, which inhibits its binding to synaptobrevin and SNAP25 (Pevsner et al., 1994). On the other hand, the majority of synaptobrevin is complexed to the synaptic vesicle protein synaptophysin, which regulates its availability for entry into the fusion complex (Edelmann et al., 1995; Washbourne et al., 1995). Synaptophysin and n-sec1 work therefore as molecular switches that regulate the interaction of synaptobrevin and syntaxin with the other members of the fusion complex . In addition to NSF, SNAPs and SNAREs, other proteins appear to play a major role in the exocytotic cycle of synaptic vesicles. Among these, the small GTP binding protein rab3 and synaptotagmin. Rab3, is probably required to maintain a normal reserve of synaptic vesicles for accelerated exocytosis during repetitive stimulation, when synaptic vesicle recycling becomes rate limiting (Sudhof, 1995). Synaptotagmin, a calcium and phospholipid binding protein, plays a prominent role in calcium-regulated neurotransmitter release probably serving as a calcium sensor that triggers synaptic vesicle fusion (Brose et al., 1992, reviewed in Kelly, 1995), even if the existence of multiple calcium sensors for exocytosis is suggested by the presence of at least two distinct components of evoked transmitter release at mature synapses activated by different concentrations of extracellular calcium (Goda and Stevens,1994). It is possible that the NSF-SNAP-SNARE complex works before synaptotagmin by priming synaptic vesicles and conferring fusion competence upon them (O'Connor, 1994), while synaptotagmin follows to act as a rapid trigger for exocytosis as soon as calcium enters the nerve terminals (Augustine, 1995). Since a molecular association between synaptotagmin and syntaxin has been demonstrated, it has been proposed that when NSF and SNAPs bind to the complex, synaptotagmin is displaced and fusion may occurr (Sollner et al., 1993b). Alternatively, the association of syntaxin and synaptotagmin may occurr after dissociation of the complex by NSF (O'Connor et al., 1994) and this interaction may trigger exocytosis. Accordingly with this idea, the interaction between syntaxin and synaptotagmin is stimulated by calcium (Li et al., 1995; Chapman et al., 1995). After exocytosis the membrane of synaptic vesicle is rapidly retrieved by endocytosis, probably involving clathrin-coated pits and coated vesicles as intermediates, and requiring the GTP-protein dynamin (Takei et al., 1995; reviewed in De Camilli et al., 1995).

In our laboratory some work has been performed during the last years to dissect the molecular events underlying the formation and the functional activation of the presynaptic compartment during synaptogenesis in cultured hippocampal neurons. Synaptic vesicles are present in hippocampal neurons even before synaptogenesis, when they first have a diffuse distribution and then become specifically sorted to the axon. Formation of synaptic contacts coincides with a further redistribution of synaptic vesicles, which become clustered at presynaptic sites (Fletcher et al., 1991; Matteoli et al., 1991, 1992). The mechanisms which trigger synaptic vesicle relocation during synaptogenesis have not been clarified yet. However, since large synaptic vesicle clusters form only when an axon makes a contact with a postsynaptic cell, one should envisage a mechanism of transduction whereby contact of some cell surface components of the axon with the postsynaptic cell triggers a conformational change at the cytosolic face of the axonal membrane, which results in a recruitment of synaptic vesicles.

Antibodies directed against the lumenal domain of the synaptic vesicle protein synaptotagmin, which become internalized in the vesicle lumen in parallel with the exo-endocytotic recycling, have been used to evaluate the dynamics of synaptic vesicles during synaptogenesis and to quantify the rate of synaptic vesicle exocytosis independently of neurotransmitter secretion. By using this method, it has been demonstrated that the property of synaptic vesicles to undergo cycles of exo-endocytosis is not dependent upon synapse formation (Matteoli et al., 1992; Kraszewski et al. 1995). Synaptic vesicles are able to undergo fusion with the plasmalemma since early stages of development in vitro, with a mechanism which is already calcium- and depolarization-dependent (Kraszewski et al. 1995). In agreement with the early activation of the basic mechanisms of synaptic vesicle recycling, all the molecular components of the fusion machinery investigated so far (synaptobrevin, syntaxin, SNAP25, synaptophysin, synaptotagmin, rab3) have been found to be expressed by hippocampal neurons already before the formation of synaptic contacts (Fletcher et al., 1991; Matteoli et al., 1991, 1992; Galli et al., 1995; Coco et al., in preparation). The property of the vesicles to undergo exo-endocytosis temporally correlates with the ability of neurons to release glutamate in the extracellular space, as assayed by analyzing glutamate-induced cytosolic calcium changes to sense presynaptically released neurotransmitter, suggesting that synaptic vesicle recycling before synaptogenesis may have in fact a functional role (Verderio et al., 1995).

However, even if synaptic vesicles seem to be functional since early stages of neuronal development, specific mechanisms controlling the exocytotic process appear to be selectively activated in parallel with synaptogenesis. First, the formation of synaptic contacts correlates with a down-regulation of basal synaptic vesicle exocytosis (Kraszewski et al. 1995), suggesting that synapse maturation may coincide with the strengthening of a mechanism which reduces the constitutive exocytosis (Debello et al., 1993; Popov and Poo, 1993; Kraszewski et al. 1995). Second, the calcium dependency of the evoked synaptic vesicle recycling appears to be different before and after the establishment of synaptic contacts. Whereas two distinct components of evoked transmitter release are active at the mature synapse of hippocampal neurons, according to the two-calcium sensor mechanism regulating the transmitter release proposed by Goda and Stevens (Goda and Stevens,1994), only the component of evoked release activated at low calcium concentration (< 200mM) has been identified in developing axons before synaptogenesis, indicating that the formation of synaptic contacts coincides with the specific activation of a low affinity calcium sensor (> 800mM) for exocytosis (Coco et al., in preparation). Finally, distinct populations of calcium channels are involved in the control of neurotransmitter release before and after synaptogenesis, ω-CTx-GVIA sensitive calcium channels playing a primary role in mediating calcium dependent glutamate release at early developmental stages and ω-Aga-IVA sensitive channels mainly mediating glutamate release from mature synaptic contacts (Verderio et al.,1995).

This study of synaptic vesicles exocytosis during synaptogenesis allowed us to conclude that synaptic vesicles distributed throughout the growing axon of developing neurons are equipped with a molecular machinery which conferes on them functional abilities. However, specific components and/or properties of the exocytotic machinery (a mechanism reducing the probability of completing a fusion event of constitutive exocytosis, a low-affinity calcium sensor, a populations of calcium channels specifically involved in the exocytotic process) appear to be selectively activated only after synaptic vesicles have reached their final destination, at the mature synaptic contact (Matteoli et al.,1995).

2 Postsynaptic Specialization

The postsynaptic site is generally represented by a portion of the dendrite or perykarial plasmamembrane which contains an almost cristalline array of ionotropic receptors, the neurotransmitter-gated ion channels, and is lined at its cytoplasmic side by specialized submembranous cytoskeleton. Most of our knowledge about the biological events underlying the differentiation of a postsynaptic compartment has derived from studies carried out using a simple and accessible experimental model, the neuromuscolar junction. At the neuromuscolar junction, cellular and molecular mechanisms responsible for aggregation of acetylcholine receptors underneath presynaptic membrane have been described in some detail. An association between the 43K protein and actin has been proposed as a mechanism for immobilizing acetylcholine receptors at the postsynaptic site (Froehner, 1993). In addition, a search for molecules that direct the accumulation of acetylcholine receptors at nerve-muscle synapses led to the discovery of agrin (McMahan, 1990). Recently, a protein complex has been identified and purified from Torpedo electric organ postsynaptic membranes, which is likely to serve as an agrin receptor (Bowe et al., 1994). The native receptor is a heteromeric complex of two membrane glycoproteins of 190 KD and 50 KD. The 190 KD subunit is sufficient to bind agrin. The 190 KD and 50 KD subunits are related to the dystrophin-associated glycoproteins a and b-dystroglycan respectively (Bowe et al., 1994). In the CNS, a functional analogue of 43K, gephyrin, has been described to promote and sustain clustering of the glycine receptor in cultured spinal neurons by mediating the interaction between receptors and cytoskeleton (Kirsh et al. 1993). This finding has suggested the general concept that a peripheral membrane protein may link neurotransmitter receptors to the submembrane cytoskeleton.

Immunocytochemical and functional studies have been carried out in cultured hippocampal neurons to define the presence and subcellular localization of ionotropic glutamate receptors at different stages of neuronal development. Antibodies directed against different subunits of the AMPA/kainate glutamate receptors were used in a classical morphological study (Craig et al., 1993) to define the distribution of these receptors in growing neurons. Moreover, the localization of Ca^{2+} transients associated with the activation of ionotropic glutamate receptors of both NMDA and non NMDA type was monitored at different stages of maturation, providing therefore a map of functional receptors distribution during neuronal development in vitro (Verderio et al., 1994). These studies provided the evidence that glutamate receptors are expressed and are already functional since early stages of neuronal development. At this stage, glutamate receptors appear to be uniformely distributed throughout the neuronal membrane. With the establishment of neuronal polarity, both NMDA and non-NMDA glutamate receptors segregate in the somatodendritic region of the neuron, independently on synapse formation. Eventually, synapse maturation promotes a further, intradendritic compartimentalization of glutamate receptors which become clustered at the postsynaptic sites (Craig et al., 1993; Verderio et al., 1994).

Both the progressive exclusion of glutamate receptors from the axon and the following clustering at synapses appeared to be independent from spontaneous electrical activity and on receptor activation. Indeed, hippocampal neurons grown in the presence of the sodium channel blocker TTX or in the presence of the glutamate receptor antagonists APV/CNQX were able to develop a mature synaptic network, with abundant synaptic boutons along dendritic spines and a normal compartmentalization of glutamate receptors at all stages of development (Verderio et al., 1994). These results indicated that neuronal electrical activity, which is known to play a major role in the acquisition of the mature anatomy and physiology of the CNS (for a review see Hockfield and Kalb, 1993), as well as in the maturation of several postsynaptic parameters in muscle cells (Hall and Sanes, 1993), is not involved in directing the clustering of glutamate receptors at postsynaptic sites.

More recently, the clustering of glutamate receptors in the CNS has been extensively studied although little is still known about the molecules that mediate receptor clustering and anchoring at synapses. By the use of the two-hybrid system, an interactions between the NMDA receptor subunit NR2 and members of the PDS-95/SAP90 family of membrane associated putative guanylate kinases have been identified from two different laboratories (Kornau et al., 1995; Niethammeret al., 1996). The interaction is mediated by binding of the cytoplasmic tails of the NMDA receptor subunits, which contain a C-terminal binding motif virtually identical to that found in the K^+ channels subunits, to the first two PDZ domains of PDS-95/SAP90, a synaptic protein associated with the cytoskeleton. These PDZ domains, also known as GLGF or DHR domains, have been described in many membrane-associated proteins. These data suggest that members of the PDS-95 family may serve to anchor NMDA receptors to the membrane cytoskeleton at postsynaptic sites (Niethammeret al., 1996).

Recently, the presence of glutamate receptor clusters has been reported in hippocampal neurons, also at presynaptic sites (Shigemoto et al., 1996). These receptors, previously postulated to serve as autoreceptors by pharmacological and physiological studies, have been identified as metabotropic glutamate receptors mGluR7. Interestingly, the level of mGluR7 expression is different when synapses are formed between pyramidal neurons or between a pyramidal cell and an interneuron. This finding raises the possibility that presynaptic neurons could modulate neurotransmitter release at individual synapse according to the postsynaptic target.

3 Neurotrasmitter Transporters

Rapid re-uptake is essential for terminating synaptic transmission by most neurotransmitters in the nervous system. Specific carriers transport neurotransmitters across the plasma membrane of neuronal cells and sorrounding glial cells, allowing the clearance of the neurotransmitter from the synaptic cleft and its further metabolic processing (reviewed in Amara and Arriza, 1993; Schloss et al.,

1994). The exception to this is acetylcholine, which is enzimatically inactivated by postsinaptic acetylcholinesterase giving choline, which is then transported by a presynaptic high-affinity transporter before re-acetylation. Once inside the cell, neurotransmitters are further transported into synaptic vesicles by distinct transport systems powered by a pH or voltage gradient across the vesicle membrane (reviewed in Edwards,1992). Plasma membrane transporters, besides terminating the postsynaptic action of neurotransmitters, are essential to replenish the neurotransmitter supply of the neurons and (in the case of glutamate) to keep the extracellular neurotransmitter concentration below neurotoxic levels. Two main families of plasma membrane uptake carriers have been cloned, and the steichiometries, physiological function and mechanisms of modulation of some of these transporters are now better understood. Uptake carriers for GABA, noradrenaline, dopamine, serotonin, glycine, proline and betaine have been shown to belong to a family of homologous proteins that derive their energy from the co-transport of Na^+ and Cl^- ions with the neurotransmitter into the cell. A second type of neurotransmitter transporters couples to inwardly directed Na^+ ion gradient an outward K^+ current. At this family belong glutamate transporters (reviewed in Attwell and Mobbs, 1994).

Glutamate is the major excitatory neurotransmitter in the brain, and both high- and low-affinity uptake of glutamate occur in neurons and glial cells. The high-affinity transporter will also transport aspartate. These transporter proteins are essential in limiting the neurotoxic effects of high synaptic glutamate concentrations. Such effects have been implicated in epilepsy, brain ischaemia and amyiotrophic lateral sclerosis, where glutamate transport is impaired (Rothstein et al. 1992). At least four different high affinity transporter subtypes are now cloned, i.e. GLAST (Storck et al., 1992), GLT1 (Pines et al., 1992), EAAC1 (Kanai and Hediger, 1992) and EAAT4 (Fairman et al., 1995a,b), which constitute a gene family with specialized brain distributions. GLT1 and GLAST are both expressed almost exclusively in astrocytes, although with differences in their respective cellular and regional contents (Rothstein et al., 1994; Torp et al., 1994; Lehre et al., 1995; Chaudhry et al., 1995). EAAC1 is neuron-specific and widely distributed throughout the brain (Rothstein et al., 1994; Kanai et al., 1995), while EAAT4 is abundant only in the cerebellum, particularly in Purkinje cells (Fairman et al., 1995a,b).

To be a good candidate for carrying out the reuptake of glutamate at synaptic cleft, a neuronal transporter should be properly located at presynaptic sites. Indeed, according to several functional and immunochemical studies, the synaptic terminals of glutamatergic neurons should possess a high affinity glutamate uptake system (uptake studies in cells and synaptosomes, Gunderesen et al., 1993). However, the localization study by Rothstein et al. (1994), indicated that EAAC1 was present in the somatodendritic compartment of glutamatergic cells, rather than in presynaptic axon terminals.Therefore, this protein may not correspond to the conventional view of a presynaptic carrier in glutamatergic synaptic function.

The expression and subcellular localization of EAAC1 during and after synaptogenesis in vitro, in comparison with a number of pre- and post-synaptic proteins belonging to the synaptic machinery, were studied by immunocytochemistry in cultured hippocampal neurons, which are characterized (see above) by the possibility to easily detect presynaptic specializations, in the form of SV clusters (Fletcher et al.,1991; Matteoli et al,1992), and postsynaptic structures, in the form of developing spines (Craig et al., 1993; Verderio et al., 1994; Papa et al.,1995). EAAC1 was found to be expressed in hippocampal neurons since very early stages of neuronal develompment, well before the formation of synaptic contacts, as well as in GABAergic interneurons, which represent about 7% of the total population of neuronal cells in culture. In all cases, EAAC1 was compartmentalized in the somatodendritic region of the cell. On the other hand, typical presynaptic markers were segregated in the developing axons (Fletcher et al.,1991; Matteoli et al,1992, see above), indicating that EAAC1 doesn't follow the route of a typical presynaptic protein during neuronal maturation. In fully differentiated neuronal cultures, rich of synaptic contacts, EAAC1 never colocalized with the stainings produced by presynaptic markers nor with the stainings produced by a typical postsynaptic marker, the subunit GluR1 of the AMPA/Kainate glutamate receptor. Thus EAAC1 is present in the somatodendritic region of the neuron but it is never enriched in the postsynaptic compartment, represented by glutamate receptor enriched spine heads (Coco S.,Verderio C., Trotti D, Rothstein JD, Volterra A. and Matteoli M, submitted).

These data indicated that EAAC1 is not to be considered a typical pre- or postsynaptic protein. The extra-synaptic localization of this high- affinity glutamate carrier tends to exclude a role of this transporter in determining the time course of glutamate inside the cleft. Indeed, spontaneous oscillations of cytosolic calcium and of membrane potential produced by glutamate released at synaptic contacts were found to be heavily influenced by the presence of glial cells, suggesting that glial transporters may play a major role in controlling the glutamate concentration in the surroundings of the synapse, leading to the repetitive excitation of the postsynaptic cell (Verderio C., Bacci A., S. Coco, Forbicini P.,G. Fumagalli, M. Matteoli in preparation). These observations, together with EAAC1 early expression and with its localization, not restricted to glutamatergic fibers, suggest an important role for this carrier in buffering ambient glutamate and in controlling the levels of tonic glutamate receptor activation during development.

4 Conclusions

By providing a source of neurons synchronized in the formation of synaptic contacts, characterized by distinguishable and accessible pre- and post-synaptic compartments, cultured hippocampal neurons represent an interesting model for dissecting the molecular events underlying the formation of pre- and post-synaptic elements and for analysing the structural and functional maturation of synaptic contacts. The results from these studies may be useful to shed a light on those mechanisms which control synaptogenesis between neurons in the central nervous system.

Acknowledgements:

These studies have been supported by grants from the Italian CNR, from Human Frontier Science Program (HFSP) and from Telethon-Italy (grant n. 672 to M.M.).

Some of the experiment reported in this review has been carried out in collaboration with the laboratories of Prof. Guido Fumagalli (University of Verona), Prof. Pietro De Camilli (Yale University) and Dr. Andrea Volterra (University of Milano).

References

Amara SG and Arriza JL (1993) *Curr Opin Neurobiol* 3, 337-344.
Attwell D and Mobbs P (1994) *Curr Opin Neurobiol* 4, 353-359.
Bartlett WPand Banker GA (1984) *J Neurosci* 4, 1954-1965.
Betz WJ, Mao F and Bewick GS (1992) *J Neurosci* 12, 363-375.
Brose N, Petrenko AG, Sudhof TC and Jahn R (1992) *Science* 256,1021-1024.
Bowe MA, Deyst KA, Leszyk JD and Fallon JR (1994) *Neuron* 12, 1173-1180.
Burgoyne RD (1991) *The neuronal cytoskeleton* (New York: Wiley Liss).
Chapman ER, Hanson PI, An S and Jahn R (1995) *J Biol Chem* 270, 23667-23771.
Chaudry FA, Lehre KP, Van Lookeren Campagne M, Ottersen OP, Danbolt NC and Storm-Mathisen J (1995) *Neuron* 15, 711-720.
Craig AM, Blackstone CD, Huganir RL and Banker GA (1993) *Neuron* 10, 1055-1068.
DeBello WM, Betz H and Augustine GJ (1993) *Cell* 74, 947-950.
De Camilli P, Takei K and McPherson PS (1995) *Curr Opin Neurobiol* 5, 559-562.
Dotti CG, Sullivan CA and Banker GA (1988) *J Neurosci* 8, 1454-1468.
Edelmann L, Hanson PI, Chapman ER and Jahn R (1995) *Embo J* 14, 224-231.
Edwards RH (1992) *Curr Opin Neurobiol* 2, 586-594.
Fairman WA, Vandenberg RJ, Arriza JL, Kavanaugh MP and Amara SG (1995a) *Nature* 375, 599-603.
Fairman WA, Vandenberg RJ, Arriza JL, Shannon EM, Murdoch GH, Kavanaugh MP and Amara SG (1995b) *Soc Neurosci Abs* 21, 1861.
Fletcher TL, Cameron PL, De Camilli P and Banker GA (1991) *J Neurosci* 11, 1617-1626.
Froehner SC (1993) *Annu Rev Neurosci* 16, 347-368.
Galli T, Garcia EP, Mundigl O, Chilcote TJ and De Camilli P (1995) *Neuropharmacology* 34, 1351-1360.
Goda Y and Stevens CF (1994) *Proc Natl Acad Sci USA* 91,12942-12946.
Gudersen V, Danbolt NC, Ottesen OP and Storm-Mathisen J (1993) *Neurosci* 57, 97-111.
Hall ZW and Sanes JR (1993) *Cell 72/Neuron 10* Rev. Suppl, 99-121.

Hokfield S and Kalb RG (1993) *Curr Opin Neurobiol* 3, 87-92.

Huttner WB (1993) *Nature* 365, 104-105.

Kanai Y and Hediger MA (1992) *Nature* 360, 467-471.

Kelly RB (1995) *Current Biology* 5,257-259.

Kirsch J, Wolters I, A Triller and H Betz (1993) *Nature* 366, 745-748.

Kornau H-C, Schenker LT, Kennedy MB and Seeburg PH (1995) *Science* 269, 1737-1740

Kraszewski K, Mundigl O, Daniell L, Verderio C, Matteoli M and De Camilli P (1995) *J Neurosci* 15, 4328-4342.

Lehre KP, Levy LM, Ottesen OP, Storm-Mathisen J and Danbolt NC (1995) *J Neurosci* 15, 1835-1853.

Li C, Ullrich B, Zhang JZ, Anderson RGW, Brose N and Sudhof TC (1995) *Nature* 375, 594-599.

Link E, et al. (1992) *Biochim Biophys Commun* 189,1017-1023.

Matteoli M, Takei K, Cameron R, Johnston PA, Hurlbut P, Jahn R, Sudhof TC and De Camilli P (1991) *J Cell Biol* 115, 625-633.

Matteoli M, Takei K, Perin MS, Sudhof TC and De Camilli P (1992) *J Cell Biol* 117, 849-861.

Matteoli M, Verderio C, Kraswzeski K, Mundigl O, Coco S, Fumagalli G and De Camilli P (1995) *J Phisiol (Paris)* 89, 51-55.

McMahan UJ (1990) In: *Cold Spring Harbor Symposium on Quantitative Biololgy* LV, 407-418.

Niemann H, Blasi J and Jahn R (1994) *Trends in Cell Biol* 4,179-185.

Niethammer M, Kim E and Sheng M (1996) *J Neurosci* 16, 2157-2163

O'Connor V, Augustine GJ and Betz H (1994) *Cell* 76, 785-787.

Papa M, Bundman MC, Greenberger V and Segal M (1995) *J Neurosci* 15, 1-11.

Pevsner J, Hsu S-C, Brown JEA, Calakos N, Ting AE, Bennet MK and Sheller RH (1994) *Neuron* 13, 353-361.

Pines G, Danbolt NC, Bioras M, Zhang Y, Bendahan A, Eide L, Koepsell H, Seeberg E, Storm-Mathisen J and Kanner BI (1992) *Nature* 360, 464-467.

Popov SV and Poo MM (1993) *Cell* 73, 1247-1249.

Rothman JE (1993a) *Nature* 362, 318-324.

Rothman JE (1994) *Nature* 372, 55-63.

Rothstein JD, Martin L, Levey AI, Dykes-Hoberg M, Jin L, Wu D, Nash N and Kuncl RW (1994) *Neuron* 13, 713-725.

Rothstein JD, Martin LJ and Kuncl RW (1992) *New Engl J Med* 326,1464-1468.

Sanes JR (1989) *Ann Rev Neurosci* 12, 491-516.

Schiavo G, Benfenati F, Poulain B, Rossetto O, Polverino de Laureto P, DasGupta BR and Montecucco C (1992a) *Nature* 359, 832-835.

Schiavo G, Poulain B, Rossetto O, Benfenati F, Tauc L and Montecucco C (1992b) *EMBO J* 11, 3577-3583.

Schloss P, Puschel AW and Betz H (1994) *Curr Opin Neurobiol* 6, 595-599.

Schweizer FE, Betz H and Augustine GJ (1995) *Neuron* 14, 689-696.

Shigemoto R, Kulik A, Roberts JDB, Ohishi H, Nusser Z, Kaneko T and Somogyi P (1996) *Nature* 381, 523-525.

Sollner T, Whiteheart SW, Brunner M, Erdjument-Bromage H, Geromanos S, Tempst P and Rothman JE (1993a) *Nature* 362, 318-324.

Sollner T, Bennett MK, Whiteheart SW, Scheller RH, and Rothman JE (1993b) *Cell* 75, 409-418.

Storck T, Schulte S, Hofmann K and Stoffel W (1992) *Proc Natl Acad Sci USA* 89, 10955-10959.

Sudhof TC (1995) *Nature* 375, 645-653.

Takei K, McPherson PS, Schmid SL and De Camilli P (1995) *Nature* 374, 186-190.

Torp R, Danbolt NC, Babaie E, Bjoras M, Seeberg E, Storm-Mathisen J and Ottersen OP (1994) *Eur J Neurosci* 6, 936-942.

Verderio C, Coco S, Fumagalli G and Matteoli M (1994) *J Cell Biol* 126,1527-1536.

Verderio C, Coco S, Fumagalli G and Matteoli M (1995) *Proc Natl Acad Sci USA* 92,6449-6453.

Washbourne P, Schiavo G. and Montecucco C. (1995) *Biochemical J* 305,721-724.

GABA$_B$ Receptor Control of Neurotransmitter Release in Mammalian Brain: Modification During Chronic Inflammation

N G Bowery, M Malcangio, H Teoh

Department of Pharmacology, Medical School, University of Birmingham, Edgbaston, Birmingham B 15 2TT, U.K.

The evidence for a GABA receptor which was distinct from the classical chloride-dependent ionotropic receptor first emerged in mammalian peripheral tissues (Bowery et al 1981; see Bowery 1989). In a variety of isolated nerve-smooth muscle and atrial muscle preparations, GABA and its analogue, β-chlorophenyl GABA (baclofen), were shown to depress the contractile response to nerve stimulation. A typical example of this is shown in Figure 1 where increasing concentrations of baclofen produced a decrease in contraction of the twitch response of the rat anoccocygeus muscle. This reduction in contraction was produced by a decrease in transmitter release from the autonomic innervation. No evidence for any effect at the postsynaptic site was observed. Thus, in general, the influence of GABA was restricted to the nerve input and could be construed as acting on the presynaptic release mechanism in peripheral tissues. The characteristics of the receptor mediating this effect showed little resemblance to the established GABA receptor but it seemed unlikely that it had any physiological significance, perhaps with the exception of the enteric nervous system where GABA neurones have been demonstrated (see Kerr & Ong 1995).

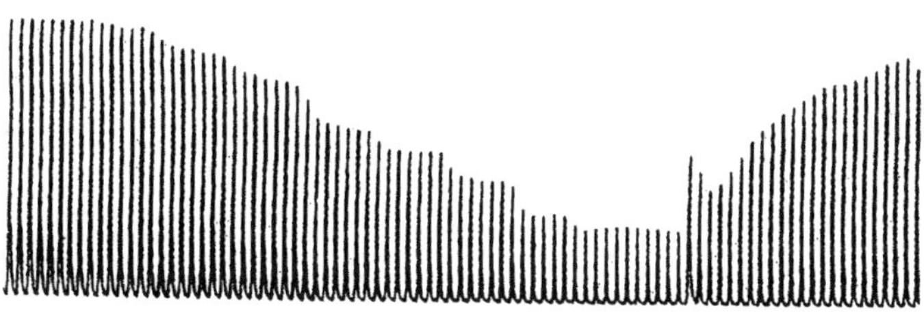

Fig 1. Effect of (-)baclofen on the electrically evoked twitch of the rat isolated anococcygeus muscle. The tissue was stimulated for 1 sec, 0.5 ms at 10 Hz every 20 s at supramaximal voltage. Increasing concentrations of (-)baclofen (to 10^{-8} - 10^{-4} M) were added cumulatively to the bathing medium at the points indicated below the record.

NATO ASI Series, Vol. H 100
Neurotransmitter Release and Uptake
Edited by Şakire Pöğün

Concomitant with these studies in peripheral tissues we were attempting to obtain evidence for this novel site within the mammalian brain. First indications of this arose in studies in which we examined the release of previously accumulated neurotransmitter amines in rat brain slices (Bowery et al 1980). Baclofen and GABA both inhibited the release of noradrenaline, 5HT and dopamine in a manner consistent with our observations in peripheral tissues. This indicated to us that this novel receptor may also be present on presynaptic fibres within the brain but a postsynaptic location could not be excluded. Subsequent to these studies a radiolabelled binding assay for the receptor was developed using tritiated baclofen and GABA and it was at this stage we were able to designate the site as the $GABA_B$ receptor to contrast with the established receptor which we termed the $GABA_A$ receptor site (Hill and Bowery 1981).

Since these early studies many characteristics of the $GABA_B$ receptor have been described and its physiological role(s) within the brain has been extensively studied (see reviews by Bowery 1993; Ong & Kerr 1995; Mott & Lewis 1994). Both pre- and post-synaptic locations for the receptor have been demonstrated in many regions of the brain. One role of the postsynaptic receptor(s) has been readily established in the generation of synaptic late hyperpolarizing potentials in, for example, hippocampal pyramidal cells and neurones of the lateral geniculate and neocortex (Crunelli & Leresche 1991; Dutar & Nicoll 1988a; Deisz et al 1993; Karlsson et al 1988). However, the role(s) of the presynaptic receptor(s) has still to be fully established, particularly within higher centres.

1 Presynaptic $GABA_B$ Receptors

Presynaptic $GABA_B$ receptors have been demonstrated in synaptosomes prepared from mammalian cerebral cortex (Pittaluga et al 1987; Raiteri et al 1990; Bonnaro & Raiteri 1993) and on presumed nerve terminals in slices prepared from brain (Bowery et al 1980; Waldmeier et al 1994). In all cases $GABA_B$ receptor activation suppresses the release of endogenous or accumulated neurotransmitter evoked by high K^+ concentrations or electrical stimulation. These receptors have been reported to influence the release of catecholamines, excitatory amino acids, neuropeptides as well as GABA itself (Bowery et al 1980; Raiteri et al 1990; Potashner 1979; Teoh et al 1996a). The concept of a physiological role for inhibitory receptors on GABAergic nerve terminals (autoreceptors) is relatively easy to accept as the endogenous activator is present. There is no requirement for any innervation. By contrast, the role of heteroreceptors on the remaining types of nerve terminal is more difficult to establish as evidence for axo-axonic innervation, outside the spinal cord, is very limited. However, in a recent study by Isaacson et al (1993), good evidence for activation by GABA released from adjacent neural inputs was provided, at least, within the hippocampus. Sufficient GABA is released from GABAergic fibres to wash-over on to terminals in close proximity at concentrations sufficient to activate $GABA_B$ receptors. These authors proposed the term 'paracrine-like' to describe this possible action of GABA on heteroreceptors.

Whether the characteristics of these presynaptic receptors are distinct from those of the postsynaptic $GABA_B$ sites remains to be fully established although there are a variety of indications that this may be so. Dutar & Nicoll (1988a) and Harnson (1990) have reported that presynaptic $GABA_B$ receptors in the hippocampus are not influenced by pertussis toxin whereas postsynaptic sites are inactivated by the toxin in common with other metabotropic G-protein-linked receptors. Raiteri and colleagues (Gemignani et al 1994) even suggest that presynaptic receptors may be subclassified on the basis of their pharmacological characteristics in association with the transmitter release process under examination. They have, so far, proposed at least 4 types of $GABA_B$ receptor 1α, 1β, 1δ and 1γ which are present on GABA, glutamate, choleocystokinin and somatostatin - releasing synaptosomes respectively. Furthermore they suggest that the $GABA_B$ receptor present in spinal cord is not sensitive to baclofen, the archetypal $GABA_B$ agonist. Unfortunately, Waldmeier et al (1994) could find no evidence to support this receptor subclassification in brain slices and we have obtained compelling evidence for the action of baclofen as a $GABA_B$ agonist in spinal cord (Malcangio & Bowery 1993;1994; Teoh et al 1996a;1996b). Nevertheless, we have obtained data using a series of $GABA_B$ antagonists to support the possibility of receptor heterogeneity, at least within the spinal cord.

2 $GABA_B$ Receptors in Spinal Cord

GABA appears to be present in the intrinsic interneurones of the spinal cord (Magoul et al.,1987; Todd and McKenzie,1989) and exerts its action via the $GABA_A$ and $GABA_B$ receptors (Hill and Bowery,1981; Bowery et al.,1983).

Autoradiographic studies using rat and human tissues have shown that, while the $GABA_A$ sites are evenly distributed throughout the laminae of the spinal cord, $GABA_B$ sites appear to be highly localised within the substantia gelatinosa (Price et al.,1984; Waldvogel et al.,1990; Malcangio et al.,1993). Neonatal capsaicin treatment and dorsal rhizotomy both reduce the density of $GABA_B$ sites in the rat substantia gelatinosa to the same extent (~ 40-50 %; Price et al.,1984; Price et al.,1987) which suggests that about half of the $GABA_B$ sites detected are located on the small-diameter primary-afferent nerve terminals.

By comparison a large proportion of the glutamate and aspartate detected in rat spinal cord is present within the terminals of the primary afferent fibres, particularly small diameter fibres. In capsaicin-treated (2 day neonatal) rats the release of glutamate, aspartate and substance P evoked by K^+, veratridine as well as dorsal root stimulation was significantly reduced in spinal cord slices. Conversely the release of endogenous GABA was unaffected by capsaicin pretreatment (Teoh et al 1996b). Thus, we can conclude that the source of released GABA is not primary afferent fibres unlike glutamate, aspartate and substance P which all appear to be present in the sensory fibres. These findings complement previously reported functional (Kangrga and Randic,1990,1991; Giovannini et al., 1991; Jeftinija et al., 1991; Skilling and Larson,1993) and anatomical (Battaglia and Rustioni,1988; De Biasi and Rustioni, 1988; Westlund et al 1989a;1989b) studies with regards to the role of the excitatory amino acids and substance P in the transmission of cutaneous information. However, our data only support the involvement of C-fibres in the release of aspartate and glutamate and do not indicate what proportion of the excitatory amino acids detected derive from other neuronal cell types. In accordance with the observations of Donnerer (1991) and Todd and McKenzie (1989), our data suggest that the increased release of GABA observed originates from the intrinsic interneurones. Hence, GABA probably plays a modulatory role at the spinal level by regulating the release of various neurotransmitters (Kangrga et al 1991; Malcangio and Bowery,1993;1994) released from both the primary afferents and the intrinsic interneurones.

$GABA_B$ receptor activation suppresses the evoked release of glutamate, substance P and GABA from rat spinal cord slices. Thus far all but one of the $GABA_B$ receptor antagonists tested prevented this inhibition in a dose-dependent manner (Malcangio & Bowery 1993; Teoh et al.,1996a). The one antagonist which contrasted with the remainder is CGP 56999. This high affinity antagonist failed to block the effect of (-) baclofen on evoked glutamate release even though it prevented the action of the $GABA_B$ agonist on GABA and substance P release (Teoh et al 1996a). This one separation, whilst limited, would support the general concept of receptor heterogeneity within the $GABA_B$ receptor family (Bonnano & Raiteri 1993) but the overall data do not support the classification proposed by Raiteri and colleagues (Gemignani et al 1994).

3 Plasticity of $GABA_B$ Receptors in Spinal Cord

Tolerance to the action of baclofen in man has long been recognised following systemic administration. Large doses are required to obtain adequate muscle relaxation which precipitates a reduced response to the drug as well as possible side effects. These have largely been overcome by intrathecal administration of small amounts of baclofen delivered by a pump inserted in the peritoneal cavity (Penn et al 1989). The tolerance to baclofen can be readily demonstrated in rats administered chronic baclofen (10 mg/kg i.p. once daily for 21 days). Spinal cord slices prepared from such animals readily release endogenous substance P, GABA and glutamate on stimulation of the dorsal roots. However, the inhibitory response to baclofen on these release systems is markedly reduced. The dose-response curve to (-) baclofen is shifted significantly to the right (Malcangio et al 1995). Conversely, rats treated with a $GABA_B$ antagonist (CGP 36742 or CGP 46381) for the same time period, demonstrated an enhanced sensitivity to baclofen shifting the dose/response curve significantly to the left (Malcangio et al 1995). These alterations in responses to baclofen were supported by a decrease and an increase, respectively, in the B_{max} for ^3H GABA binding to $GABA_B$ sites in the dorsal horn of rats from the same groups, (Malcangio et al 1993). Thus, prolonged activation or antagonism of the $GABA_B$ receptor in spinal cord produces an inhibition or enhancement of receptor activation. This apparent plasticity of the receptor was also manifest in the olfactory cortex of the same animals (Malcangio et al 1995) but was not manifest in all parts of the brain (Pratt & Bowery 1993). In the olfactory cortex, as in the spinal cord, changes in the receptor produced alterations in the functional response to baclofen and synaptically mediated $GABA_B$ receptor responses (Malcangio et al 1995) indicating that synaptic, as well as possible non-synaptic, receptors can be modified by chronically-administered receptor ligands. This may have significant consequences under abnormal conditions of altered GABA release such as in chronic inflammation.

4 Chronic Inflammation: Influence on GABAergic Function in Spinal Cord

It has been reported that adjuvant arthritic rats show an increase in the number of γ-aminobutyric acid (GABA)-immunoreactive cells due to an enhanced concentration of intracellular GABA in lamina I-III of the dorsal horn (L3-LS) (Castro-Lopes et al:,1992). This increase in the amount of GABA occurs within the zone of projection of afferent nerves emanating from the inflamed foot, supporting the possibility that GABA modulates the nociceptive input at the segmental level (CastroLopes et al.,1992).

The transmission of afferent impulses within the spinal cord is possibly mediated, in part, via peptidergic mechanisms (see Sluka et al.,1992) and among the peptide candidates proposed is the undecapeptide, substance P (SP). This is thought to be released from the central end of sensory fibres as well as from the peripheral terminals where it has been implicated in the genesis of neurogenic inflammation. Increased biosynthesis of SP in the dorsal root ganglia of adjuvant arthritic rats is reflected in an increase in the mRNA levels of preprotachykinin A, the precursor of SP (Minami et al.,1989) and the number of SP-immunoreactive fibres increases in the dorsal horn of the lumbar spinal cord (Kar et al.,1991). In polyarthritic rats, spontaneous release of SP from the spinal dorsal horn exceeds that in control rats (Oku et al.,1987). Movement of the ankle joint which is innocuous and does not affect the release of SP in non-inflamed control rats, is converted into a noxious stimulus producing a significant increase in SP release in polyarthritic rats (Oku et al.,1987).

Since GABA and baclofen can inhibit the electrically-evoked release of substance P from the rat spinal cord (Malcangio & Bowery 1993) it seemed possible that the increase in GABA within the spinal cord of monoarthritic rats may act to reduce any enhancement in the output of nociceptive peptide. To examine this question rats were injected into the left hind paw (intradermal) with 5OOμg heat-killed Mycobacterium tuberculosis in 0.1 ml mineral oil (complete Freud's adjuvant) under halothane anaesthesia. Control animals were injected with 0.1 ml mineral oil alone (incomplete Freud's adjuvant). Spinal cord slices were obtained at 7,14, 21 and 28 days after injection. In addition thermal hyperalgesia to radiant heat was assessed in each rat by using the paw - withdrawal test where the latency to withdraw was taken as an index of the nociceptive index (Hargreaves et al 1988).

Electrical simulation of the dorsal roots attached to control rat isolated spinal cord slices induced a significant increase (174 ± 18% of basal outflow which was 30.4 ± 3.1 fmol per fraction 8 ml^{-1} in SP-Ll release (Figure 2, open column).
One week after injection of CFA no differences in the amount of SP-L1 released from the spinal cord of IFA and CFA rats were detected (Figure 2) whereas after two weeks the spinal cord from CFA-treated rats tended to release more SP-L1 than IFA cords.

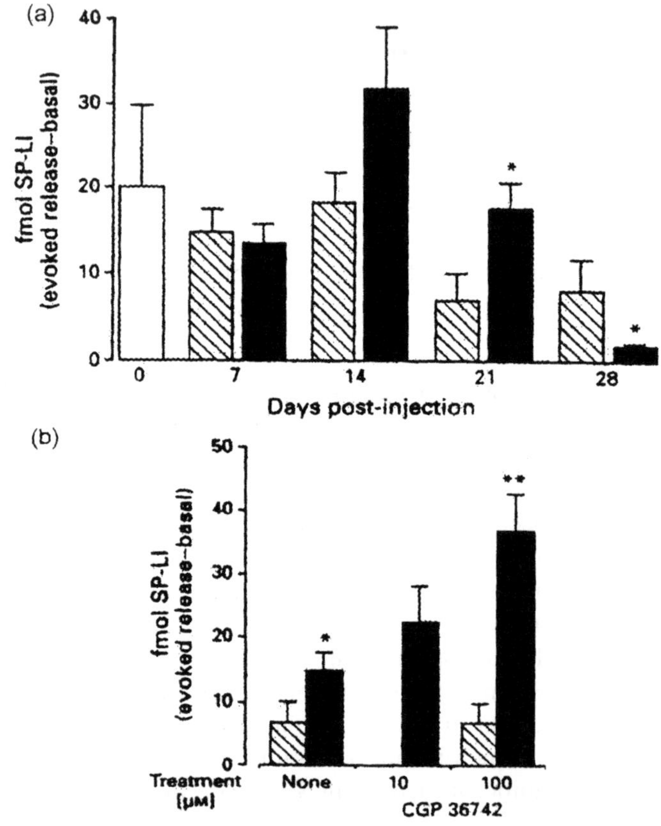

Fig 2a. Effect of monoarthritis on substance P-like immunoreativity (SP-LI) from Lewis rat spinal cord slices. Basal outflow of SP-L1 from the spinal cord of saline-injected (open column) IFA (widely hatched column) and CFA (solid black column) rats was 36.7 ± 3.0 fmol 8 ml^{-1} per fraction, n=61 (pooled means). SP-LI was released by electrical stimulation of the dorsal roots at 20 V, 0.5 ms, 1 Hz for 8 min. Values are mean ± s.e.mean. n=5 spinal cords for open column, 9 spinal cords for IFA except at 28 days when they were 3; 8 spinal cords for CFA at 7 days; 12 spinal cords for CFA at 14 and 21 days; 3 spinal cords for CFA at 28 days. *P<0.05 CFA vs IFA (Student's t test). **Fig 2 (b).** Effect of CGP 36742, on electrically-evoked substance P-like immunoreactivity (SP-LI) release from the spinal cord of IFA and CFA rats. Basal outflow of SP-LI from IFA (widely hatched column) and CFA (solid black column) rat spinal cord slices was 6.5 ± 1.5 fmol 8ml^{-1} per fraction, n=26. drug was perfused 5 min prior to and during stimulation (20V, 0.5 ms, 1Hz for 8 min). Values are mean ± s.e.mean of 3-4 slices for each group. *P<0.01 CFA in absence vs presence of drug (Student's t test). *(With permission from* Br. *J. Pharmacol).*

Twentyone days after injection, the spinal cord of CFA rats released significantly more peptide compared to IFA rats (Figure 2). By contrast, four weeks after injection, the release of SP-Ll from the spinal cord of CFA rats was almost absent (Figure 2). Basal outflow of SP-L1, total content of SP-Ll and SP-Ll release induced by capsaicin were unaffected in CFA and IFA rats throughout the 28 days period. Evoked release of SPLI from the spinal cord of CFA rats was greatly increased by perfusion with the $GABA_B$ antagonist, CGP 36742 (10 and 100μM) which did not alter the release from IFA spinal cord. The effect of this drug was dose-dependent, achieving statistical significance at 100μM. Intraperitoneal injection of the $GABA_B$ antagonist, significantly decreased the withdrawal latency of both inflamed and contralateral paw of CFA rats up to 45 min after treatment CGP 36742 was not hyperalgesic in IFA rats. The $GABA_B$ agonist (-)-baclofen, could reverse the hyperalgesia in the flamed paw but a high dose of 3 mg kg^{-1} was necessary to raise the paw withdrawal latency from 5.2 ± 0.8 s to 8.2 ± 0.5 s ($n = 6$, $P<0.05$, Student's t test).

Concomitant with this apparent increase in GABA - mediated inhibition in substance P release, was a decrease in $GABA_B$ receptor binding within the dorsal horn of the spinal cord (Castro-Lopes et al 1995). This could result from the increase in sustained receptor activation arising from the raised level of GABA in the number of GABA-immunoreactive cells in laminae I-III following CFA injection (Castro-Lopes et al 1992). This would be equivalent to the chronic administration of baclofen reported above which similarly diminished the number of $GABA_B$ binding sites.

Thus, the induction of chronic inflammation appears to activate the GABAergic system within the spinal cord. This increases presynaptic inhibition of primary afferent inputs to the dorsal spinal cord via $GABA_B$ receptors to reduce nociception. Plasticity of the $GABA_B$ receptors, however, may limit the long term response to activation.

References

Battaglia, G., Rustioni, A. (1988). Coexistence of glutamate and substance P in dorsal root ganglion neurons of the rat and monkey. J. Comp. Neurol., 277, 302.

Bonanno, G., Raiteri, M. (1993). Multiple $GABA_B$ receptors. Trends Pharmacol. Sci, 14: 259-261.

Bowery, N.G. (1989). $GABA_B$ receptors and their significance in mammalian pharmacology. Trends Pharmacol. Sci.10, 401-407.

Bowery, N.G., (1993). $GABA_B$ Receptor Pharmacology. Annu. Rev. Pharmacol. Toxicol.109-147.

Bowery, N.G., Doble, A., Hill, D.R., Hudson, A.L., Shaw, J.S., Turnbull, M.J. and Warnngton, R. (1981). Bicuculline-insenstive GABA receptor on peripheral autonomic nerve terminals. Eur. J. Pharmacol. 71: 53-70.

Bowery, N.G., Hill, D.R., and Hudson A.L. (1983). Characteristics of $GABA_B$ receptor binding sites on rat whole brain synaptic membranes, Br. J. Pharmacol. 78, 191.

Bowery, N.G., Hill, D.R., Hudson, A.L., Doble, A., Middlemiss, D.N., Shaw J. and Turnbull, M. (1980) (-) Baclofen decreases neurotransmitter release in the mammalian CNS by an action at a novel GABA receptor. Nature 283: 92-94.

Castro-Lopes, J.M., Malcangio, M., Pan, B.H., Bowery, N.G. (1995). Complex changes of $GABA_A$ and $GABA_B$ receptor binding in the spinal cord dorsal horn following peripheral inflammation or neurectomy. Brain Research 679, 289-297.

Castro-Lopes, J.M., Tavares, I., Tolle, T.R., Coito, A., Coimbra, A. (1992). Increase in GABAergic cells and GABA levels in the spinal cord in unilateral inflammation of the hindlimb of the rat. Eur. J. Neurosci, 4, 296-301.

Crunelli, V., Leresche, N. (1991). A role for $GABA_B$ receptors in excitation and inhibition of thalamocortical cells. Trends Neurosci.14:16-21.

De Biasi, S., Rustioni, A. (1988). Glutamate and substance P co-exist in primary afferent terminals in the superficial laminae of spinal cord. Proc. Natl. Acad. Sci. USA 85, 7820 - 7824.

Deisz, R. A., Billard, J.M. and Zieglgansberger, W. (1993). Pre and postsynaptic $GABA_B$ receptors of rat neocortical neurones differ in their pharmacological properties. Neuroscience Letters 154, 209-212.

Donnerer, J., (1991). Depolarization-evoked release of glutamate, aspartate and γ-aminobutyric acid from rat dorsal spinal cord slices does not originate from capsaicinsenstive neurons, Brain Res. 555, 332.

Dutar, P., Nicoll, R.A. (1988a). A physiological role for $GABA_B$ receptors in the central nervous system. Nature 332:156-158.

Dutar, P., Nicoll, R.A. (1988b). Pre-and postsynaptic $GABA_B$ receptors in the hippocampus have different pharmacological properties. Neuron 1: 585-598.

Gemignani, A., Paudic, P., Bonanno, G. and Raiteri M. (1994). Pharmacological Discrimination Between γ-Aminobutyric Acid Type B Receptors Regulating Cholecystokinin and Somatostatin Release from Rat Neocortex Synaptosomes. Mol. Pharmacol., 45, 558-562.

Giovannini, G., Lacey M.G., Pepeu G., Nistri, A. (1991). Release of endorgenous glutamate and aspartate from frog spinal cord in vitri. Eur. J. Pharmacol.,195, 47.

Hargreaves, K. Dubner, R., Brown, F., Flores, C., Joris, J. (1988). A new and sensitive method for measuring thermal nmociception in cutaneous hyperalgesia. Pain, 32, 77-88.

Harrison, N.L. (1990). On the presynaptic action of baclofen at inhibitory synapses between cultured rat hippocampal neurones. J. Physiol. 422: 433-446.

Hill, D.R., Bowery, N.G. (1981) ^{3}H-Baclofen and ^{3}H-GABA bind to bicucullineinsensitive $GABA_B$ sites in rat brain. Nature 290:149-152.

Isaacson, J.S., Solis, J.M. and Nicoll, R.A. (1993). Local and diffuse synaptic actions of GABA in the hippocampus. Neuron 10:165-175.

Kangrga, I., Jiang, M. Randic, M. (1991). Action of (-)baclofen on rat dorsal horn neurones. Brain Res. 562, 265.

Kangrga, I., Larew J.S.A., Randic, M. (1990). The effects of substance P and calcitonin-gene-related peptide on the efflux of endogenous glutamate and aspartate from rat spinal dorsal horn in-vitro. Neurosci. Lett,108,155.

Kangrga, I., Randic, M. (1991). Outflow of endogenous amino acids from the rat spinal dorsal horn slice in response to low- and high-threshold primary afferent fibres. Modulation of μ-opioids, Brain Res. 553, 347.

Kar, S., Gibson, S.J., Rees, R.G., Jura, W.G.Z.O., Brewerton, D.A., Polak, J.M. (1991). Increased calcitonin-gene-related peptide (CGRP), substance P and enkephalin immunoreactivities in dorsal spinal cord and loss of CGRP-immunoreactive motoneurons in arthritic rats depend on intact peripheral nerve supply. J. Mol. Neurosci. 3, 7-18.

Karlsson, G., Pozza, M. and Olpe, H.-R. (1988). Phaclofen: a $GABA_B$ blocker reduces long-duration inhibition in the neocortex. Eur. J. Pharmacol. 148: 485-486.

Kerr, D.I.B. and Ong, J. (1995). $GABA_B$ receptors. Pharmac. Ther., 67,187-246.

Magoul, R. Onteniente, B., Geffard, M. & Calas, S. (1987). Anatomical distribution and ultrastructural organization of the GABAergic system in the rat spinal cord. An immunocytochemical study using anti-GABA antibodies. Neuroscience, 20,1001- 1009.

Malcangio, M., Bowery, N.G. (1993). γ-Aminobutyric $acid_B$, but not γ-aminobutyric $acid_A$ receptor activation, inhibits electrically-evoked substance P-like immunoreactivity from rat spinal cord in vitro. J. Pharmacol. Exp. Ther., 266,1490- 1496.

Malcangio, M., Bowery, N.G. (1993). Gamma-aminobutyric $acid_B$ but not gammaamino butyric and $acid_A$ receptor activation inhibits electrically evoked substance Plike immunoreactivity release from the rat spinal cord in Vitro. J. Pharmac. Exp. Ther., 266,1490-1496.

Malcangio, M. and Bowery, N.G. (1994). Spinal cord SP release and hyperalgesia in monoarthritic rats: involvement of the $GABA_B$ receptor system. Br. J. Pharmacol. 113:1561-1566.

Malcangio, M., and Bowery, N.G.1994. Spinal cord SP release and hyperalgesia in monoarthritic rats: involvement of the $GABA_B$ receptor system. Br. J. Pharmacol. 113,1561.

Malcangio, M., Libri, V., Teoh, H., Constanti, A. and Bowery, N.G. (1995). Chronic (-) baclofen or CGP 36742 alters $GABA_B$ receptor senstivity in rat brain and spinal cord. Neuroreport, 6: 339-403.

Malcangio, M., Libri, V., Teoh, H., Constanti, A., Bowery, N.G. (1995). Chronic (-) baclofen or CGP 36742 alters $GABA_B$ receptor sensitivity in rat brain and spinal cord. Neuroreport 6, 399-403.

Malcangio, M., Silva Da H., Bowery N.G. (1993). Plasticity of $GABA_B$ receptor in rat spinal cord detected by autoradiography, Eur. J. Pharmacol. 250,153.

Minami, M., Kuraishi, Y., Kawamura, M., Yamaghuchi, T., Masu, Y., Nakanishi, S., Satoh, M. (1989). Enhancement of preprotachykinin in the rat spinal cord: possible involvement of substance P-containing spinal neurons in nociception. Neurosci. Lett., 98,105-110.

Mott, D.D., Lewis, D.L. The pharmacology and function of $GABA_B$ receptors. Int. Rev. Neurobiol. 36: 97-223,1994.

Oku, R., Satoh, M. & Tagaki, H., (1987). Release of substance P from the spinal dorsal horn is enhanced in polyarthritic rats. Neurosci. Lett., 74, 315-319.

Penn, R.D., Savoy, S.M., Corcos, D., Latash, M., Gottlieb, G., et al.1989. Intrathecal baclofen for severe spinal spasticity. N. Engl. J. Med. 320:1517-21.

Penn, R.R. and Mangieri, E.A. (1993). Stiff-man syndrome treated with intrathecal baclofen. Neurology 43: 2412.

Pittaluga, A., Asaro, D., Pettegrinni, G. and Raiteri, M. (1987). Studies of ^{3}H-GABA and endogenous GABA release in rat cerebral cortex suggest the presence of autoreceptors of the $GABA_B$ type. Eur. J. Pharmacol.144: 45-52.

Pratt, G.D., Bowery, N.G. (1993). Repeated administration of desipramine and a $GABA_B$ receptor antagonist, CGP 36742, discretely up-regulates rat frontal cortical $GABA_B$ receptor binding sites. Br. J. Pharmacol.,110, 724-735.

Price, G.W. Kelly, J.S.., Bowery, N.G. (1987). The location of $GABA_B$ receptor binding sites in mammalian spinal cord. Synapse 1:530-38.

Price, G.W., Blackburn, T.P., Hudson, A.L. Bowery, N.G.1984. Presynaptic $GABA_B$ sites in the interpeduncular nucleus. Neuropharmacology 23:861-62.

Raiteri, M., Giralt, M.T., Bonanno, G., Pittaluga, A., Fedele, E. and Fontana, G. (1990). Release-regulating GABA autoreceptors in human and rat central nervous system. In: $GABA_B$ receptors in mammalian function. Eds: Bowery, N.G., Bittiger, H. & Olpe H.-R. Publ. Wiley, Chichester pp 81-98.

Skilling, S.R., Larson, A.A. (1993). Capsaicin inhibits whereas rhizotomy potentiates substance P-induced release of excitatory amino acids in the rat spinal cord in vivo. Neurosci. Lett.150,107.

Sluka, K.A., Dougherty, P.M., Okin, L.S., Willis, W.D., Westlund, K.N. (1992). Neuronal changes in acute arthritis in monkey III. Changes in subtance P, CGRP and glutamate in the dorsal horn of the spinal cord. Brain Res. Rev.,17, 29-38.

Teoh, H., Malcangio, M., and Bowery N.G. (1996a). The effects of novel $GABA_B$ antagonists on the release of amino acids from spinal cord of the rat. Br. J. Pharmacol. (in press).

Teoh, H., Malcangio, M., Fowler, L.J. & Bowery, N.G. (1996b). Evidence for release of glutamic acid, aspartic acid and substance P but not γ-aminobutyric acid from primary afferent fibres in rat spinal cord. Eur. J. Pharmacol (in press).

Todd, A.J. & McKenzie, J. (1989). GABA-immunoreactive neurons in the dorsal horn of the rat spinal cord. Neuroscience, 3 l, 799-806.

Waldmeier, P.C., Wicki, P., Feldtrauer, J.-J., Mickel, S.J. Bittiger, H. and Baumann, P.A. (1994). GABA and glutamate release affected by $GABA_B$ receptor antagonists with similar potency: no evidence for pharmacologically different presynaptic receptors. Br. J. Pharmacol.113,151-1521.

Waldovogel, H.J., Faull, R.L.M., Jansen, K.L.R., Dragunow, J., Richards, J.G., Mohler, H., Streit, P., (1990). GABA. GABA receptors and benzodiazepine receptors in the human spinal cord: an autoradiographic and immunohistochemical study at the light and electron microscopic levels. Neuroscience 39, 361.

Westlund, K.N., McNeill, D.L., Patterson, J.T. Coggeshall, R.E. (1989b). Aspartate immunoreactive axons in normal rat L4 dorsal roots. Brain Res. 489, 347.

Phosphoproteins and the Regulation of Vesicular Neurotransmitter Release

Jacques J. H. Hens[¶], Govert Hoogland, Willem-Hendrik Gispen, and Pierre N.E. De Graan

Department of Medical Pharmacology, Rudolf Magnus Institute for Neurosciences, Utrecht University, Universiteitsweg 100, 3584 CG Utrecht, The Netherlands.
¶ To whom correspondence should be addressed

1 Introduction

Neurotransmission requires rapid docking, fusion and recycling of neurotransmitter vesicles in the nerve terminal. Many of the proteins involved in this complex mechanism have now been identified (reviewed by Scheller,1995; Sudhof,1995) and are present in presynaptic nerve terminals, where they constitute the molecular apparatus for the storage and secretion of neurotransmitters and termination of their effects. Presynaptic nerve terminals store a variety of different neurotransmitters in vesicles which are secreted into the extracellular space by means of regulated exocytosis i.e. upon a specific signal, usually a depolarization-induced Ca^{2+} influx. Neurotransmitter-containing vesicles firmly dock at the presynaptic plasma membrane prior to exocytosis and are rapidly reclaimed and recycled after fusion and secretion. In mammalian brain nerve terminals two classes of secretory neurotransmitter vesicles can be distinguished on an ultrastructural basis, namely small clear-cored vesicles and large dense-cored vesicles (reviewed by Kelly,1993). Small clear-cored vesicles, which store neurotransmitters such as glutamate, GABA, glycine, acetylcholine and possibly catecholamines, are released swiftly at the active synaptic zone and undergo multiple cycles of exo- and endocytosis within the nerve terminal. Large dense-cored vesicles, which contain catecholamines or a variety of neuropeptides, on the other hand, cannot refill after exocytosis and therefore need to be recycled via the trans Golgi network. The extreme rapidity of small clear-cored vesicle fusion with the presynaptic plasma membrane (within 200-300 μs) requires that at the resting state a number of synaptic vesicles is docked as pre-assembled complexes at their sites of fusion. The assembly of these fusion complexes occurs preferably in the close vicinity of Ca^{2+} channels in order to allow Ca^{2+} to bind to its receptor protein, the Ca^{2+}-sensor, without delay and to instantly mediate vesicle fusion.

Although Ca^{2+} clearly induces neurotransmitter release, it is certainly not the sole regulator of this process. Indeed, many studies that focussed on the role of Ca^{2+} dependent protein kinases in the regulation of neurotransmitter release, notably

NATO ASI Series, Vol. H 100
Neurotransmitter Release and Uptake
Edited by Şakire Pöğün

protein kinase C (PKC)[1] and Ca^{2+}/calmodulin-dependent protein kinase II (CaMKII), demonstrated that several proteins undergo Ca^{2+}-dependent phosphorylation or dephosphorylation events in conjunction with neuronal depolarization and secretion. Presently, an increasing body of evidence indicates that second messenger-regulated changes in the activity of several types of protein kinases and protein phosphatases affect the ability and extent of neurotransmitter release, implying that these protein kinases and phosphatases and their substrates are intimately involved in the mediation and regulation of Ca^{2+}-induced, vesicular neurotransmitter release. Phosphoproteins observed in nerve terminals and convincingly been implicated in the molecular mechanism of neurotransmitter release include for instance the synapsins (Greengard et al.,1993), synaptotagmin (Bennett et al., 1993; Davletov et al., 1993), rabphylin-3A (Fykse et al., 1995), synaptobrevin/VAMP (Nielander et al.,1995), dynamin I (Robinson et al.,1993), MARCKS (Aderem, 1992) and B-50/GAP-43 (Dekker et al., 1989b; Hens et al., 1995).

In this chapter we discuss the implications of protein phosphorylation and dephosphorylation processes in the regulation of neurotransmitter release. Firstly, the main protein kinases and phosphatases will be discussed, that so far have been implicated in the regulation of neurotransmitter release. In the second part of this chapter we will focus on the nervous tissue-specific PKC-substrate B-50, and in particular on how it affects vesicular neurotransmitter release in a phosphorylation dependent fashion.

2 Protein Kinases and Release

Protein phosphorylation occurs when a protein kinase mediates the transfer from the terminal phosphate of ATP to a substrate protein. Depending on the type of enzyme, protein kinases only phosphorylate serine (Ser) and threonine (Thr) or tyrosine (Tyr) residues in a restricted group of substrates. Protein kinase activity, protein phosphorylation and neurotransmitter release are found to be highly correlated in the brain. Pharmacological approaches using protein kinase inhibitors and activators have established more directly that protein kinases are indeed involved in the regulation of neurotransmitter release. In particular, the Ca^{2+}-dependent protein kinases CaMKII and PKC have been shown to affect the extent of neurotransmitter release. Therefore, we will shortly discuss these two protein kinases.

---

[1] The abbreviations used are: AKAP79 A kinase anchor rotein-79; CaM, calmodulin; CaMKII, calcium/calmodulin-dependent protein kinase type II; CasKII, casein kinase II; CCK8, cholecystokinin-8; DAG, diacylglycerol; DARPP-32, dopamine- and cAMP-regulated phosphoprotein of 32 kDa; GAP-43, growth-associated protein-43; IgG, immunoglobin antibody; MARCKS, myristoylated alanine-rich C kinase substrate; PIP, phosphatidylinositol 3-phosphate; PKC, protein kinase C; Ser, serine; Thr, threonine; Tyr, tyrosine.

2.1 Calcium/Calmodulin-Dependent Protein Kinase II and Synapsin I

In presynaptic nerve terminals, CaMKII is tightly associated with synaptic vesicles and the plasma membrane, where it functions as a multimer of α- (50 kDa) and β/β' subunits (60 kDa) (reviewed by Colbran and Soderling, 1990; Greengard et al., 1993). In the absence of Ca^{2+} and calmodulin (CaM), native CaMKII is mactive due to an interaction of its autoinhibitory domain with both the ATP- and substrate binding element of the catalytic domain. In the presence of Ca^{2+} and CaM this interaction is disrupted and CaMKII rapidly autophosphorylates Thr-residues in its α and β-subunit, and is now capable to phosphorylate exogenous substrates. Protein phosphatases 1, 2A (Lai et al.,1986; Schworer et al.,1986) and 2C (Fukanaga et al., 1993) dephosphorylate autophosphorylated CaMKII. Autophosphorylation transforms CaMKII into a Ca^{2+}-independent form and results in prolonged kinase activity even after intracellular $[Ca^{2+}]$ has returned to a basal level. In addition, Ca^{2+} independent CaMKII might be required for the phosphorylation of substrates such as calcineurin (see also paragraph 3), because it was observed that binding of Ca^{2+}/CaM to calcineurin blocks the CaMKII phosphorylation site in calcineurin.

The involvement of CaMKII in neurotransmitter release was demonstrated by CaMKII micro-injection into squid giant synapses, in which it enhanced depolarization-induced neurotransmitter release. Similarly, K^+-induced noradrenaline and glutamate release from synaptosomes were facilitated after introduction of activated CaMKII (reviewed by Greengard et al.,1993). For the facilitatory action of CaMKII on release a small group of phosphoproteins, the synapsins, is thought to be important. The synapsins (Ia, Ib, IIa and IIb) are a family of neuron-specific phosphoproteins specifically associated with the cytoplasmic site of small clear-cored vesicle membranes. Virually all nerve terminals contain synapsins, where they have been implicated in the phosphorylation-dependent linkage of small clear-cored vesicles to the actin-based cytoskeleton. Synapsin I and II are phosphorylated at the same Ser residue at the NH2-terminal domain, by either protein kinase A or by Ca^{2+}/CaM-dependent protein kinase I. CaMKII-mediated phosphorylation of two additional sites at the COOH-terminal domain of synapsin I results in a dissociation of synapsin I from small clear-cored vesicles. In vitro, calcineurin (see also paragraph 3) was found to dephosphorylate synapsin I (King et al.,1984). Dephosphorylated synapsin I, either by itself or bound to a vesicle, triggers the nucleation of actin monomers and thus speeds up the actin polymerization process. This nucleation activity of synapsin I as well as its ability to bundle actin filaments, were lost after CaMKII-mediated synapsin I phosphorylation. Introduction of dephospho-synapsin I into squid giant synapses or synaptosomes inhibited evoked neurotransmitter release, whereas CaMKII-phosphorylated synapsin I was without effect. Moreover, presynaptic injection of anti-synapsin I IgGs depressed neurotransmitter release evoked by high-frequency (20 Hz) but not by low-frequency (0.2 Hz) stimulation in lamprey neurons (Pieribone et al.,1995). Based on these observations a function of synapsin I was postulated in synaptic vesicle recruitment from the so-called reserve (distal) pool to the releasable (proximal) pool. The liberation of synaptic vesicles from the reserve pool is assumed to replenish the releasable pool, and thus regulates the availability of neurotransmitter vesicles for exocytosis. These data are in accordance with recent studies in mice lacking either CaMKII-substrate synapsin I (Rosahl et al.,1993; Li et al.,1995) or both synapsin I and II (Rosahl et al.,1995). The data obtained in these studies indicate that synapsin I is not essential for neurotransmitter release per se, but is essential for the accelerated recruitment of vesicles to the active zone during repetitive stimulation, thus preventing depletion during conditions of sustained neurotransmission.

Summarizing, CaMKII is a Ca^{2+}/CaM-regulated kinase with a clear implication in neurotransmitter release. Synapsin I is a major presynaptic CaMKII substrate, for which a large body of evidence indicates that it regulates the size of the releasable pool of vesicles in a phosphorylation-dependent fashion.

2.2 Protein Kinase C

Protein kinase C (PKC) is a family of monomeric enzymes that are activated in the presence of Ca^{2+} and membrane phospholipids (reviewed by Tanaka and Nishizuka, 1994). Diacylglycerol (DAG) activates PKC by increasing its affinity for Ca^{2+}, an action also performed by tumour promoting β-type phorbol esters. Four Ca^{2+} dependent PKC isoforms are known; α (type III), β1, β2 (type II) and γ (type I), which share a conserved C2 domain that is absent in Ca^{2+} -independent PKC isoforms and thought to be important in the Ca^{2+}-dependent interaction with membranes. A similar C2 domain is also present in other phosphoproteins implicated in secretion, such as synaptotagmin and rabphylin-3A (Sutton et al.,1995; and references therein). All four Ca^{2+} -dependent PKC isoforms can be autophosphorylated, but the physiological relevance of this autophosphorylation is still unclear (Ohno et al., 1990).

The involvement of PKC in neurotransmitter release was demonstrated by the observation that phorbol esters potentiated neurotransmitter release in Aplysia neurons, rat brain synaptosomes and permeated chromaffin cells and PC12 cells (reviewed by Dekker et al.,1991a; Robinson, 1991). Micro-injection of PKC into different types of neurons was also found to enhance synaptic activity. These studies did not establish whether PKC is a prerequisite for secretion or has a modulatory function. In an attempt to resolve this issue it was observed that specific PKC inhibitors did not affect the degree of Ca^{2+}-induced release from permeated chromaffin cells ('Terbush and Holz,1990) and permeated synaptosomes (Hens et al., 1993a,b), whereas the phosphorylation of PKC-substrate B-50 (see also paragraph 4) was significantly reduced by these inhibitors in the permeated synaptosomes. These data thus indicate that PKC has a modulatory function rather than a mediating function in exocytosis. Indeed, a facilatory metabotropic glutamate receptor pathway has been described that activates PKC and enhances the K^{+}-channel blocker 4 - aminopyridine - induced Ca^{2+} entry and glutamate release in synaptosomes (Herrero et al., 1992; Coffey et al., 1994). The list of PKC substrates implicated in neurotransmitter release is still growing, and includes e.g.: MARCKS (Aderem,

1992), dynamin I and B-50. The phosphorylation-dependent implications of the latter two substrates in neurotransmission are discussed later in this chapter.

3 Protein Phosphatases, Dephosphorylation and Release

Protein phosphatases reverse the actions of protein kinases and are therefore essential components in processes regulated in a phosphorylation-dependent fashion including neurotransmission. At least four different classes of phospho-Ser/phospho-Thr phosphatases have been identified, usually classified in type 1 and type 2 according to their substrate specificity and regulation (Ingebritsen et al., 1983). Protein phosphatases 1 and 2A are most abundantly present in brain (reviewed by Cohen, 1989). Large amounts of the Ca^{2+}/CaM-dependent protein phosphatase 2B, calcineurin, have also been detected in several brain regions (Klee,1988; Steiner et al.,1992). The Mg^{2+}-dependent protein phosphatase 2C is enriched in cerebellum and hippocampus (Abe et al.,1992). Protein phosphatase 1 interacts with proteins such as inhibitor-l, -2 and DARPP-32 (**d**opamine- and **cAMP-r**egulated **p**hospho**p**rotein of **32** kDa), which regulate the protein phosphatase 1 activity in a phosphorylation-dependent manner (reviewed by Girault,1993), and can be regarded as inhibitory subunits. Protein phosphatase 2A is specifically inhibited by protein kinase G-phosphorylated G-substrate, which is mainly present in cerebellum Purkinje cells (Detre et al.,1984). Moreover in neurons an association between protein kinase A, AKAP79 (A kinase anchor protein) and calcineurin was shown, targeting both the kinase and phosphatase to a subcellular site where they may regulate the phosphorylation state of key neuronal substrates (Coghlan et al., 1995). In vitro calcineuin is phosphorylated by PKC and autophosphorylated CaMKII, which partially inactivate its phosphatase activity. Calcineurin exhibits very slow autodephosphorylation and is rapidly dephosphorylated by protein phosphatase 2A (Hashimoto et al.,1989). All these data thus illustrate that protein phosphorylation processes also regulate the activity of protein phosphatases, and vice versa.

The recent discovery of more or less specific phosphatase inhibitors, such as okadaic acid, calyculin A and the cyclosporin A derivatives, has resulted in several reports demonstrating a role of protein phosphatases in the regulation of neurotransmitter release. At crayfish and frog neuromuscularjunctions, application of okadaic acid, a relatively specific membrane-permeable inhibitor of protein phosphatases 1 and 2A, has been reported to act presynaptically to increase transmitter release (Swain et al.,1991). In cultured cell systems (Wu and Wagner, 1991; Yanagihara et al.,1991) and rat synaptosomes (Sim et al.,1993; Verhage et al., 1995), however, okadaic acid was found to inhibit stimulated exocytosis of several different types of transmitters, suggesting that intrinsically Ca^{2+}-independent phosphatases 1 and 2A are involved in this process. Addition of cyclosporin A or FK506, which inhibit calcineurin activity by binding to the cytosolic immunophilins cyclophilin and FKBP-12, respectively, was found to potentiate the 1-mM 4-aminopyridine-evoked release of glutamate but not the K^+-evoked release from synaptosomes (Nichols et al.,1994; Sihra et al.,1995). These observations, however, are likely to result from a suppression of voltage-gated calcium-channel activity and probably do not relate to a direct action on exocytosis itself. In other systems, such as mast cells (Hultsch et al.,1991) and pancreatic acinar cells (Groblewski et al.,1994) cyclosporin A derivatives were found to inhibit secretion. Another more direct implication of calcineurin in stimulus-secretion coupling occurring after the Ca^{2+} influx was obtained using specific anti-calcineurin antibodies (IgGs) and purified protein phosphatases. Micro-injection of purified calcineurin into protozoa *Paramecium* cells triggered exocytosis, whereas application of anti-calcineurin IgGs inhibited exocytosis (Momayezi et al., 1987). Similarly, introduction of anticalcineurin IgGs into permeated rat brain synaptosomes inhibited Ca^{2+}-induced vesicular noradrenaline release (De Graan et al., 1992). These data indicate that calcineurin-mediated dephosphorylation is an essential event in the molecular mechanism of neurotransmitter release occurring after the Ca^{2+} influx. However, with the exception of dynamin I, the relevant calcineurin substrate(s) are still largely unidentified. The implications of calcineurin-mediated dynamin I dephosphorylation for vesicle recycling will be discussed below in more detail and further illustrate the importance of protein phosphatases in the process of neurotransmission.

3.1 Dynamin I and Endocytosis

Dynamin I is a neuronal PKC-substrate that is rapidly (within 2 seconds) dephosphorylated by calcineurin upon a depolarization-induced Ca^{2+} influx and has been implicated in endocytosis, the process of membrane retrieval after exocytosis (reviewed by Robinson et al.,1994). Dynamin I dephosphorylation in synaptosomes is fully reversible for several cycles of depolarization and repolarization. Peptide sequencing revealed four isoforms of dynamin I, possesing microtuble-stimulated GTPase activity. PKC-mediated phosphorylation at the COOH-terminal domain enhances the GTPase activity of dynamin I, which is confined to the NH_2-terminal domain (Robinson et al.,1993). Phosphorylation of dynamin I by casein kinase II (CasKII) prevents phosphorylation by PKC but does not affect its GTPase activity per se. The COOH-terminal domain of dynamin I was suggested to interact with other endocytotic components and phospholipid membranes (Herskovits et al., 1993). Recent information on the function and the mechanism of action of dynamin I was obtained in hypotonically-lysed synaptosomes, which were incubated with brain cytosol, ATP and GTP-γ-S. Electron microscopical evaluation of these preparations revealed tubular invaginations of clathrin-coated vesicle membrane which were surrounded by stacks of rings that consisted of dynamin I (reviewed by De Camilli, 1995). Based on these data the following model for dynamin I in endocytosis has been proposed. Ca^{2+}-dependent stimulation of neurotransmitter release leads to a rapid calcineurin-mediated dephosphorylation of dynamin I which correlates with a decrease in GTPase activity (thereby increasing the concentration of activated, GTP bound dynamin I) and an increased rearrangement of dynamin I at the plasma membrane. Dephosphorylation may thus facilitate assembly of dynamin I rings and retrieval of clathrin-coated vesicle membrane during endocytosis. Another PKC substrate implicated in vesicle fusion and recycling is B-50, which will be discussed next.

4 B-50 (GAP-43) and Release

B-50 [also known as growth-associated protein-43 (GAP-43), neuromodulin and F1] is a nervous tissue-specific protein, that is highly expressed in neurons during development and nerve regeneration, and has been implicated in neurite outgrowth, long-term potentiation, neuronal signal transduction and neurotransmitter release (reviewed by Gispen et al.,1991; Liu and Storm, 1990; Strittmatter et al., 1992; Pasinelli et al.,1995). In mature, non-damaged nervous tissue B-50 is expressed in most neurons, but at lower levels; expression levels are highest in human associative brain areas and rat hippocampal and olfactory areas. In mature neurons B-50 is predominantly found in presynaptic plasma membranes and not in dendrites. B-50 has also been identified in PC12 cells and chromaffin cells. The biochemistry and function of B-50 in neurotransmitter release have been studied by several labs and will be discussed below.

4.1 Biochemistry of B-50

B-50 (226-amino acids in rat; PAGE 43-48 kDa) is attached to the presynaptic membrane through two cysteine residues at positions 3 and 4, either by thioester linkage to palmitic acid or by some other, still unresolved, mechanism. In vitro, B-50 has been reported to bind calmodulin (Alexander et al.,1988; De Graan et al.,1990) and filamentous actin (Strittmatter et al.,1992; Hens et al.,1993c), to serve as a substrate ,for several kinases and phosphatases, to affect polyphosphoinositol metabolism by inhibiting PIP-kinase (Van Hooff et al., 1988), and to augment the binding of GTP-γ-S to trimeric G proteins G_o and G_i (Strittmatter et al..,1990). In situ, PKC phosphorylates B-50 at a single site (Ser-41 in rat). B-50 dephosphorylation can be mediated by type 1 and type 2A protein phosphatases and by calcineurin (Hens et al., 1995; and references therein). B-50 is an atypical calmodulin-binding protein, as it has a higher affinity for calmodulin in the absence than in the presence of Ca^{2+}. The phosphorylation state of B-50 may be a regulatory factor in calmodulin binding, because phosphorylation of Ser-41 within the presumed calmodulin-binding domain (residues 39-51 in rat), reduces its affinity for calmodulin. In vitro, CasKII phosphorylates rat B-50 at Ser-192 and Ser-193 (major phospho-sites), and Thr-88, Thr-89 and Thr-95 (minor phospho-sites). B-50 phosphorylation by CasKII did not affect the ability of B-50 to bind CaM, whereas CaM binding to B-50 inhibited CasKII-mediated phosphorylation of B-50. This indicates that CaM regulates CasKII-mediated B-50 phosphorylation by binding to B50 and thus sterically hinders the interaction of CasKII with its phosphorylation sites on B-50. Because PKC phosphorylation prevents CaM binding to B-50, CasKII phosphorylation may indirectly be controlled by PKC-mediated B-50 phosphorylation (Pisano et al.,1988; Apel et al.,1991).

4.2 B-50 Phosphorylation and Release

Based on its presynaptic localisation, its proposed role in signal transduction, and a high degree of correlation between the degree of PKC-mediated B-50 phosphorylation and neurotransmitter release (Dekker et al., 1989a), we have proposed that B-50 is involved in the molecular mechanism of neurotransmitter release. Direct evidence supporting this hypothesis was obtained in experiments with synaptosomes, which were permeated with the bacterial toxin Streptolysin-O. In these permeated synaptosomes we showed that affinity-purified anti-B-50 IgGs, which interfere with PKC-mediated B-50 phosphorylation, inhibit Ca^{2+}-induced release of noradrenaline, thus establishing a causal relation between B-50 and neurotransmitter release (Dekker et al.,1989b;1991b; Hens et al.,1993a). Studies in PC12 cells also observed an inhibition of catecholamine release by introduction of anti-B-50 IgGs (Norden et al.,1991) or by expression of antisense B-50 mRNA (Ivins et al.,1993).

4.3 B-50 and Release from Chromaffin Cells

In contrast to brain and PC12 cells, B-50 in chromaffin cells is found mainly in the cytosolic compartment rather than attached to the plasma membrane. Because purified, depalmitoylated B-50 was found to activate heterotrimeric G_o proteins in vitro (Strittmatter et al., 1990), Bader c.s. decided to examine whether cytosolic, depalmitoylated B-50 could act as an endogenous regulator for the G_o proteins in chromaffin cells, which in these cells are associated with the secretory organelles and block exocytosis upon their activation. Indeed, depalmitoylated B-50 introduction into permeated chromaffin cells blocked the Ca^{2+} -induced catecholamine release, and this effect could be reversed with anti-G_o IgGs (Vitale al., 1994). These data indicate that cytosolic, depalmitoylated B-50 can block Ca^{2+} -induced exocytosis in chromaffin cells by activating a protein G_o family. Palmitoylated B-50, assumed to be membrane-associated, lacked this ability. Although B-50 seems not to be present in all cells of the adrenal gland (Grant et al.,1992), these observations may reflect seperate functions of B-50 in brain and adrenal medulla. Moreover, the role of B-50 phosphorylation in exocytosis as well as the mechanism by which activated G_o proteins prevent vesicle fusion in chromaffin cells remain to be further elaborated on in future.

4.4 The NH_2-Terminal Domain of B-50 Is Important in Ca^{2+}-Induced Release from Nerve Terminals

Although the exact role of B-50 in neurotransmitter release from nerve terminals has not been resolved yet, introduction of site-specific anti-B-50 IgGs and specific PKC inhibitors into Streptolysin-O-permeated synaptosomes (Hens et al.,1993b;1995) convincingly demonstrated that the NH_2-terminal domain of B-50 plays an essential role after the Ca^{2+}trigger in this mechanism (Fig.1). In this series of experiments we have introduced two different monoclonal anti-B-50 antibodies, NM2 and NM6, into permeated synaptosomes to study their effect on Ca^{2+}-induced release. The NH2-terminal-directed anti-B-50 IgGs NM2 were selected for their ability to recognize an epitope (amino acid residues 39-43) in the presumed CaM-binding domain of B-50 (Oestreicher et al., 1994), and inhibited B-50 phosphorylation, B-50 dephosphorylation and B-50/CaM binding, as well as the Ca^{2+}-induced release of noradrenaline and cholecystokinin-8 (CCK-8). The COOH-terminal-directed anti-B50 IgGs NM6, which recognize an epitope between residue 132 and 213 in rat B-50, have little or no effect on these B-50 parameters and do not affect Ca^{2+}-induced release (Fig.1). Because several rather selective PKC inhibitors, which inhibited B50 phosphorylation in permeated synaptosomes, had no effect on the Ca^{2+}-induced release from permeated synaptosomes, we concluded that the phosphorylation state of B-50, rather than the activity of PKC per se, might be an important factor in neurotransmitter release. Interestingly, the PKC phosphorylation site of B-50 is located within the CaM-binding domain of B-50. It is therefore not surprising that monoclonal NM2 does not distinguish between phosphorylation, dephosphorylation and CaM-binding properties of B-50. In fact, under physiological conditions these properties are also thought to be tightly associated, because not only the Ca^{2+} concentration, but also the degree of phosphorylation of B-50 determines the CaMbinding capacity of B-50 at the membrane. Therefore, we hypothesised that a local CaM store function of B-50 at the plasma membrane might be important for the mechanism of Ca^{2+}-induced release. The size of the CaM store is determined by the degree of phosphorylation of the B-50 proteins. This degree of phosphorylation is determined by the balance between PKC activity and phosphatase activity. Upon Ca^{2+} entry CaM dissociates from the B-50 and locally activates Ca^{2+} /CaM dependent processes, for instance CaMKII, which is thought to be involved in recruitment of vesicles for release (see paragraph 2.1; Greengard et al.,1993), and Ca^{2+}/CaM -dependent phosphatase calcineurin, one of the phosphatases dephosphorylating B-50. Because of the mutually exclusive nature of phosphorylation and CaM binding, i.e.: in its phosphorylated form B-50 does not bind CaM, and when bound to CaM, B-50 cannot be phosphorylated, presumably both the degree of B-50 phosphorylation and CaM binding are important regulatory properties of B-50 in the mechanism of vesicular neurotransmitter release. To elaborate further on this hypothesis we have studied the role of CaM in Ca^{2+}-induced release by introducing anti-CaM IgGs, exogenous CaM and several CaM antagonist into permeated synaptosomes and were able to obtain evidence for a role of CaM in Ca^{2+}-induced noradenaline release (Hens et al.,1996).

4.5 Does B-50 serve as a CaM store in release?

Previously, the use of CaM antibodies and antagonists had indicated a role of CaM in Ca^{2+}-induced neurotransmitter release from nerve terminals, although conclusive evidence was still lacking, because:1) CaM antagonists that inhibited secretion from intact and permeated cells have limited specificity (Hidaka et al,1981; Schatzman et al.,1983); And 2) a reduction in secretion from intact cells after manipulation of CaM activity using either CaM antagonists (Sasakawa et al, 1983; Clapham and Neher, 1984; Sitges and Talamo, 1993) or intracellular application of anti-CaM IgGs (Kenigsberg and Trifaró, 1985) was partially due to inhibition of the Ca^{2+} entry (reviewed by: Augustine et al.,1987; Trifaró et al.,1992). This latter problem can be avoided by the use of Ca^{2+} ionophores or cell permeation techniques, and indeed, using the former approach, an inhibition of secretion that could not be attributed to a block of Ca^{2+} entry was reported for CaM antagonist trifluoperazine in chromaffin cells (Clapham and Neher, 1984; Kenigsberg et al., 1982).

Kinetic analysis of exocytosis in permeated chromaffin cells has revealed defined sequential steps along the secretory pathway of large dense-cored vesicles, namely vesicle recruitment, priming, docking and fusion. These sequential steps were

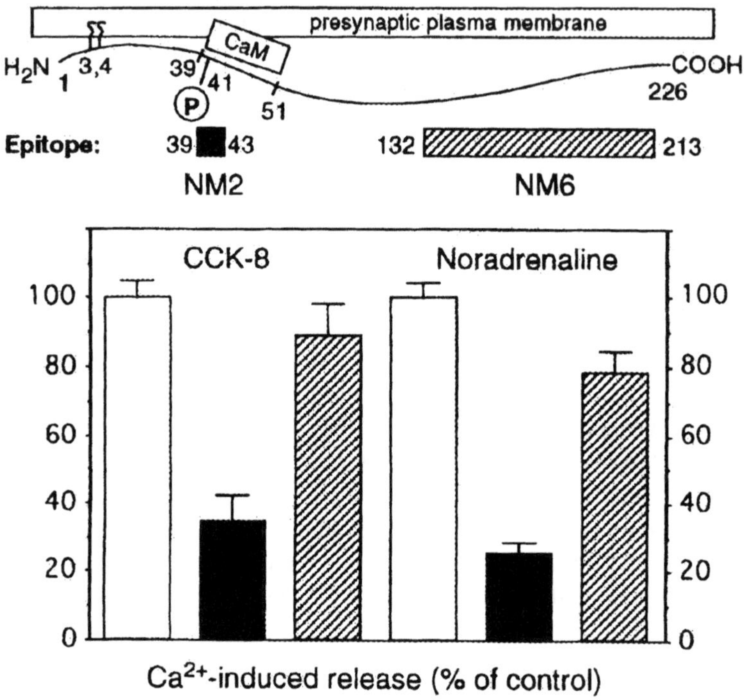

Fig 1. Effect of monoclonal anti-B-50 IgGs NM2 and NM6 on the release of noradrenaline and neuropeptide cholecystokinin-8 (CCK-8) from Streptolysin-O-permeated synaptosomes. At the top the PKC phosphorylation site (Ser-41), CaM-binding domain (residues 39-51), and membrane association (cysteines 3 and 4) of rat B-50 are schemattically indicated. The blocks underneath indicate the epitopes in rat B-50 recognized by NM2 (filled block) and NM6 (hatched block). At the bottom the effect of NM2 (filled symbols) and NM6 (hatched symbols) anti-B-50 IgGs on the Ca^{2+} -induced release are shown Ca^{2+} -induced release was obtained from permeated synaptosomes by increasing the free $[Ca^{2+}]$ from 10^{-8} M to 10^{-5} M for 5 min at 25 °C. Ca^{2+}-induced release in the absence of IgGs (open symbols) was set at 100 % and represented 12 % of the total synaptosomal noradrenaline content and 6 % of the total synaptosomal CCK-8 content. Anti-B-50 IgGs were present throughout the experiment at 50 µg of IgG/ml. Data were calculated from previously published experiments (Hens et al., 1993b; 1995) and represent the means ± SEM (bars) of 3-5 independent experiments.

characterized and defined by their dependency for Ca^{2+}, ATP, pH and temperature sensitivity (Bittner and Holz, 1992a,b). The ultimate triggering steps leading to exocytosis were found to be Ca^{2+} -dependent but did not need ATP. The preceding step, resulting in primed vesicles, was also Ca^{2+}-dependent but did also require ATP.

Using this permeated chromaffin cell model it was found that CaM was involved in triggering of the Ca^{2+}-induced exocytosis (Chamberlain et al.,1995). Moreover, in patch-clamped chromaffin cells membrane capacitance measurements indicated that CaM is involved in rapid endocytosis (Artalejo et al., 1996). Using permeated synaptosomes, we have demonstrated that anti-CaM IgGs partially inhibit Ca^{2+} induced noradrenaline release (Hens et al., 1996). Altogether, these data present convincing evidence for an important role of CaM in vesicle fusion and recycling, and lend further support to the hypothesis that B-50 serves as a local CaM store underneath the presynaptic plasma membrane in the mechanism of neurotransmitter release from nerve terminals, which is regulated in a Ca^{2+}- and phosphorylation dependent fashion.

5 Concluding Remarks

The process of neurotransmitter release consists of several discrete steps, each of which is mediated or regulated by a variety of different proteins. In this chapter we have reviewed a large body of evidence convincingly demonstrating the involvement of several protein kinases, phosphatases and their neuronal substrates in the molecular mechanism of vesicle fusion and recycling. It is still a matter of debate to what extent enzyme reactions like protein phosphorylation and dephosphorylation reactions are actually involved in the mediation of vesicle fusion itself. The rapidity of the fusion process of docked small clear-cored vesicles (200 - 300 μs between Ca^{2+} entry and vesicle fusion) is used as an argument against the involvement of protein phosphorylation reactions in this process (Jahn and Sudhof, 1993). Ca^{2+} was suggested to trigger only a few conformational changes that ultimately result in vesicle fusion. Phosphorylation reactions take hundreds of milliseconds to seconds, and would therefore be too slow to mediate vesicle fusion. Kinetic analysis of protein (de)phosphorylation reactions, however, has only been performed in vitro under conditions in which enzyme and substrate were completely solubilized. It might, therefore, very well be that the reaction time is significantly reduced when the protein kinases and phosphatases are also present together with their substrates in preassembled vesicle fusion complexes. Moreover, kinetic analysis of neurotransmitter release in chromaffin cells (see also paragraph 4.5; Bittner and Holz,1992a,b; Neher and Zucker,1993; Von Rüden and Neher,1993), PC12 cells (Hay and Martin,1992) and pituitary melanotrophs (Thomas et al.,1993) revealed defined sequential steps along the secretory pathway of large dense-cored vesicles, namely vesicle recruitment, priming, docking and fusion. As mentioned before (se paragraph 4.5) the ultimate steps leading to exocytosis were found to be Ca^{2+}-dependent but did not need ATP, whereas the preceding step, resulting in primed vesicles, was also Ca^{2+} dependent but also required ATP. Similarly, for *Paramecium* cells it was reported that ATP was required to prime the release system but was not required for the fusion event itself (Vilmart-Seuwen et al., 1986). These data indicate that the ultimate MgATP-independent steps of exocytosis from large dense-cored vesicles occur without the need for the activation of protein kinases and thus without the need for (ATP-dependent) protein phosphorylation reactions. This model, however, does not exclude the involvement of protein phosphatases (see paragraph 3) in the ultimate MgATP-independent steps.

Summarizing, we have reviewed compelling evidence for a role of the Ca^{2+} dependent protein kinases CaMKII and PKC and protein phosphatase calcineurin in neurotransmitter release. CaMKII presumably mediates its facilatory effect on secretion via the phosphorylation of synapsin I, which links synaptic vesicles to the actin-based cytoskeleton. Dynamin I and B-50 are two major PKC substrates implicated in neurotransmission. Calcineurin-mediated dynamin I dephosphorylation is thought to initiate vesicle membrane retrieval during endocytosis. In nerve terminals B-50 may serve as a local CaM store at sites of vesicle fusion, which is regulated in a Ca^{2+} - and phosphorylation-dependent fashion. For other proteins implicated in neurotransmitter release, such as for instance synaptotagmin, rabphylin3A, synaptobrevin/VAMP, and MARCKS, protein phosphorylation has been reported although the exact role of these phosphorylations in the regulation of the release process still remains to be elucidated. The observation that activation of PKC and CaMKII are essential in the induction of long-term potentiation, a well-known experimental model for studying the activity-dependent enhancement of synaptic plasticity, suggests that the phosphorylation state of certain release proteins may regulate the extent of neurotransmitter release (see e.g.: Greengard et al.,1993). As a matter of fact, temporal changes in the phosphorylation state of proteins such as B-50 (Gianotti et al.,1992; Ramakers et al.,1995) have been observed during long-term potentiation. Therefore the regulation of the neurotransmitter release process by Ca^{2+}-dependent protein phosphorylation and dephosphorylation events may be essential for learning and memory processes in the brain.

Acknowledgements

J.J.H.H. is supported by grant 903-53-091 of the Netherlands Organization for Scientific Research (MW-NWO) and G.H. by grant A91 of the Dutch Epilepsy Foundation.

References

Abe H., Tamura S., and Kondo H. (1992) Localization of mRNA for protein phosphatase 2C in the brain of adult rats. *Mol. Brain* Res.13, 238-288

Aderem A. (1992). The MARCKS brothers: a family of protein kinase C substrates. Cell 71, 713-716.

Aigner L., Arber S., Kapfhammer J.P., Laux T., Schneider C., Botteri F., Brenner H.R., and Caroni P. (1995) Overexpression of the neural growth-associated protein GAP-43 induces nerve sprouting in the adult nervous system of transgenic mice. Cell 83, 269-278.

Alexander K.A., Wakim B.T., Doyle G.S., Walsh K.A. and Storm D.R. (1988) Identification and characterization of the calmodulin-binding domain of neuromodulin. J. *Biol. Chem.* 263, 7544-7549.

Apel E.D., Litchfield D.W., Clark R.H., Krebs E.G. and Storm D.R. (1991) Phosphorylation of neuromodulin (GAP-43) by casein kinase II. Identification of phosphorylation sites and regulation by calmodulin. J. *Biol. Chem.* 266,10544-10551.

Artalejo C.R., Elhamdani A., and Palfrey H.C. (1996) Calmodulin is the divalent cation receptor for rapid endocytosis, but not exocytosis, in adrenal chromaffin cells. *Neuron* 16,195-205.

Augustine G.J., Charlton M.P., and Smith S.J. (1987) Calcium action in synaptic transmitter release. *Ann.* Rev. Neurosci.10, 633-693.

Bennett M.K., Miller K.G., and Scheller R.H. (1993). Casein kinase II phosphorylates the synaptic vesicle protein p65. *J.* Neurosci.13,1701-1707.

Bittner M.A., and Holz R.W. (1992a) Kinetic analysis of secretion from permeabilized adrenal chromaffin cells reveals distinct components. J. *Biol. Chem.* 267,16219-16225.

Bittner M.A., and Holz R.W. (1992b) A temperature-sensitive step in exocytosis. J. *Biol. Chem.* 267,16226-16229.

Chamberlain L.H., Roth D., Morgan A., and Burgoyne R.D. (1995) Distinct effects of α-SNAP,14-3-3 proteins, and calmodulin on priming and triggering of regulated exocytosis. J. *Cell* Biol.130,1-8.

Clapham D.E., and Neher E. (1984) Trifluoperazine reduces inward ionic currents and secretion by separate mechanisms in bovine chromaffin cells. J. *Physiol. 353,* 541-564.

Coffey E.T., Herrero I., Sihra T.S., Sanchez-Prieto J., and Nicholls D.G. (1994) Glutamate exocytosis and MARCKS phosphorylation are enhanced by a metabotropic glutamate receptor coupled to a protein kinase synergistically activated by diacylglycerol and arachidonic acid. *J. Neurochem.* 63,1303-1310.

Coghlan V.M., Pernno B.A., Howard M., Langeberg L.K., Hicks J.B., Gallatin, and Scott J.D. (1995) Association of protein kinase A and protein phosphatase 2B with a common anchoring protein. *Science* 267,108-111.

Cohen P. (1989) The structure and regulation of protein phosphatases. *Ann. Rev. Biochem.* 58, 453-508.

Colbran R.J., and Soderling T.R. (1990) Calcium/calmodulin-dependent protein kinase II. Cur. *Top. Cell. Reg.* 31,181-221.

Davletov B., Sontag J.-M., Hata Y., Petrenko A.G., Fyske E.M., Jahn R., and Südhof T.C. (1993). Phosphorylation of synaptotagmin I by casein kinase II. J. *Biol. Chem.* 268, 6816-6822.

De Camilli P. (1995) Molecular mechanisms in synaptic vesicle recycling. FEBS *Lett.* 369, 3-12.

De Graan P.N.E., Oestreicher A.B., De Wit M., Kroef M., Schrama L.H., and Gispen W.H. (1990) Evidence for the binding of calmodulin to endogenous B-50 (GAP*43)* in native synaptosomal plasma membranes. *J. Neurochem.* 55, 2139-2141.

De Graan P.N.E., Hens J.J.H., Oestreicher A.B., and Gispen W.H. (1992) Antibodies against B-50/GAP-43 and calcineurin inhibit Ca^{2+}-induced catecholamine release. *Soc. Neurosci.* Abstr.18, 634 #273.3.

Dekker L.V., De Graan P.N.E., Versteeg D.H.G., Oestreicher A.B. and Gispen W.H. (1989a) Phosphorylation of B-50 (GAP-43) is correlated with neurotransmitter release in rat hippocampal slices. J. *Neurochem.* 52, 24-30.

Dekker L.V., De Graan P.N.E., Oestreicher A.B., Versteeg D.H.G., and Gispen W.H. (1989b) Inhibition of noradrenaline release by antibodies to B-50 (GAP-43). *Nature 342,* 74-76.

Dekker L.V., De Graan P.N.E., and Gispen W.H. (1991a) Transmitter release: target of regulation by protein kinase C? *Prog. Brain Res.* 89, 209-233.

Dekker L.V., De Graan P.N.E., Pijnappel P., Oestreicher A.B. and Gispen W.H. (1991b) Noradrenaline release from Streptolysin O-permeated rat cortical synaptosomes: Effects of calcium, phorbol esters, protein kinase inhibitors, and antibodies to the neuron-specific protein kinase C substrate B-50 (GAP-43). J. *Neurochem.* 56,1146-1153.

Dent E.W. and Meiri K.F. (1992) GAP-43 phosphorylation is dynamically regulated in individual growth cones. J. *Neurobiol.* 23,1037-1053.

Detre J.A., Nairn A.C., Aswad D.W. and Greengard P. (1984) Localization in mammalian brain of G-substrate, a specific substrate for guanosine 3',5'-cyclic menophosphate-dependent protein kinase. J. *Neurosci.* 4, 2843-2849.

Fukunaga K., Kobayashi T., Tamura S. and Miyamoto E. (1993) Dephosphorylation of autophosphorylated Ca^{2+}/calmodulin-dependent protein kinase II by protein phosphatase 2C. J. *Biol. Chem.* 268,133-137.

Fykse E.M., Li C., and Südhof T.C. (1995). Phosphorylation of rabphylin-3A by Ca^{2+}/calmodulin- and cAMP-dependent protein kinases in vitro. J. *Neuroscience* 15, 2385-2395.

Gianotti C., Nunzi M.G., Gispen W.H., and Corradetti R. (1992) Phosphorylation of the presynaptic protein B-50 (GAP-43) is increased during electrically induced long-term potentiation. *Neuron 8,* 843-848.

Girault J.-A. (1993) Protein phosphorylation and dephosphorylation in mammalian central nervous system. *Neurochem. Int.* 23,1-25.

Gispen W.H., Nielander H.B., De Graan P.N.E., Oestreicher A.B., Schrama L.H., and Schotman P. (1991) Role of the growth-associated protein B-50/GAP-43 in neuronal plasticity. *Mol. Neurobiol.* 5, 61-85.

Grant N.J., Konig F., Deloulme J.C., Aunis D., and Langley K. (1992) Noradrenergic, but not adrenergic chromaffin cells in the adrenal gland express neuromodulin (GAP-43). Eur. *J. Neurosci.* 4,1257-1263.

Greengard P., Valtorta F., Czernik A.J., and Benfenati F. (1993) Synaptic vesicle phosphoproteins and regulation of synaptic function. *Science* 259, 780-785.

Groblewski G.E., Wagner A.C.C., and Williams J.A. (1994) Cyclosporin A inhibits Ca^{2+}/calmodulin-dependent protein phosphatase and secretion in pancreatic acinar cells. J. *Biol. Chem.* 269,15111-15117.

Hay J.C. and Martin T.F.J. (1992) Resolution of regulated secretion into sequential MgATP-dependent and calcium-dependent stages mediated by distinct cytosolic proteins. J. *Cell* Biol.119,139-151.

Hashimoto Y. and Soderling T.R. (1989) Regulation of calcineurin phosphorylation. Identification of the regulatory site phosphorylated by Ca^{2+}/calmodulin-dependent protein kinase II and protein kinase C. J. Biol. *Chem.* 264,16524-16529.

Hens J.J.H., De Wit M., Dekker L.V., Boomsma F., Oestreicher A.B., Margolis F., Gispen W.H., and De Graan P.N.E. (1993a) Studies on the role of B-50 (GAP-43) in the mechanism of Ca^{2+}-induced noradrenaline release: Lack of involvement of protein kinase C after the Ca^{2+} trigger. J. *Neurochern.* 60,1264-1273

Hens J.J.H., Ghijsen W.E.J.M., Dimjati W., Wiegant V.M., Oestreicher A.B., Gispen W.H. and De Graan P.N.E. (1993b) Evidence for a role of protein kinase C substrate B-50 (GAP-43) in Ca^{2+}-induced neuropeptide cholecystokinin-8 release from permeated synaptosomes. J. *Neurochem.* 61, 602-609.

Hens J.J.H., Benfenati F., Nielander H., Valtorta F., Gispen W.H., and De Graan P.N.E. (1993c) B-50/GAP-43 binds to actin filaments without affecting actin polymerization and filament organization. *J. Neurochem.* 61,1530-1533.

Hens J.J.H., De Wit M., Boomsma F., Mercken M., Oestreicher A.B., Gispen W.H., and De Graan P.N.E. (1995) N-terminal-specific anti-B-50 (GAP-43) antibodies inhibit Ca^{2+}-induced noradrenaline release, B-50 phosphorylation and dephosphorylation, and calmodulin binding. J. *Neurochem.* 64,1127-1136.

Hens J.J.H., Oestreicher A.B., De Wit M., Marquart A., Gispen W.H. and De Graan P.N.E. (1996) Evidence for a role of calmodulin in calcium-induced noradrenaline release from permeated synaptosomes: effects of calmodulin antibodies and antagonists. J. *Neurochem.* 66, in press.

Herrero I., Miras-Portugal M.T., and Sánchez-Prieto J. (1992) Positive feedback of glutamate exocytosis by metabotropic presynaptic receptor stimulation. *Nature* 360,163-166.

Herskovits J.S., Burgess C.C., Obar R.A. and Vallee R.B. (1993) Effects of mutant rat dynamin on endocytosis. J. *Cell* Biol.122, 565-578.

Hidaka H., Asano M., and Tanaka T. (1981) Activity-structure relationship of calmodulin antagonists: Naphthalenesulfonamide derivitives. *Mol. Pharmacol.* 20, 571-578.

Hultsch T., Albers M.W., Schreiber S.L., and Hohman R.J. (1991) Immunophilin ligands demonstrate common features of signal transduction leading to exocytosis or transcription. *Proc. Natl. Acad* Sci. *USA* 88, 6229-6233.

Ingebritsen T.S., Steward A.A., and Cohen P. (1983) The protein phosphatases involved in cellular regulation. 6. Measurement of type 1 and type 2 protein phosphatases in extracts of mammalian tissues: an assessment of their physiological roles. *Eur. J.* Biochem.132, 297-307.

Ivins K.J., Neve K.A., Feller D.J., Fidel S.A., and Neve R.L. (1993) Antisense GAP43 inhibits the evoked release of dopamine from PCl2 cells. J. *Neurochem.* 60, 626-63 3.

Kelly R.B. (1993) Storage and release of neurotransmitters. *Cell/Neuron* 72/10 (Suppl.), 43-53.

Kenigsberg R.L. and Trifaró J.M. (1985) Micro-injection of calmodulin antibodies into cultured chromaffin cells block catecholamine release in response to stimulation. *Neuroscience* 14, 335-347.

Kenigsberg R.L., Côté A., and Trifarb J.M. (1982) Trifluoperazine, a calmodulin inhibitor, blocks secretion in cultured chromaffm cells at a step distal from calcium entry. *Neuroscience* 7, 2277-2281.

King M.M., Huang C.Y., Chock P.B., Nairn A.C., Hemmings H.C., Chan K.F., and Greengard P. (1984) Mammalian brain phosphoproteins as substrates for calcineurin. J. *Biol. Chem.* 259, 8080-8083.

Klee C.B., Draetta G.F., and Hubbard M.J. (1988) Calcineurin. *Adv. Enzymol.* 61, 149-200.

Lai Y., Nairn A.C., and Greengard P. (1986) Autophosphorylation reversibly regulates the Ca^{2+}/calmodulin-dependent protein kinase II. *Proc. Natl. Acad. Sci.* USA 83, 4253-4257.

Li L., Chin L.S., Shupliakov O., Brodin L., Sihra T.S., Hvalby Ø., Jensen V., Zheng D., McNamara J.O., Greengard P., and Andersen P. (1995) Impairment of synaptic vesicle clustering and of synaptic transmission, and increased seizure propensity, in synapsin I-deficient mice. *Proc. Natl. Acad. Sci. USA* 92, 9235-9239.

Liu Y. and Storm D.R. (1990) Regulation of free calmodulin levels by neuromodulin: neuron growth and regeneration. *Trends Pharmacol.* Sci.11,107-111.

Momayezi M., Lumpert C.J., Kersken H., Gras U., Plattner H., Krinks M.H. and Klee C.B. (1987) Exocytosis induction of Paramecium tetraurelia cells by exogenous phosphoprotein phosphatase in vivo and in vitro: possible involvement of calcineurin in exocytotic membrane fusion. J. *Cell* Biol.105,181-189.

Neher E. and Zucker R.S. (1993) Ca^{2+}-dependent steps in chromaffin cell secretion. *Neuron* 10, 21-30.

Nichols R.A., Suplick G.R., and Brown J.M. (1994) Calcineurin-mediated protein dephosphorylation in brain nerve terminals regulates the release of glutamate. J. Biol. *Chem.* 269, 23817-23823.

Nielander H.B., Onofri F., Valtorta F., Schiavo G., Montecucco C., Greengard P., and Benfenati, F. (1995). Phosphorylation of VAMP/synaptobrevin in synaptic vesicles by endogenous protein kinases. J. *Neurochem.* 65,1712-1720.

Norden J.J., Lettes A., Costello B., Lin L., Wouters B., Bock S., and Freeman J.A. (1991) Possible role of GAP-43 in calcium regulation/neurotransmitter release. *Ann.* N. Y. *Acad. Sci.* 627, 75-93.

Oestreicher A.B., Hens J.J.H., Marquart A.B., Mercken M., De Graan P.N.E., Zwiers H. and Gispen W.H. (1994) Monoclonal antibody NM2 recognizes the protein kinase C phosphorylation site in B-50 (GAP-43) and neurogranin (BICKS). J. *Neurochem.* 62, 881-889.

Ohno S., Konno U., Akita Y., Yano A., and Suzuki K. (1990) A point mutation at the putative ATP-binding site of protein kinase Cα abolishes the kinase activity and renders it down-regulation-insensitive. A molecular link between autophosphorylation and down-regulation. *J. Biol. Chem.* 265, 6296-6300.

Pasinelli P., Ramakers G.M.J., Urban I.J.A., Hens J.J.H., Oestreicher A.B., De Graan P.N.E., and Gispen W.H. (1995) Long-term potentiation and synaptic protein phosphorylation. *Behav. Brain* Res. 66, 53-59.

Pieribone V.A., Shupliakov O., Brodin L., Hilfiker-Rothenfluh S., Czernik A.J., and Greengard P. (1995) Distinct pools of synaptic vesicles in neurotransmitter release. *Nature 375,* 493-497.

Pisano M.R., Hegazy M.G., Reimann E.M., and Dokas L.A. (1988) Phosphorylation of protein B-50 (GAP-43) from adult rat brain cortex by casein kinase II. *Biochem. Biophys.* Res. Comm.155,1207-1212.

Ramakers G.M.J., De Graan P.N.E., Urban I.J.A., Kraay D., Tang T., Pasinelli P., Oestreicher A.B., and Gispen W.H. (1995) Temporal differences in the phosphorylation state of pre- and postsynaptic protein kinase C substrates B50/GAP-43 and neurogranin during long term potentiation. J. Biol. *Chem.* 270, 13892-13898.

Robinson P.J. (1991). The role of protein kinase C and its neuronal substrates dephosphin, B-50, and MARCKS in neurotransmitter release. *Mol. Neurobiol. 5,* 87-130.

Robinson P.J., Sontag J.-M., Liu J.-P., Fykse E.M., Slaughter C., McMahon H., and Südhof T.C. (1993). Dynamin GTPase regulated by protein kinase C phosphorylation in nerve terminals. *Nature* 365,163-166.

Robinson P.J., Liu J.-P., Powell K.A., Fykse E.M., and Südhof T.C. (1994). Phosphorylation of dynamin I and synaptic-vesicle recycling. *Trends* Neurosci.17, 348-353.

Rosahl, T.W., Geppert, M., Spillane, D., Herz, J., Hammer, R.E., Malenka, R.C., and Sudhof, T.C. (1993). Short-term synaptic plasticity is altered in mice lacking synapsin I. *Cell* 75, 661-670.

Rosahl, T.W., Spillane, D., Missler, M., Herz, J., Selig, D.K., Wolff, J.R., Hammer, R.E., Malenka, R.C., and Südhof, T.C. (1995). Essential functions of synapsin I and II in synaptic vesicle regulation. *Nature* 375, 488-493.

Sasakawa N., Kumakura K., Yammaooto S., and Kator S. (1983) Effects of W-7 on catecholamine release and $^{45}Ca^{2+}$ uptake in cultured adrenal chromaffin cells. Life *Sci. 33,* 2017-2024.

Schatzman R.C., Raynor R.L., and Kuo J.F. (1983) N-(6-aminohexyl)-5-chloro-1-naphthalenesulfonamide (W-7), a calmodulin antagonist, also inhibits phospholipid-sensitive calcium-dependent protein kinase. *Biochim. Biophys. Acta* 755,144-147.

Scheller, R.H. (1995). Membrane trafficking in the presynaptic nerve terminal. *Neuron* 14, 893-897.

Schrama L.H., Heemskerk F.M.J., and De Graan P.N.E. (1989) Dephosphorylation of protein kinase C-phosphorylated B-50/GAP-43 by the calmodulin-dependent phosphatase calcineurin. *Neurosci. Res. Commun.* 5,141-147.

Schworer C.M., Colbran R.J., and Soderling T.R. (1986) Reversible generation of a Ca^{2+}/calmodulin-dependent protein kinase II by a autophosphorylation mechanism. J. *Biol. Chem.* 261, 8581-8584.

Sihra T.S., and Nichols R.A. (1993) Mechanisms in the regulation of neurotransmitter release from brain terminals: current hypotheses. *Neurochem.* Res. 18, 47-58

Sihra T.S., Nairn A.C., Kloppenburg P., Lin Z., and Pouzat C. (1995) A role for calcineurin (protein phosphatase-2B) in the regulation of glutamate release. *Biochem. Biophys.* Res. *Comrnun.* 212, 609-616.

Sim A.T.R., Lloyd H.G.E., Jarvie P.E., Morrison M., Rostas J.A.P. and Dunkley P.R. (1993) Synaptosomal amino acid release. Effect of inhibiting protein phosphatases with okadaic acid. *Neurosci.* Lett.160,181-184.

Sitges M., and Talamo B.R. (1993) Sphingosine, W-7, and trifluoperazine inhibit the elevation in cytosolic calcium induced by high K^+ depolarization in synaptosomes. *J. Neurochem.* 61, 443-450.

Steiner J.P., Dawson T.M., Fotuhi M., Glatt C.E., Snowman A.M., Cohen N., and Snyder S.H. (1992) High brain densities of the immunophilin FKBP colocalized with calcineurin. *Nature* 358, 584-587.

Strittmatter S.M., Valenzuela D., Kennedy T.E., Neer E.J. and Fishman M.C. (1990) Go is a major growth cone protein subject to regulation by GAP-43. *Nature 344,* 836-841.

Strittmatter S.M., Vartanian T., and Fishman M.C. (1992) GAP-43 as a plasticity protein in neuronal form and repair. *J. Neurobiol.* 23, 507-520.

Südhof, T.C. (1995). The synaptic vesicle cycle: a cascade of protein-protein interactions. *Nature* 375, 645-653.

Sutton R.B., Davletov B.A., Berghuis A.M., Südhof T.C., and Sprang S.R. (1995) Structure of the first C_2 domain of synaptotagmin I: a novel Ca^{2+} /phospholipidbinding fold. Cell 80, 929-938.

Swain J.E., Robitaille R., Dass G.R., and Charlton M.P. (1991) Phosphatases modulate transmission and serotonin facilitation at synapses: studies with the inhibitor okadaic acid. J. *Neurobiol.* 22, 855-864.

Tanaka C., and Nishizuka Y. (1994) The protein kinase C family for neuronal signaling. *Annu.* Rev. Neurosci.17, 551-567.

Terbush D.R., and Holz R.W. (1990) Activation of protein kinase C is not required for exocytosis from bovine adrenal chromaffin cells. J. *Biol. Chem.* 265, 21179-21184.

Thomas P., Wong J.G., Lee A.K. and Almers W. (1993) A low affinity Ca^{2+} receptor controls the final steps in peptide secretion from pituitary melanotrophs. *Neuron* 11, 93-104.

Trifarb J.-M., Vitale M.L., and Rodriguez Del Castillo A. (1992) Cytoskeleton an molecular mechanisms in neurotransmitter release by neurosecretory cells. Eur. J. *Pharmacol.* 225, 83-104.

Van Hooff C.O.M., De Graan P.N.E., Oestreicher A.B., and Gispen W.H. (1988) B50 phosphorylation and polyphosphoinositide metabolism in nerve growth cone membranes. *J. Neurosci.* 8,1789-1795.

Verhage M., Hens J.J.H., De Graan P.N.E., Boomsma F., Wiegant V.M., Lopes da Silva F., Gispen W.H., and Ghijsen W.E.J.M. (1995) Ba^{2+} replaces Ca^{2+}/calmodulin in the activation of protein phosphatases and in exocytosis of all major transmitters. *Eur. J. Pharmacol.* 291, 387-398.

Vitale N., Deloulme J.-C., Thiersé D., Aunis D., and Bader M.-F. (1994) GAP-43 controls the availability of secretory chromaffin granules for regulated exocytosis by stimulating a granule-associated Go. J. *Biol. Chem.* 269, 30293-20298.

Von Rüden L. and Neher E. (1993) A Ca^{2+}-dependent early step in the release of catecholamines from adrenal chromaffin cells. *Science* 262,1061-1065.

Wu Y.N., and Wagner P.D. (1991) Effects of phosphatase inhibitors and a protein phosphatase on norepinephrine secretion by permeabilized bovine chromaffm cells. *Biochim. Biophys. Acta* 1092, 384-390.

Yanagihara N., Toyohira Y., Koda Y., Wada A. and Izumi F. (1991) Inhibitory effect of okadaic acid on carbachol-evoked secretion in cultured bovine adrenal medullary cells. *Biochem. Biophys.* Res. Comm.174, 77-83.

Acetylcholine Release from Mouse Neuroblastoma Cells Co-Transfected with Mediatophore and Choline Acetyltransferase cDNAs

Alain Bloc, Emmanuelle Roulet, Françoise Loctin and Yves Dunant.

Pharmacologie, Centre Médical Universitaire, CH-1211 Genève 4, Switzerland

1 Introduction

A family of 16 kDa proteins has been isolated from a wide variety of species and appeared to be an universal highly conserved membrane proteolipid (Holzenburg *et al.*, 1993). Comparison of the amino acid sequences showed a high degree of homology between these proteins which are components of various translocator systems.

First, a 16 kDa protein has been isolated as the major subunit of the hydrophobic membrane sector (V0) of the vacuolar H+-ATPase (Mandel et al., 1988). The vacuolar H+-ATPase is an oligomeric complex which functions as a proton pump, directing limited vacuolar acidification of plant and fungal vacuoles, lysosomes, synaptic vesicles, coated vesicles, etc... The proton gradient generated by ATP hydrolysis drives transport of solutes such as ions and neurotransmitters. The 16 kDa protein has been identified as the membrane binding proteolipid subunit "c" of the V0 sector of the vacuolar-ATPase complex and proposed to form the proton channel.

Second, a 16 kDa proteolipid, called ductin, has been described as the principal protein component of gap junctions isolated from the hepatopancreas of the lobster Nephrops norvegicus (Norway lobster) and from other different preparations (see Finbow et al., 1995).

Third, Israël and co-workers have isolated from presynaptic plasma membrane of the *Torpedo* electric lobe a protein which was called the ***mediatophore*** since it was characterised by its ability to release acetylcholine (ACh) in a calcium-dependent manner when incorporated into liposomes. ACh release by the mediatophore mimics several typical features of synaptic transmission (Israël *et al.*, 1986; see Dunant *et al.*, this symposium), including the production of pulsatile and quantal jets of transmitter. In its native form the mediatophore has an apparent molecular weight of 200 kDa and is a homo-oligomer made of the 16 kDa subunits with a close homology (80%) to the proteins described above (Birman *et al.*, 1990). Electron microscope analyses show that this oligomeric complex has a structure similar to those found in the gap junction containing a stain-penetrable pore (Israël and Dunant, 1993). Immuno-histochemistry in the *Torpedo* electric organ has revealed that the mediatophore oligomer is localised at active zones of the presynaptic membranes (Brochier *et al.*, 1993); additionally, it has to be addressed to the plasma membrane

NATO ASI Series, Vol. H 100
Neurotransmitter Release and Uptake
Edited by Şakire Pöğün

to support transmitter release as demonstrated by reconstitution experiments in *Xenopus* oocytes (Leroy and Meunier, 1995).

2 Results

Previous works from our laboratory (Falk-Vairant et al., 1996) have demonstrated that quantal ACh release can be restored in the mouse neuroblastoma N18TG-2 cells by transfection of a cDNA encoding the Torpedo mediatophore subunit. N18TG-2 cells are normally deficient for both synthesis and release of ACh (Israël et al., 1994; Kimura et al., 1992; Zhong et al., 1995), but can be loaded with the neurotransmitter by incubation in a choline-free culture medium containing high ACh concentration and acetylcholinesterase inhibitor (Israël et al., 1994). After this treatment, only transfected N18TG-2 cells can release ACh upon electrical depolarisation or calcium influx.

2.1 Is the property to release ACh specific to the Torpedo 16 kDa protein ?

Using the same strategy, we examined whether transfection of a 16 kDa protein other than the *Torpedo* mediatophore subunit can also restore calcium-dependent ACh secretion in the release-deficient N18TG-2 cell line. To answer this question N18TG-2 cells were transfected with cDNA encoding the *Nephrops norvegicus* 16kDa protein (cloned under the control of the CMV promoter; kindly provided by M. Finbow).

After ACh loading, release from control and transfected cells was examined using both the choline oxydase chemiluminescent assay (see Israël *et al.*, 1994) and an electrophysiological technique (Falk-Vairant *et al.*, 1996) adapted from Evers *et al.* (1989; and see also M-m. Poo, this symposium). In the biochemical assay, release was elicited by successive addition of calcium ionophore (A23187, 4μM) and calcium (5-10μM), and ACh secretion was monitored by recording of the light emission. In the electrophysiological assay, release was triggered by electrical depolarisation of a single cell brought in close contact to a cultured embryonic *Xenopus* myocyte voltage-clamped in the whole-cell configuration (holding potential: -70mV). Real time ACh release from N18TG-2 cell was then detected as nicotinic inward currents recorded on the adjoining myocyte. Results show that, as in the case of the *Torpedo* protein, transfection of the *Nephrops* 16 kDa protein cDNA is able to restore evoked ACh release in N18TG-2 previously loaded with this neurotransmitter. Typical recordings of evoked release from transfected cells are shown in Fig. 1. Both *Nephrops* and *Torpedo* 16 kDa protein transfected cells generated, on the adjoined myocyte, inward currents which can be blocked by the nicotinic receptor antagonist (+)-Tubocurarine and display step-like amplitude levels and a similar time course (see Fig. 3C).

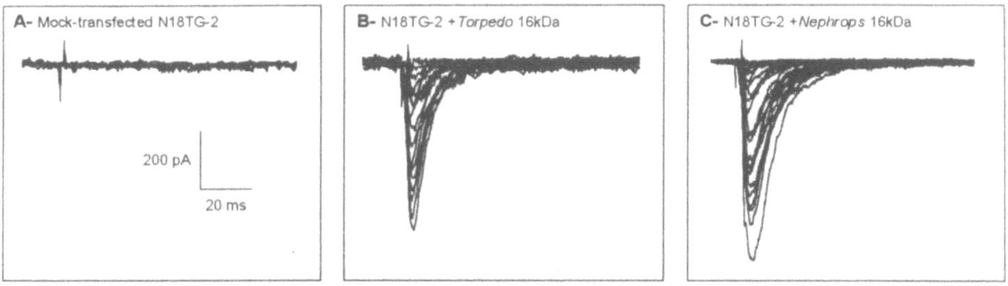

Fig 1. Typical recordings of ACh release from N18TG-2 transfected cells using the electrophysiological procedure. A: mock-transfected N18TG-2 cells (transfected with pREP4 vector alone); B: N18TG-2 cells transfected with the Torpedo 16 kDa cDNA inserted into pREP4 vector. C: N18TG-2 cells transfected with the Nephrops 16 kDa cDNA inserted into pcDNA3 vector. After filling with ACh, cells were tested for release by electrical stimulation (1V, 2ms at 0.25 Hz) using a voltage-clamped (Hp=-70mV) Xenopus myoball as ACh detector. Each graph shows successive recordings superimposed on the stimulus artefact and obtained from a single transfected N18TG-2 cell.

2.2 Is ACh release after mediatophore transfection dependent on the source of neurotransmitter ?

Kimura *et al.* (1992, and see Zhong *et al..*, 1995) have shown that, in contrast to other cell lines, N18TG-2 cells are unable to release the ACh after transfection with choline acetyltransferase (ChAT). We then examined whether N18TG-2 cell expressing both 16 kDa subunit and ChAT can use the synthesized neurotransmitter in the release process supported by the mediatophore. In this purpose, cells were co-transfected with cDNAs encoding the *Torpedo* 16 kDa protein (cloned into the plasmid pREP4) together with the rat ChAT cDNA (cloned into the plasmid pSRNEO) and selected for their resistance to both geneticin and hygromycin. Several monoclonal cell lines were generated and tested for ChAT activity and ACh release. Cells were not artificially filled with ACh this time. Fig. 2 shows the results obtained from control cells and clones expressing consistent ChAT activities. Only cells expressing both ChAT and mediatophore subunit were able to release ACh when tested using both the luminescent and the electrophysiological assays.

ACh secretion by co-transfected clones examined with the electrophysiological assay was shown to be sensitive to the non selective voltage-dependent calcium channel blocker cadmium and to (+) -Tubocurarine (Fig. 3B). The evoked currents recorded on voltage-clamped *Xenopus* myocytes displayed again discrete amplitude levels and time course characteristics similar to those obtained from N18TG-2 cells transfected with 16 kDa protein alone and loaded with ACh (Fig. 3C).

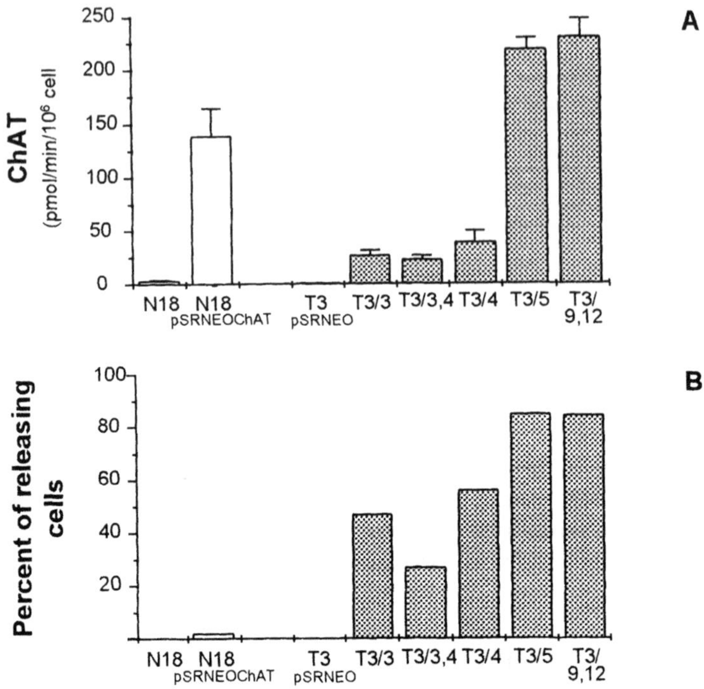

Fig 2. Comparison of ChAT activities (**A**) and percent of releasing cells determined using the electrophysiological assay (**B**). Measurements were performed on different monoclonal cell lines generated after co-transfection of both *Torpedo* 16 kDa protein and rat ChAT cDNAs. N18: native N18TG-2 cells; N18pSRNEOChAT: N18TG-2 cells transfected with rat ChAT cDNA; T3pSRNEO: N18TG-2 co-transfected with both 16 kDa cDNA and pSRNEO vector alone; T3/3, ..., T3/9,12: 16 kDa protein and ChAT co-transfected clones. Cells were not artificially loaded with Ach

2.3 Is a change in vacuolar H^+-ATPase activity involved in the restoration of ACh release after 16 kDa transfection?

The high degree of homology between primary structure of the mediatophore 16 kDa subunit and the "c" subunit of the V_o sector of the vacuolar H^+-ATPase allows an alternative interpretation of our data not involving an increase of mediatophore oligomer complexes after transfection of the 16 kDa subunit in N18TG-2 cells. Restoration of release in this case might result from an improvement of vesicular ACh secretion arising from an increased vesicular H^+ gradient due to an enhancement in H^+-ATPase activity by formation of hybrid H^+-ATPase complexes comprising endogenous catalytic subunits and *Torpedo* integral membrane component. The vacuolar H^+-ATPase activity have been shown indeed to be restored by expressing the *Nephrops* 16 kDa protein cDNA in a haploid *Saccharomyces cerevisiae* strain in which the VMA3 gene coding for the 16 kDa proteolipid has been disrupted (Harrison *et al.*, 1994).

To test this hypothesis we determined the amount of ACh released from N18TG-2 expressing the *Torpedo* mediatophore 16 kDa subunit after pre-incubation of the cells with 1μM of the vacuolar H^{+}-ATPase Bafilomycin A_1 (Fig. 4). No difference between control and pre-incubated cells were measured on both choline and ACh released determined biochemically.

A.

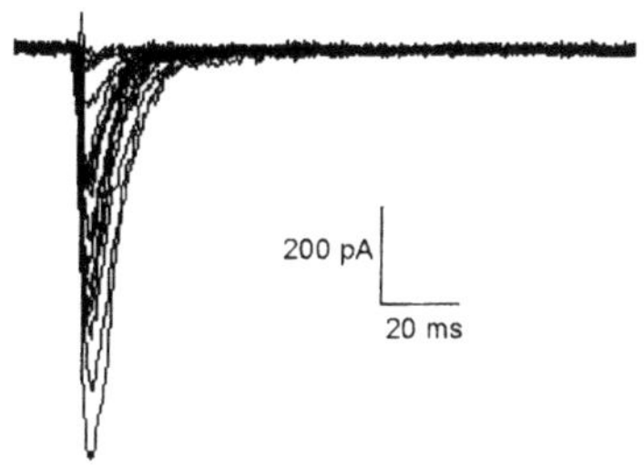

B.

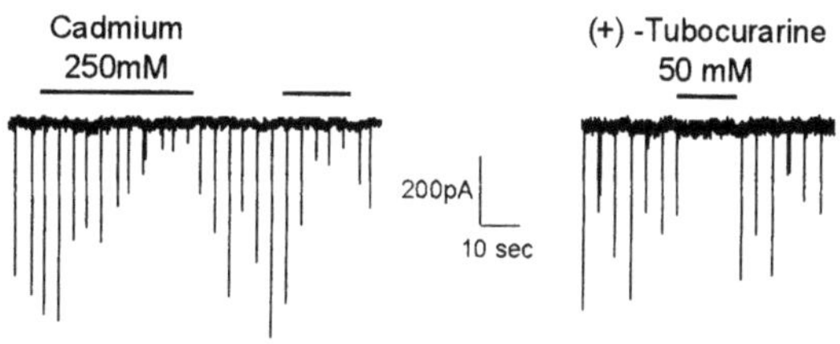

C.

	N18TG-2 + *Torpedo* 16kDa	N18TG-2 + *Nephrops* 16kDa	N18TG-2 + *Torpedo* 16kDa & rat ChAT
Rise Time Half-decay Time	1.49 ms ± 0.08 5.18 ms ± 0.56	1.93 ms ± 0.07 6.18 ms ± 0.33	1.48 ms ± 0.1 6.13 ms ± 0.25

Fig 3. ACh release from N18TG-2 cells transfected with both the *Torpedo* 16 kDa cDNA inserted into pREP4 vector and the rat ChAT inserted into pSRNEO vector. **A.** Successive recordings from a single co-transfected cell superimposed on the stimulus artefact (same stimulation protocol as in Fig.1). **B**. Continuous traces showing the effects of bath application of cadmium or (+) -Tubocurarine on release evoked from a co-transfected cell. **C**. Comparison of time course characteristics observed for ACh release from N18TG-2 cells transfected with the *Torpedo* or the *Nephrops* 16 kDa protein and loaded with the neurotransmitter or from co-transfected N18TG-2. Value are means ± sem of at least 50 recordings. Rise time: time taken for the signals to rise from 10 to 90% of peak amplitude (measured from the origin of the stimulation pulse); Half-decay Time: time taken for the signals to fall by 50% of peak amplitude.

3 Conclusions

The fact that, as previously shown for the *Torpedo* mediatophore subunit (Falk-Vairant *et al.*, 1996), transfection of the 16 kDa protein from *Nephrops* was also effective in re-establishing evoked release of ACh from N18TG-2 cells allows us to propose that this property is not specific of the *Torpedo* protein. Thus, the involvement of the 16 kDa proteolipid in various functions can not be explained only by slight differences in primary structure of the protein. As proposed by Finbow *et al.* (1995), it seems more likely that the various 16 kDa subunit functions result from differential addressing and orientation into plasma or vesicular membranes. Results from expression experiments in *Xenopus* oocytes (Leroy and Meunier, 1995) tend to indicate furthermore that another protein might be involved in the specific targeting of mediatophore subunits to the plasma membrane. Over-expression of the *Torpedo* 16 kDa proteolipid in N18TG-2 cells, which naturally contain very low levels of this protein in their plasma membrane (Israël *et al.*, 1994), might overcome a deficient targeting process, allowing mediatophore complexes to reach the membrane and mediate ACh release.

Data obtained from N18TG-2 cells expressing both *Torpedo* 16 kDa protein and rat ChAT clearly indicate that the release process supported by the mediatophore can be fuelled by ACh synthesized by the cells. This finding together with the characteristics of the mediatophore-mediated ACh release, such as calcium

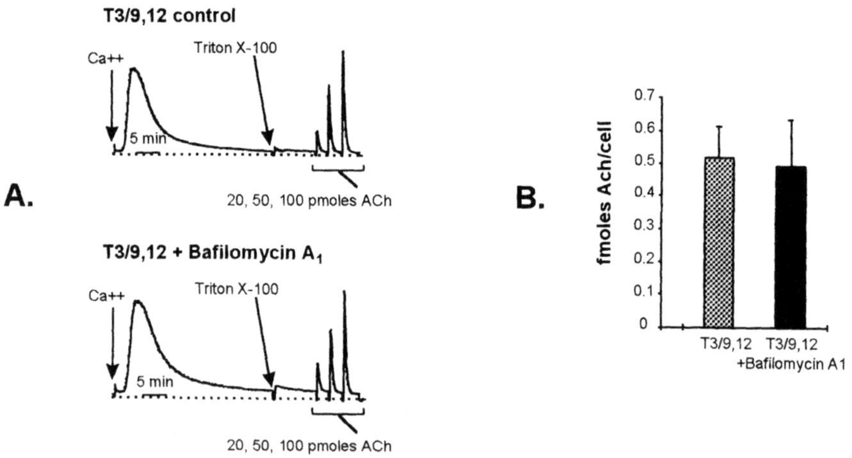

Fig 4. Effects of Bafilomycin A_1 pre-incubation on ACh release from N18TG-2 cells transfected with both the *Torpedo* 16 kDa cDNA inserted into pREP4 vector and the rat ChAT inserted into pSRNEO vector. **A.** Typical recordings using the bioluminescent choline oxydase procedure of release from co-transfected cells (T3/9,12) with or without pre-incubation with Bafilomycin A_1 (1 mM, 2 hours, at 37°C). Release (from a suspension containing 0.18×10^6 cells in both conditions) was trigger by calcium (10mM) addition after treatment with the calcium ionophore (A23187, 4mM). Secretion of both choline and ACh was elicited by the calcium influx. Triton-x100 (0.1%) was then added to quantify the remaining ACh and choline contents, and subsequent applications of standard doses of ACh were performed to calibrate the responses. **B**. Histogram showing mean values ± sem (from at least three independent experiments) of ACh release from N18TG-2 co-transfected cells (T3/9,12) with or without Bafilomycin A_1 pre-incubation. Specific ACh secretion was determined in parallel assays by adding acetylcholinesterase in the reaction mixture after calcium, allowing distinction between choline and ACh released.

dependence, quantal nature, and time courses similar to those observed on newly formed *Xenopus* neuromuscular junctions (Evers *et al.*, 1989 ; Girod *et al.*, 1992), argue for a physiological role of the mediatophore in ACh release process. Furthermore, the similarities between the releases observed from cells artificially filled with ACh and cells transfected with ChAT validate the procedure of neurotransmitter loading by pre-incubation. Since N18TG-2 cells, even over-expressing ChAT, are unable to release ACh when co-cultured with striated muscle cells (Kimura *et al.*, 1992; Zhong *et al.*, 1995), it remains now to determine whether these cells, co-transfected with both mediatophore subunit and ChAT cDNAs, are able to form functionally operating synapses in this kind of co-culture.

We found also that Bafilomycin A_1 pre-incubation has no effect on ACh release from N18TG-2 expressing the *Torpedo* 16 kDa protein. This finding is to be compared with previous data showing that this cell line is deficient in ACh vesicular transporter (Varoqui, 1995) and in other proteins involved in the vesicular secretion process such as synaptophysin (Suburo *et al.*, 1992). It is thus hardly to be expected that restoration of ACh release in N18TG-2, after transfection of the *Torpedo* 16 kDa protein, result from an increased vacuolar H^+-ATPase activity allowing more neurotransmitter to be taken up into synaptic-like vesicles and subsequently released. Taken together, these data lead us to consider that, in 16 kDa protein-transfected N18TG-2 cells, ACh is probably released from a cytoplasmic pool thanks to a enhancement of mediatophore oligomeric complexes expressed in the plasma membrane of the cells.

Acknowledgements

This work was supported by the Fonds National Suisse pour la Recherche Scientifique (grant Nr 31-37703.93) and has been performed in collaboration with the group of M. Israël, Laboratoire de Neurobiologie Cellulaire & Moléculaire, CNRS, Gif-sur-Yvette, France. We are grateful to M. Finbow, CRC Beatson Laboratories, Glasgow, Scotland, for providing us with the *Nephrops* 16 kDa cDNA.

Bibliography

Birman, S., Meunier, F.M., Lesbats, B., LeCaer, J.P., Rossier, J. and Israël, M. (1990) A 15 kD proteolipid found in mediatophore preparations from Torpedo presents high sequence homology with the bovine chromaffin granule protonophore. FEBS Lett. 261, 303-306.

Brochier, G., Israël, M. and Lesbats, B. (1993) Immunolabelling of the presynaptic membrane of Torpedo electric organ nerve terminals with an antiserum towards the acetylcholine releasing protein mediatophore. Biol.Cell 78, 145-154.

Evers, J., Laser, M., Sun, Y., Xie, Z. and Poo, M.M. (1989) Studies of nerve-muscle interactions in Xenopus cell culture: Analysis of early synaptic currents. J. Neurosci. 9, 1523-1539.

Falk-Vairant, J., Corrèges, P., Eder-Colli, L., Salem, N., Roulet, E., Bloc, A., Meunier, F., Lesbats, B., Loctin, F., Synguelakis, M., Israël, M. and Dunant, Y.

(1996) Quantal acetylcholine release induced by mediatophore transfection. Proc.Natl.Acad.Sci.USA 93, 5203-5207.

Finbow, M. E., Harrison, M., and Jones, P. (1995) Ductin - a proton pump component, a gap junction channel and a neurotransmitter release channel. Bioessays. 17, 247-255.

Girod, R., Eder-Colli, L., Medilanski, J., Dunant, Y., Tabti, N. and Poo, M.M. (1992) Pulsatile release of acetylcholine by isolated nerve terminals (synaptosomes) isolated from the Torpedo electric organ. J. Physiol. Lond. 450, 325-340.

Harrison, M., Jones, P., Kim, Y-I., Finbow, M. and Findlay, J. (1994) Functional properties of a hybrid vacuolar H+-ATPase in Saccharomyces cells expressing the Nephrops 16-kDa proteolipid. Eur. J. Biochem. 221, 111-120.

Holzenburg, A., Jones, P. C., Franklin, T., Pali, T., Heimburg, T., Marsh, D., Findlay, J. B., and Finbow, M. E. (1993) Evidence for a common structure for a class of membrane channel. Eur. J. Biochem. 213, 21-29.

Israël, M. and Dunant, Y. (1993) Acetylcholine release, from molecule to function. Prog. Brain. Res. 98, 219-233.

Israël, M., Lesbats, B., Synguelakis, M. and Joliot, A. (1994) Acetylcholine accumulation and release by hybrid NG108-15, glioma and neuroblastoma cells - Role of a 16 kDa membrane protein in release. Neurochem. Int. 25, 103-109.

Israël, M., Morel, N., Lesbats, B., Birman, S. and Manaranche, R. (1986) Purification of a presynaptic membrane protein that mediates a calcium-dependent translocation of acetylcholine. Proc.Natl.Acad.Sci.USA 83, 9226-9230.

Kimura, Y., Oda, Y., Deguchi, T. and Higashida, H. (1992) Enhanced acetylcholine secretion in neuroblastoma x glioma hybrid NG108-15 transfected with rat choline acetyltransferase cDNA. FEBS 314, 409-412.

Leroy, C. and Meunier, F-C. (1995) Differential targeting to the plasma membrane of the Torpedo 15-kDa proteolipid expressed in oocytes. J. Neurochem. 65, 1789-1797.

Mandel, M., Moriyama, Y., Hulmes, J.D., Pan, Y.C.E., Nelson, H. and Nelson, N. (1988) cDNA sequence encoding the 16-kDa proteolipid of chromaffin granule implies gene duplication in the evolution of H+-ATPases. Proc.Natl.Acad.Sci.USA 85, 5521-5524.

Suburo, A., Wheatley, S., Horn, D., Gibson, S., Jahn, R., Fischer-Colbrie, R., Wood, J., Latchman, D. and Polack J. (1992) Intracellular distribution of neuropeptides and secretory proteins during differentiation of neuronal cell lines. Neuroscience 46, 881-889.

Varoqui, H. (1995) Ph.D. thesis (INAPG, Paris).

Zhong, Z.G., Misawa, H., Furuya, S., Kimura, Y., Noda, M., Yokoyama, S. and Higashida, H. (1995) Overexpression of choline acetyltransferase reconstitutes discrete acetylcholine release in some but not all synapse formation-defective neuroblastoma cells. J. Physiol. (Paris) 89, 137-145.

Neurotransmitter Release: An Overview

George J. Augustine

Department of Neurobiology, Duke University Medical Center Box 3209, Durham, NC 27710, U.S.A.

1 Introduction

During this course we heard many excellent talks that provided a nice survey of our rapidly advancing understanding of the process of neurotransmitter release. These talks described experiments that used diverse experimental approaches to yield interesting, though sometimes contradictory, conclusions. I have been assigned the not-so-simple task of summarizing what we have learned over the course. One goal of my summary is to highlight what I perceive to be the central themes of our lectures. My second goal is to look to the future by pointing out questions whose answers we do not yet understand.

2 Organization of the Active Zone

From Tom Reese we learned about the highly stereotyped structure of the active zone that is seen at all fast-releasing nerve terminals (Heuser & Reese, 1977). However, many questions still remain about the general structure of active zones. Of paramount importance is the question of how the active zone is organized; that is, how do the organelles of the active zone become properly positioned relative to each other and maintained in a precise alignment over the postsynaptic part of the active zone? While we have no firm answer to this question, the proposal of Mu-Ming Poo that there are physical links between pre- and postsynaptic proteins is a likely explanation that should be tested critically.

3 Synaptic Vesicle Cycling

Regarding the details of the synaptic vesicle cycle, we heard from Tom Reese about the sequential progression of synaptic vesicle components through a series

NATO ASI Series, Vol. H 100
Neurotransmitter Release and Uptake
Edited by Şakire Pöğün

membranes are composed of both proteins and lipids, it is perhaps likely that both of these constituents are important. Whatever the source of these pores, we still need to understand how calcium regulates their opening to yield exocytosis at nerve terminals.

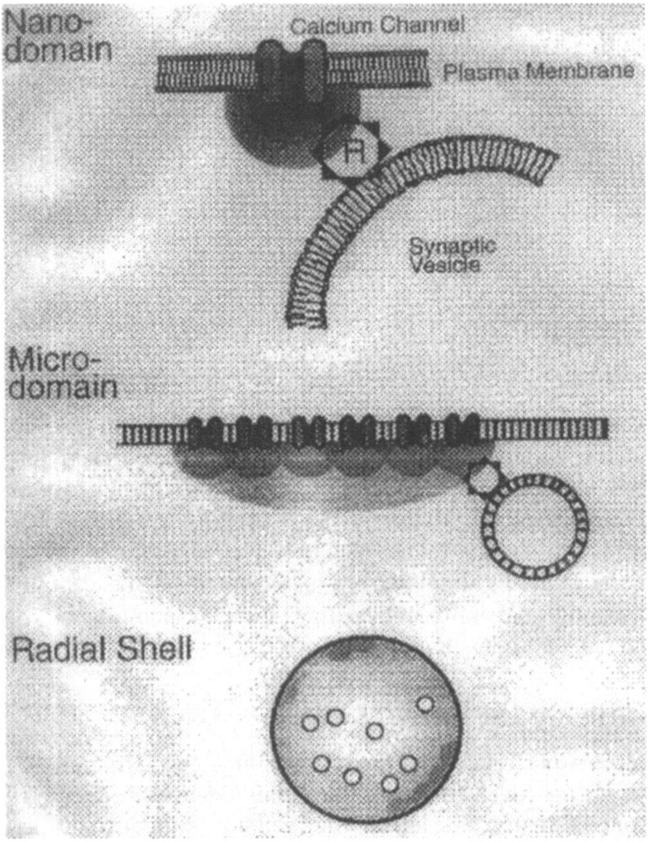

Fig. 1. Spatial ranges of calcium action in transmitter release. **Top-** Calcium entering through single open channels create nanodomains; a calcium receptor (R) within this range will sense calcium concentration (shading) that rapidly rises to tens to hundreds of micromolar. **Middle -** Clustering of calcium channels can lead to overlapping nanodomains and result in a microdomain. **Bottom -** If calcium receptors are far from calcium channels, they will sense a slowly-rising radial diffusion of calcium throughout the cell. Modified from Schweizer et al. (1995).

We also heard from Julio that amines are released from mast cell vesicles only after being displaced from the vesicular matrix via a process of ion exchange. Another question for the future is to determine whether similar processes are involved in the rapid release of neurotransmitters from synaptic vesicles.

6 Non-neuronal Secretion

We heard from Mu-Ming Poo and Phil Haydon that non-neuronal cells, such as fibroblasts and glial cells, also appear capable of releasing neurotransmitters in a calcium-regulated and quantal fashion (Parpura et al., 1994; Girod et al., 1995). This then raises the question of what makes neurons unique. According to the SNARE hypothesis, the interaction of synaptotagmin and SNARE proteins confers calcium regulation at the end of a sequence of relatively slow reactions. This sort of modularity is capable of explaining how diverse signals can regulate exocytosis in different types of cells, but this hypothesis requires more direct testing.

7 Calcium Domains

Rami Rahamimoff and I described the roles of ion channels in neurotransmitter release, with special emphasis on the voltage-gated ion channels that serve as the source of calcium ions that triggers release. One idea that emerged from these discussions was the concept of a local domain of calcium action centered on individual ion channels. As I discussed, the spatial arrangement of calcium channels - relative to each other and relative to calcium receptors that trigger exocytosis - is a critical parameter and is likely to vary between different types of nerve terminals. At one extreme is the squid giant synapse and the frog neuromuscular junction, where single nanodomains appear sufficient to trigger neurotransmitter release (Figure 2). At the other extreme are peptide-secreting terminals, where the distance between calcium channels and calcium receptors appears to be quite large in a type of shell arrangement. It is likely that many other nerve terminals may exhibit spatial arrangements intermediate between these two extremes. One important goal for the future will be to determine what the spacing is between open calcium channels and calcium receptors for each nerve terminal of interest.

A major challenge toward achieving this goal is to develop experimental methods that will allow these parameters to be determined reliably. From Yusuf Tan, Phil Haydon, Julio Fernandez and others, we heard about various optical methods - including video microscopy, confocal microscopy, two-photon microscopy, atomic force microscopy, and near-field microscopy - which offer some possibilities for making such determinations. However, to date these methods have met with limited success. The primary limitation of all these methods is that while they can look at calcium channels or an intracellular calcium signal, they cannot yet define the spatial relationship between these entities and the calcium receptors. To date, most success has come from bioassays, such as were described by me and by David Nichols (Adler et al., 1991; Verhage et al., 1991). It is clear that designing experimental paradigms to define these spatial relationships remains as a key challenge for the future.

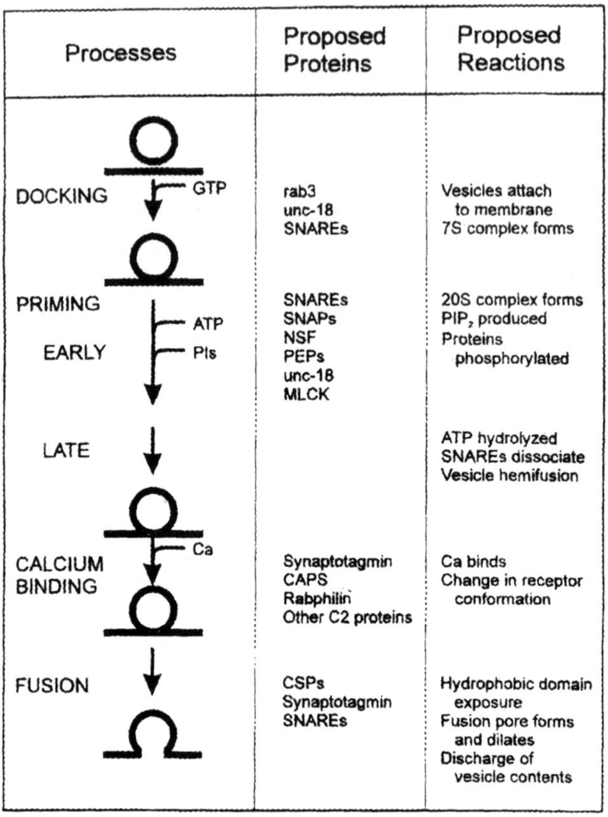

Fig. 2. A model of calcium-regulated neurotransmitter release. Processes known to be involved in transmitter release (or other forms of calcium-regulated exocytosis) are shown on the left, while the proteins and reactions proposed to underlie these processes are shown on the middle and on the right, respectively. Modified from Augustine et al. (1996).

8 Mechanisms of Endocytosis

As very elegantly described by Tom Reese in his talk, the process of endocytosis is responsible for retrieving synaptic vesicle membrane from the plasma membrane following exocytosis. Tom presented very good evidence that, at least in very active nerve terminals, endocytosis occurs via clathrin coated vesicles (Heuser, 1989). However, there are still those who believe that kiss-and-run exo/endocytosis may be occurring under more physiological conditions (Fesce et al., 1994). Particularly relevant in this regard are the experiments described by Julio Fernandez showing us that transient fusion is a common occurrence in mast cells loaded with GTP-γ-S (Alvarez de Toledo et al., 1993).

There appear to be several questions that need to be answered before we will understand the mechanism of endocytosis. First we need to know where endocytosis occurs under physiological conditions. The work of Tom Reese and colleagues shows clearly that clathrin-coated endocytosis occurs in presynaptic areas interposed between active zones, whereas (by definition) kiss-and-run exocytosis must occur at the active zone. Thus, one means of distinguishing between these two mechanisms is to define where exocytosis occurs under physiological conditions.

Second, as we heard from Jacques Hens, *Drosophila* genetics has demonstrated that dynamin, a protein that pinches off clathrin-coated vesicles to form coded vesicles, is essential for endocytosis. Shabire, a *Drosophila* mutant with a temperature-sensitive form of dynamin, is paralyzed within one to two minutes at temperature that prevents dynamin from working (Koenig & Ikeda, 1983). This rapid paralysis shows that synapses stop quickly without dynamin, implying that coated vesicles are the main mechanism of synaptic vesicle exocytosis under physiological conditions. However, even this evidence leaves open the possibility that kiss-and-run exocytosis might be mediated by dynamin, so another pressing question for the future is to determine whether or not dynamin can mediate kiss-and-run exocytosis.

Finally, one of the most remarkable characteristics of endocytosis is that it occurs in nerve terminals only following exocytosis. Thus, an unknown mechanism must regulate endocytosis to achieve this coupling to exocytosis. How this coupling is achieved is an interesting but unanswered regulatory question.

9 Lingering Problems with the Conventional Wisdom

Although the model upon which my talk is based is generally accepted, there are still some skeptics. These alternative points of view were well represented at the course and help emphasize still other things that we do not understand about neurotransmitter release. For example, although most do not put much faith in the calcium-voltage hypothesis that we heard from Itzchak Parnas, Itzchak is quite correct in pointing out that calcium diffusion cannot be rate-limiting for transmitter release (Parnas & Parnas, 1994). Thus, we need to identify the molecular processes that terminate neurotransmitter release. Likewise, both Yves Dunant and Tom Reese have pointed out that, under physiological rates of release, it is almost impossible to see exocytosis in the electron microscope, even following rapid-freeze fixation. The reason for this anomaly is not really well established and is worthy of further consideration. Likewise, Yves' provocative experiments with expression of mediatophore protein certainly are interesting, though there are clear alternatives to his interpretation that this protein is responsible for calcium-regulated transmitter release (Falk-Vairant et al., 1996). So the role of mediataphore in release remains an open and important question.

10 Synaptic Plasticity Mechanisms

Although the ability of synapses to release neurotransmitter in such a rapid and tightly-regulated fashion certainly is remarkable, even more remarkable is the fact that synaptic transmitter release is modifiable via a number of forms of plasticity.

One of the clear areas of future exploration will be to explain these forms of plasticity in mechanistic terms.

We heard from Itzchak Parnas, David Nichols and Mu-Ming Poo that a one form of regulation occurs because a number of neurotransmitters can activate presynaptic receptors that modify transmitter release. These receptors are likely to be important mediators of plasticity, so we need to understand their mechanism of action. As a starting point, we need to know whether these presynaptic modulators modify transmitter release by altering the opening of ion channels - to modify the amount of calcium available for transmitter release - or whether they work downstream of calcium signaling. Such an understanding of the locus of synaptic plasticity is clearly a tractable, yet important, question.

As we heard from Jacques Hens and others, nearly all proteins implicated in neurotransmitter release can be post-translationally modified by mechanisms such as phosphorylation and fatty acylation (Greengard et al., 1993; Sihra & Nichols, 1993). It is very likely that such processes contribute to the plasticity of neurotransmitter release, though it has not yet been possible to explain any form of synaptic plasticity in molecular terms. Certainly, identifying the roles of posttranslational modifications - as well as transcriptional and translational mechanisms - in synaptic plasticity will clearly be a priority for future work.

11 Conclusions

In conclusion, it is clear that this course has covered a lot of territory in a few days, and all of the participants are to be commended for making this a remarkable event. However, I hope that my presentation makes it clear that the new facts we have learned have only led to more questions. Thus, I hope that the goal of all participants now will be to go forth and to seek answers to these questions and then pose the next generation of questions that will follow.

Acknowledgements

I thank Don Garrett and Iris Oliver for typing this manuscript and Sakire Pogun for inviting me to participate in the NATO course in Kusadasi. Supported by NIH grant NS-21624.

References

Alvarez de Toledo, G, R Gernandez-Chacon and JM Fernandez (1993) Release of secretory products during transient vesicle fusion. *Nature*. 363, 554-557.

Augustine, GJ, ME Burns, WM DeBello, DL Pettit and FE Schweizer (1996) Exocytosis: proteins and perturbations. *Annu.* Rev. *Pharmacol. Toxicol.* 36, 659-701.

Betz WJ and GS Bewick (1992) Optical analysis of synaptic vesicle recycling at the frog neuromuscular junction. *Science*. 255, 200-203.

Del Castillo, J and B Katz (1954) Quantal components of the end-plate potential. *J. Physiol.* 124, 560-573.

Falk-Wairant, J, P corréges, L Eder-Colli, N Salem, F Meunier, B Lesbats, F Loctin, M Synguelakis, M Isräel and Y Dunant. (1996) Evoked acetylcholine release expressed by transfection of mediatophore cDNA. *J. Neurochem.* 66, (in press).

Fesce, R, F Grohovaz, F Valtorta and J Meldolesi (1994) Neurotransmitter release: fusion of "kiss-and-run." *Trends in Cell Biol.* 4, 1-4.

Girod, R, S Popov, J Alder, J Zheng, A Lohof and M-m Poo (1995) Quantal transmitter release from myocytes and fibroblasts: comparison with neuronal secretion. J. Neurosci.15, 2826-2838.

Greengard, P, F Valtorta, AJ Czernik and F Benfenati (1993) Synaptic vesicle phosphoproteins and regulation of synaptic function. *Science.* 259, 780-785.

Heuser JE and TS Reese (1973) Evidence for recycling of synaptic vesicle membrane during transmitter release at the frog neuromuscular junction. *J. Cell Biol.* 57, 315-344.

Koenig, JH and K Ikeda (1983) Evidence for a presynaptic blockage of transmission in a temperature-sensitive mutant of Drosophila. J. *Neurobiol.* 14, 411-419.

Liscovitch, M and LC Cantley (1995) Signal transduction and membrane traffic: the PI-TP/phosphoinositide connection. Cell. 81, 659.

Monck, JR and JM Fernandez (1994) The exocytotic fusion pore and neurotransmitter release. Neuron.12, 707-716.

Parnas, H and I Parnas (1994) Neurotransmitter release at fast synapses. J. *Membrane* Biol.142, 267-279.

Parpura, V, TB Basarsky, F Lui, S Jeftinija, S Jeftinija and PG Haydon (1994) Glutamate-mediated astrocyte-neuron signaling. *Nature.* 369, 744-747.

Rothman, JE (1994) Mechanisms of intracellular protein transport. *Nature.* 372, 55-63.

Ryan TA and SJ Smith (1995) Vesicle pool mobilization during action potential firing at hippocampal synapses. Neuron.14, 983-989.

Scheller, RH (1995) Membrane trafficking in the presynaptic nerve terminal. Neuron.14, 893-897.

Schweizer, FE, H Betz and GJ Augustine (1995) From vesicle docking to endocytosis: intermediate reactions of exocytosis. Neuron.14, 689-696.

Sihra, TS, and RA Nichols (1993) Mechanisms in the regulation of neurotransmitter release from brain terminals: current hypotheses. *Neurochem.* Res.18, 47-58.

Sudhof, TC (1996) The synaptic vesicle cycle: a cascade of protein-protein interactions. *Nature, 375,* 645-653.

Verhage, M, HT McMahon, WEJM Ghijsen, F Boomsma, VM Wiegant and DG Nicholls. (1991) Differential release of amino acids, neuropeptides and catecholamines from isolated nerve terminals. *Neuron.* 4, 577-524.

PART II: Neurotransmitter Uptake

The Vesicular and Plasmalemmal Dopamine Transporters: Ultrastructural Localization in Nigrostriatal Dopaminergic Neurons

Melissa J. Nirenberg and Virginia M. Pickel

Division of Neurobiology, Department of Neurology and Neuroscience, Cornell University Medical College, New York, NY 10021

Abstract

Dopaminergic neurotransmission is regulated by two distinct membrane-bound transporters: the vesicular monoamine transporter-2 (VMAT2), and the plasmalemmal dopamine transporter (DAT). These transporters play important roles in the initiation and termination of dopaminergic neurotransmission, respectively. In the present report, we review our recent electron microscopic studies showing the immunocytochemical localization of VMAT2 and DAT in rat brain. We have focused on the nigrostriatal dopaminergic neurons, which are known to release dopamine both from their somata and/or dendrites in the substantia nigra (SN) and from their axonal processes in the dorsal striatum. Our studies show that VMAT2 is selectively localized to small synaptic vesicles (SSVs) and large dense core vesicles (DCVs) in dopaminergic axons and terminals in the striatum, and to tubulovesicular organelles and rare SSVs or DCVs within dopaminergic somata, dendrites, and dendritic spines in the SN. These findings support the proposal that the axonal storage and release of dopamine are primarily vesicle-mediated, but suggest that somatodendritic dopamine storage and release may also involve vesicle-independent mechanisms. The subcellular localization of DAT was strikingly different from that of VMAT2, with extensive labeling of extrasynaptic portions of the plasma membranes of dopaminergic axons and terminals in the striatum, and of dopaminergic dendrites and dendritic spines in the SN. Dopaminergic perikarya in the SN, however, almost never contained detectable plasmalemmal labeling for DAT, suggesting that extrasynaptic portions of the distal dendritic and axonal processes are the primary functional sites for DAT-mediated dopamine reuptake. Together, these studies have provided important new information regarding the potential sites for release and reuptake of dopamine within nigrostriatal dopaminergic neurons.

1 Introduction

Dopaminergic neurotransmission involves the regulated storage, release, and reuptake of dopamine. These functions involve two distinct types of transporters: a vesicular dopamine transporter to permit dopamine uptake into synaptic vesicles or

NATO ASI Series, Vol. H 100
Neurotransmitter Release and Uptake
Edited by Şakire Pöğün

other intracellular organelles, and a plasmalemmal dopamine transporter to reduce extracellular dopamine concentrations through active reuptake of dopamine from the extracellular space (for reviews see Edwards, 1992; Rudnick and Clark, 1993). In the rat, the uptake of dopamine into vesicles or other subcellular organelles is mediated by a reserpine-sensitive vesicular monoamine transporter (VMAT2). In contrast, the uptake of dopamine across the plasma membrane is mediated by a structurally and functionally distinct cocaine-sensitive dopamine transporter (DAT). The recent molecular cloning and sequencing of the genes encoding VMAT2 (Erickson et al., 1992; Liu et al., 1992) and DAT (Giros et al., 1991; Kilty et al., 1991; Shimada et al., 1991) has led to the production of selective polyclonal antisera directed against synthetic peptide sequences unique to each of these transporters (Freed et al., 1995; Peter et al., 1995). These anti-peptide antisera have enabled us to perform the first electron microscopic immunocytochemical studies demonstrating the precise subcellular localization of VMAT2 and DAT within central monoaminergic neurons (Nirenberg et al., 1995; Nirenberg et al., 1996a; Nirenberg et al., 1996b). The results of these studies have helped to clarify the basic mechanisms of normal monoaminergic neurotransmission, and to determine the potential sites and mechanisms of responsiveness to drugs such as cocaine and amphetamines.

2 Nigrostriatal Dopaminergic Neurons

We have focused this review on the nigrostriatal dopaminergic neurons, which play a critical role in the central regulation of motor functions (reviewed in Björklund and Lindvall, 1984; Roth and Elsworth, 1995). These neurons have been extensively characterized with regard to their distribution, synaptic relationships, and morphological features (for reviews, see Pickel and Sesack, 1995; Roth and Elsworth, 1995). The nigrostriatal dopaminergic neurons release dopamine both from their axonal processes located in the patch compartments of the striatum, and from their somata and/or dendrites in the ventral tier of the SN (Gerfen et al., 1987). In primates and other susceptible species, these neurons are particularly vulnerable to neurotoxin-induced parkinsonism; and in humans they selectively degenerate in idiopathic Parkinson's disease (Manaye et al., 1989; Kaakkola and Teräväinen, 1990; Pickel et al., 1992). Since VMAT2 and DAT have both been shown to transport parkinsonism-inducing neurotoxins such as 1-methyl-4-phenylpyridinium (MPP^+), an examination of the subcellular localization of these transporters within nigrostriatal dopaminergic neurons might potentially help to clarify the cellular basis for this selective vulnerability (reviewed in Edwards, 1993; Bannon et al., 1995).

3 The Vesicular Monoamine Transporter-2 (VMAT2)

Like most classical neurotransmitters, dopamine is synthesized in the cytoplasm. Newly synthesized dopamine, as well as dopamine that has been recycled by plasmalemmal reuptake, is then actively transported into an intracellular compartment. This process is mediated by a reserpine-sensitive vesicular monoamine carrier, which uses the proton electrochemical gradient to drive the transport of dopamine and other monoamines into vesicles and/or other acidic intracellular organelles (reviewed in Edwards, 1992; Schuldiner, 1994).

The genes encoding two closely related rat vesicular monoamine transporters, VMAT1 (also known as CGAT) which is present in monoamine-containing endocrine cells but absent from neurons, and VMAT2 (also known as SVAT) which is expressed in neurons and other cells, have recently been cloned and sequenced (Erickson et al., 1992; Liu et al., 1992). VMAT2 has been shown to confer all of the known properties of the neuronal vesicular monoamine carrier, including sensitivity to reserpine and tetrabenazine, which potently inhibit vesicular monoamine uptake (Edwards, 1992; Peter et al., 1994).

Aside from their role in the uptake and intracellular storage of monoamines, the VMATs also share certain structural and functional similarities with the bacterial drug resistance transporters (Schuldiner et al., 1994; Yelin and Schuldiner, 1995). In specific, the VMATs have been shown to sequester parkinsonism-inducing neurotoxins such as MPP^+ (Liu et al., 1992; Stern-Bach et al., 1992). This has led to the proposal that the VMATs and the multidrug resistance transporters are members of a large family of toxin extruding antiporters (TEXANs), whose function is to protect the cells in which they are expressed by transporting toxins out of the cytoplasm (Yelin and Schuldiner, 1995). Thus, it is possible that differential expression of VMAT2 may partially account for the differential susceptibilities of monoaminergic neurons to neurotoxic insult (reviewed in Edwards, 1993; Bannon et al., 1995).

3.1 Regional distribution of VMAT2

The regional distribution of the vesicular monoamine carrier in central neurons has been examined by a number of different techniques, including *in situ* hybridization (Erickson et al., 1992; Liu et al., 1992), and ligand-binding autoradiography (Henry, 1989; Vander Borght et al., 1995). Together, these studies have shown that the localization of VMAT-containing sites closely resembles the known distribution of monoamine-containing cell bodies and processes. High levels of VMAT2 have been detected in the nigrostriatal dopaminergic neurons, with abundant expression of VMAT2 mRNA in the SN (Liu et al., 1992), and of VMAT binding sites in both the SN and the striatum (Henry, 1989). Light microscopic immunocytochemistry with anti-peptide antisera directed against VMAT2 has also demonstrated that the localization of this transporter parallels the known distribution of monoaminergic neurons, with the possible exception of the olfactory bulbs, in which VMAT2 immunolabeling has not been detected (Peter et al., 1995).

3.2 Subcellular Localization of VMAT2 in the Striatum

The release of dopamine from axon terminals located in the dorsolateral striatum is calcium-dependent and reserpine-sensitive, suggesting that it results from vesicular exocytosis (Llinás, 1979). This may involve one or both types of regulated secretory vesicles that are present in central monoaminergic neurons: small synaptic vesicles (SSVs) and large dense core vesicles (DCVs) (Pickel and Milner, 1987; Maley et al., 1990). SSVs typically contain classical neurotransmitters, and have been implicated in fast synaptic transmission, while DCVs also contain neuropeptides, and are involved in neuromodulation (reviewed in Matteoli et al., 1991). Most of the catecholaminergic (tyrosine hydroxylase (TH)-immunoreactive) terminals in the dorsal striatum are dopaminergic (Pickel and Sesack, 1995). Electron microscopic examination of these dopaminergic terminals has shown that they contain numerous SSVs and occasional DCVs (reviewed in Pickel and Sesack, 1995).

The subcellular sites for monoamine storage in the striatum have been examined by electron microscopy after the uptake of false neurotransmitters such as 5-hydroxydopamine (Tennyson et al., 1974; Tennyson et al., 1975). These experiments have shown that a characteristic electron dense precipitate representing the presence of the false neurotransmitter is localized to the lumens of SSVs and DCVs within striatal axons and terminals. This indicates that both SSVs and DCVs are potential sites for monoamine storage and release in the striatum. These studies, however, are limited by the potential for (1) diffusion, or (2) uptake of false neurotransmitters into cells or organelles that do not normally store monoamines (Kelly et al., 1985; Groves et al., 1994). Autoradiographic ligand binding studies have the theoretical advantage of localizing the membrane-bound vesicular monoamine carrier, and thus are not subject to the possibility of diffusion (Henry, 1989). Unfortunately, these studies have not provided sufficient resolution to determine the specific subcellular organelles with which VMAT2 is associated (reviewed in Henry, 1989).

More recently, we used higher resolution dual labeling immunocytochemical techniques to elucidate the precise subcellular localization of VMAT2, a specific membrane-bound marker for intracellular monoamine storage, within identified dopaminergic (TH-immunoreactive) neurons in the rat dorsolateral striatum. In these axons and terminals, immunogold-silver labeling for VMAT2 was selectively localized to vesicular membranes (Figure 1). Most of the VMAT2-containing vesicles were 40-60 nm in diameter and electron-lucent, and therefore were identified as SSVs (Figure 1). DCVs, identified by their larger (80-120 nm) diameter and the presence of a characteristic electron-dense core, were only infrequently observed within dopaminergic terminals in this region. When they were present, however, the DCVs were often heavily decorated with immunogold particles for VMAT2. These observations suggest that dopamine can be stored in and released from both SSVs and DCVs in striatal dopaminergic terminals. The results of these studies therefore provide direct evidence that VMAT2 is involved in the uptake of dopamine into two types of regulated secretory vesicles in striatal dopaminergic terminals.

3.3 Subcellular Localization of VMAT2 in the Substantia Nigra

Nigrostriatal dopaminergic neurons can also store and release dopamine from somata and/or dendrites located in the SN (Björklund and Lindvall, 1975; Geffen et al., 1976; Korf et al., 1976; reviewed in Cuello and Iversen, 1978; Cheramy et al., 1981). This somatodendritic dopamine release is reserpine-sensitive (Björklund and Lindvall, 1975; Elverfors and Nissbrandt, 1991; Heeringa and Abercrombie, 1995), and largely calcium-dependent, suggesting that it is mediated by vesicular exocytosis (Geffen et al., 1976; Cheramy et al., 1981). Surprisingly, however, electron microscopic studies have shown that dopaminergic dendrites in the SN contain few synaptic vesicles (Wilson et al., 1977; Wassef et al., 1981; Groves and Linder, 1983; Nirenberg et al., 1996b), and that the uptake of false or radiolabeled monoamines shows no specific association with somatodendritic vesicles (Sotelo, 1971; Cuello and Iversen, 1978). Instead, these exogenous monoamines are primarily transported into larger tubulovesicular organelles that resemble smooth endoplasmic reticulum (SER), suggesting that the SER is the major site of dendritic storage and release of dopamine in the SN (Hattori et al., 1979; Mercer et al., 1979; Wassef et al., 1981).

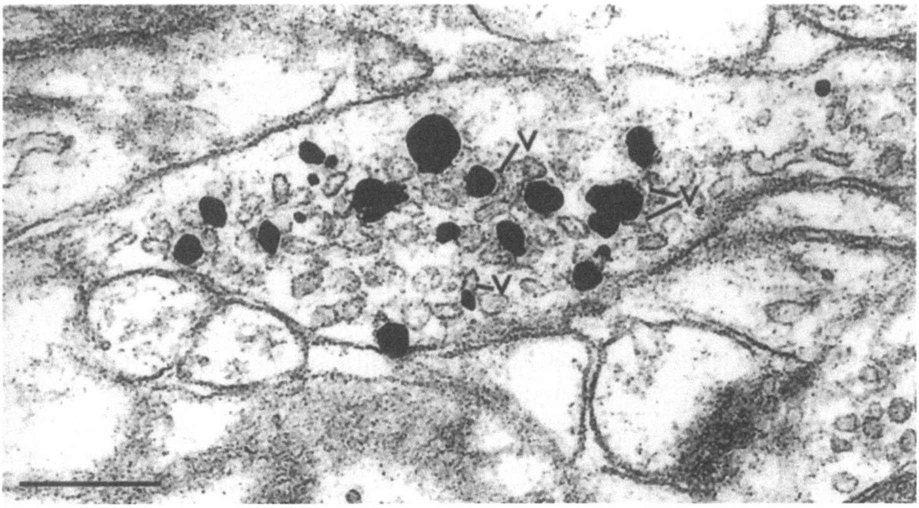

Fig 1. VMAT2 is localized to SSVs in a striatal axon. Electron micrograph showing an axonal varicosity in the dorsolateral striatum that has been immunogold-silver labeled for VMAT2. Numerous gold particles, indicating the presence of VMAT2, are localized to the membranes of pleomorphic electron-lucent synaptic vesicles (v). Bar=0.25 μm.

We recently used dual labeling immunocytochemistry with an antiserum directed against VMAT2 as a selective marker for determining the potential sites of somatodendritic storage and release of dopamine within identified dopaminergic (TH-immunoreactive) neurons in the SN (Nirenberg et al., 1996a). In these studies, we showed that VMAT2 is primarily associated with large tubulovesicular structures that resemble SER (Figure 2A-B), and is only rarely associated with SSVs (Figure 2B) or DCVs (not shown). In combination with the previous evidence from autoradiographic and uptake studies, these findings strongly suggest that the somatodendritic storage and release of dopamine principally involve tubulovesicular organelles, rather than typical synaptic vesicles.

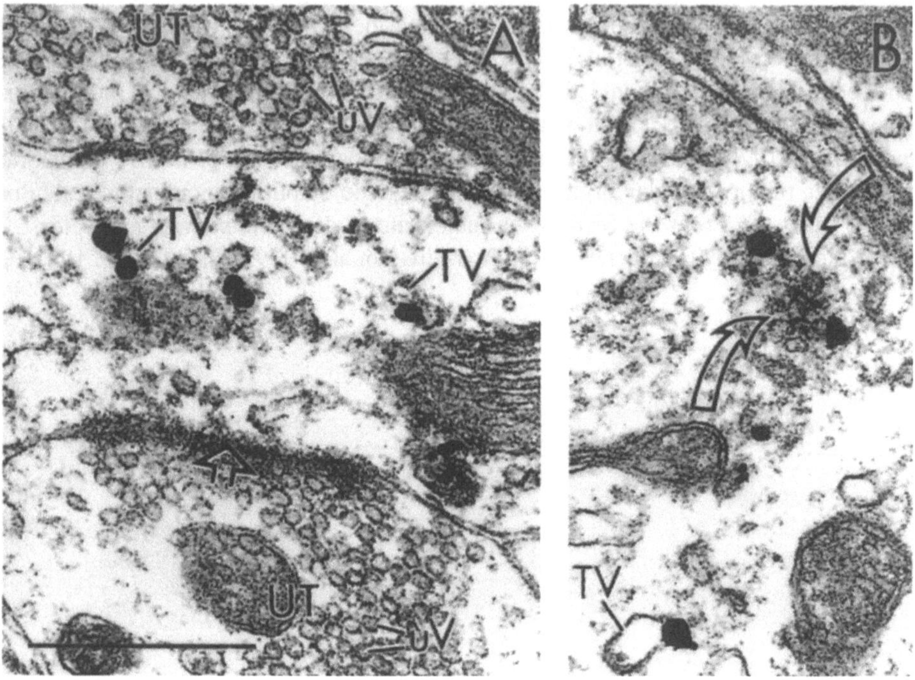

Fig 2. VMAT2 is localized to cytoplasmic tubulovesicles within dopaminergic dendrites in the substantia nigra. Electron micrographs showing dendrites in the SN that have been immunogold-silver labeled for VMAT2. In A, the tissue has been dually labeled with immunogold for VMAT2 and immunoperoxidase for TH. Immunogold particles for VMAT2 are localized to cytoplasmic tubulovesicles (TV) in a dendrite that also contains diffuse peroxidase labeling for TH. The VMAT2-containing tubulovesicles are larger than the unlabeled synaptic vesicles (uV) seen in adjacent unlabeled terminals (UT). The unlabeled terminal (UT) at the bottom of the panel forms a synaptic junction (open arrowhead) onto the dually labeled dendrite. In B, the tissue has been singly immunogold-labeled for VMAT2. VMAT2 is localized to a large tubulovesicle (TV), and to a cluster of smaller synaptic-like vesicles (curved arrows) in the cytoplasm of a dendrite. Bar=0.5 μm. *(Reproduced with permission from Nirenberg et al., 1996a).*

Since the dendritic release of dopamine in the SN is largely calcium-dependent, the release of dopamine from these tubulovesicular organelles might potentially involve a stimulus-secretion coupling mechanism similar to vesicular exocytosis (Llinás, 1979). Alternatively, the presence of calcium might cause the efflux of dopamine from tubulovesicles into the dendritic cytoplasm, leading to a local reversal of the plasmalemmal dopamine concentration gradient, and thus the release of dopamine via the plasmalemmal dopamine transporter (Bannon et al., 1995; see below).

4 The Plasmalemmal Dopamine Transporter (DAT)

Dopaminergic neurotransmission is terminated as a result of the reduction of extracellular dopamine concentrations. This is principally mediated by the active reuptake of dopamine into the presynaptic neuron by the sodium- and chloride-dependent DAT (Cass et al., 1993; Smith and Justice, 1994; Giros et al., 1996; for reviews see Edwards, 1992; Uhl and Hartig, 1992; Rudnick and Clark, 1993; Boja et al., 1994; Hitri et al., 1994). Under conditions in which the concentration gradients are reversed, this dopamine carrier can also facilitate the calcium-independent release of dopamine through reversal of the normal direction of dopamine transport. This has been demonstrated under pharmacological conditions in which the plasmalemmal dopamine gradient has been reversed, as with the application of amphetamines (Bernardini et al., 1991; Sulzer et al., 1993), or a reduction of the extracellular sodium concentration (Hurd and Ungerstedt, 1989), but might potentially also occur under physiological conditions (Bannon et al., 1995).

A single gene encoding the rat dopamine transporter (DAT) has recently been identified and cloned (Giros et al., 1991; Kilty et al., 1991; Shimada et al., 1991). Expression of the cloned DAT confers all of the properties of the pharmacologically characterized dopamine uptake carrier, including the capacity to bind cocaine and other psychostimulant drugs that are potent inhibitors of dopamine reuptake (Ritz et al., 1987; Kuhar et al., 1991). Disruption of the gene encoding the mouse DAT has been shown to lead to dramatically elevated extracellular dopamine concentrations and the absence of a detectable behavioral response to cocaine, providing further evidence of the critical importance of DAT for these functions (Giros et al., 1996).

4.1 Regional distribution of DAT

The distribution of dopamine uptake sites has been investigated by autoradiography with numerous different DAT ligands (Doucet et al., 1986; Graybiel and Moratalla, 1989; Cass et al., 1992; Cline et al., 1992), as well as by external imaging with PET and SPECT (reviewed in Boja et al., 1994). The regional distribution of DAT mRNA has also been examined by *in situ* hybridization (Shimada et al., 1992; Cerruti et al., 1993; Blanchard et al., 1994; Lorang et al., 1994); and the distribution of DAT protein has been shown using immunocytochemistry with specific anti-peptide antisera directed against this transporter (Ciliax et al., 1995; Freed et al., 1995; Revay et al., 1996). Together, these studies have shown that DAT is selectively expressed in most dopamine-containing cell groups, and is present at particularly high levels in the nigrostriatal dopaminergic neurons.

4.2 Subcellular localization of DAT in the Striatum

The subcellular localization of DAT in the striatum has previously been examined by *in vitro* light microscopic autoradiography with specific DAT ligands. In primary cultures of nigral dopaminergic neurons, DAT binding sites have been shown to be present along the processes of dopamine-immunoreactive neurons (Cerruti et al., 1991). More recently, the use of higher resolution dual labeling electron microscopic immunocytochemical techniques has enabled us to determine the precise sites at which DAT is localized within dopaminergic neurons in fixed tissue sections from the rat dorsolateral striatum (Nirenberg et al., 1996b).

In these studies, we showed that immunogold labeling for DAT is prominently localized to plasma membranes, and less frequently associated with intracellular vesicles or tubulovesicles within axonal processes and terminals in the dorsolateral striatum (Figure 3A-B). The labeled plasma membranes presumably represent the functional sites of DAT-mediated dopamine reuptake. In contrast, the labeled intracellular membranes indicate potential sites for transport of DAT-containing membranes between the cell body and the axonal plasma membrane. This pattern of labeling was observed in sections that were singly immunogold labeled for DAT (Figure 3A), and in dually labeled tissue, where most of the DAT-labeled processes also contained a dense peroxidase reaction product for TH (Figure 3B). In both singly and dually labeled tissue, the DAT-labeled axons and terminals had all of the known morphological characteristics of the dopaminergic neurons in this region, including the infrequent presence of synaptic junctions, which were small and symmetrical in appearance, and often occurred *en passant* (Figure 3A).

The plasmalemmal labeling for DAT was observed along intervaricose axonal segments as well as axonal varicosities in the striatum, where it was present both near and distant from recognized synaptic junctions (Figure 3A-B). The extensive localization of DAT at sites lacking the typical morphological characteristics of axon terminals suggests that the uptake, and thus the "spatial buffering" of dopamine is not necessarily restricted to the region of the synaptic cleft, but instead may occur predominantly at extrasynaptic sites (Nicholson and Rice, 1991). These findings suggest a role for striatal dopamine in "parasynaptic" transmission (Schmitt, 1984; Herkenham, 1987). Other studies, however, have suggested that extensive diffusion of dopamine is likely to be more prominent in the nucleus accumbens (Stamford et al., 1988), and in extrastriatal dopamine-containing regions such as the amygdala and prefrontal cortex (Garris and Wightman, 1994), than in the dorsal striatum.

4.3 Subcellular localization of DAT in the Substantia Nigra

The subcellular localization of DAT in the SN has been examined using *in vitro* light microscopic autoradiography with DAT ligands (Cerruti et al., 1991). These studies have shown that in nigral cultures, DAT is expressed along both the processes and the somata of dopamine-immunoreactive neurons. More recently, we have used electron microscopic immunogold-silver techniques to elucidate the precise subcellular localization of DAT in fixed tissue sections from the rat SN (Nirenberg et al., 1996b). In these studies, we showed that DAT is present in dopaminergic somata and dendrites in this region. Within dopaminergic perikarya (not shown), DAT was almost exclusively localized to internal membranes, including saccules of Golgi and tubulovesicular structures resembling SER. Many of the DAT-containing dendrites also contained considerable labeling of internal membranes, including cytoplasmic tubulovesicles. These DAT-labeled perikaryal and dendritic membranes presumably represent sites of synthesis, transport, and/or recycling of DAT.

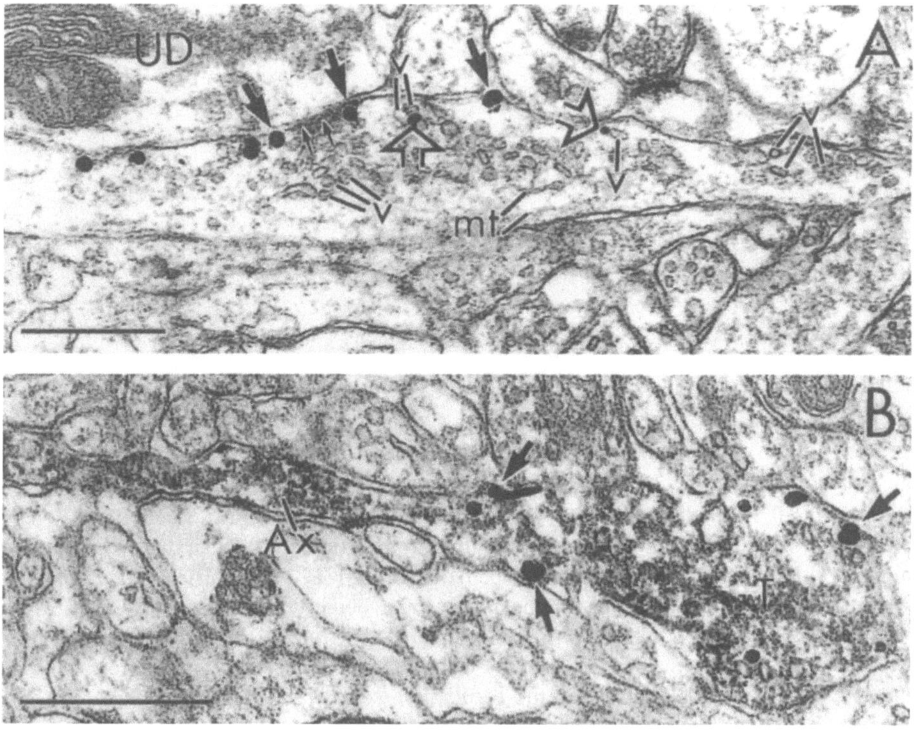

Fig 3. DAT is localized to plasma membranes of striatal axons. Electron micrographs showing immunogold particles for DAT along the plasma membranes (large closed arrows) of varicose axons in the striatum. In A, a gold particle for DAT (open arrowhead) is also localized to a small vesicle (v). The labeled axon forms a small, symmetrical synapse *en passant* (small closed arrows) with an unlabeled dendrite (UD). In B, the tissue has been dually labeled with immunoperoxidase for TH. The DAT-labeled axon (Ax) contains a diffuse, electron dense peroxidase reaction product indicating the presence of TH. Bars=0.5 μm. *(Reproduced with permission from Nirenberg et al., 1996b).*

In dopaminergic dendrites, labeling for DAT was also frequently localized to the plasma membrane (Figure 4). This plasmalemmal labeling was particularly prominent along distal dendrites and dendritic spines. The abundant labeling for DAT along distal dendrites, and its relative absence from the plasma membranes of dopaminergic somata suggest that the distal dendrites are the most likely site of uptake of dopamine and related neurotoxins in the SN. In addition, these findings suggest that in the SN, the distal dendrites of dopaminergic neurons are likely to be the primary sites of action of cocaine and other DAT-binding psychostimulants.

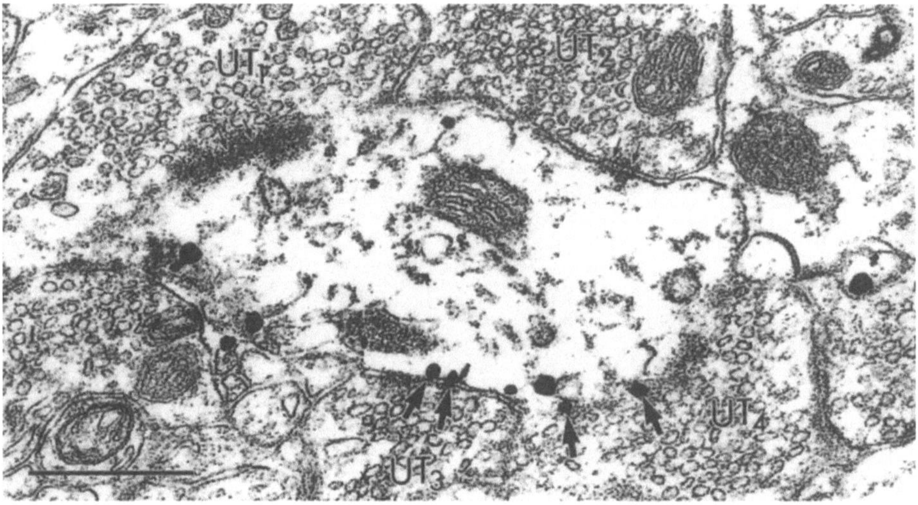

Fig 4. DAT is localized to the plasma membrane of a nigral dendrite. Electron micrograph showing a section from the rat substantia nigra that has been dually labeled with immunogold-silver for DAT and immunoperoxidase for TH. Gold particles for DAT (large arrows) are seen along the plasma membrane of a dendrite that also contains a diffuse peroxidase reaction product for TH. Several unlabeled terminals (UT) are apposed to (UT_3, UT_4) or form synaptic junctions (UT_1,UT_2) with the labeled dendrite. Small arrows=synaptic junction; Bar=0.5 μm. *(Reproduced with permission from Nirenberg et al., 1996b).*

Summary

The recent development of selective antisera directed against VMAT2 and DAT has enabled us to directly determine the immunocytochemical localization of these membrane-bound transporters within nigrostriatal dopaminergic neurons. These studies have provided important new information regarding the potential cellular sites for (1) intracellular storage and plasmalemmal reuptake of dopamine, (2) binding of psychoactive drugs such as cocaine and amphetamines, and (3) uptake or sequestration of exogenous (and putative endogenous) parkinsonism-inducing neurotoxins.

Acknowledgments

We thank Drs. Michael J. Kuhar, George R. Uhl, and Roxanne A. Vaughan (DAT) and Drs. Robert H. Edwards and Yongjian Liu (VMAT2) for generously providing the antisera that made these experiments possible. We also thank June Chan for expert technical assistance, and Drs. Adena L. Svingos and Carrie T. Drake for helpful critical commentary. VMP receives salary support from the National Institute of Mental Health (NIMH; Grant MH00078) and research support for this work from NIMH (Grant MH40342) and the National Institute on Drug Abuse (NIDA; Grant DA04600). MJN is supported by NIMH Grant MH40342.

References

Bannon MJ, Granneman JG, Kapatos G (1995) The dopamine transporter: potential involvement in neuropsychiatric disorders. In: Psychopharmacology: The Fourth Generation of Progress (Bloom FE, Kupfer DJ), pp 179-188. New York: Raven Press, Ltd.

Bernardini GL, Gu X, Viscardi E, German DC (1991) Amphetamine-induced and spontaneous release of dopamine from A9 and A10 dendrites: an *in vitro* electrophysiological study in the mouse. J Neural Transm Gen Sect (Austria) 84:183-193.

Björklund A, Lindvall O (1975) Dopamine in dendrites of substantia nigra neurons: suggestions for a role in dendritic terminals. Brain Res 83:531-537.

Björklund A, Lindvall O (1984) Dopamine-containing systems in the central nervous system. In: Handbook of Chemical Neuroanatomy. Volume II: Classical Transmitters in the Central Nervous System (Björklund A, Hökfelt T), pp 55-121. Amsterdam: Elsevier Science Publishers.

Blanchard V, Raisman-Vozari R, Vyas S, Michel PP, Javoy-Agid F, Uhl G, Agid Y (1994) Differential expression of tyrosine hydroxylase and membrane dopamine transporter genes in subpopulations of dopaminergic neurons of the rat mesencephalon. Brain Res Mol Brain Res 22:29-38.

Boja JW, Vaughan R, Patel A, Shaya EK, Kuhar MJ (1994) The dopamine transporter. In: Dopamine receptors and Transporters: Pharmacology, Structure, and Function (Niznik HB), pp 611-644.

Cass WA, Gerhardt GA, Mayfield RD, Curella P, Zahniser NR (1992) Differences in dopamine clearance and diffusion in rat striatum and nucleus accumbens following systemic cocaine administration. J Neurochem 59:259-266.

Cass WA, Zahniser NR, Flach KA, Gerhardt GA (1993) Clearance of exogenous dopamine in rat dorsal striatum and nucleus accumbens: role of metabolism and effects of locally applied uptake inhibitors. J Neurochem 61:2269-2278.

Cerruti C, Drian MJ, Kamenka JM, Privat A (1991) Localization of dopamine carriers by BTCP, a dopamine uptake inhibitor, on nigral cells cultured in vitro. Brain Res 555:51-57.

Cerruti C, Walther DM, Kuhar MJ, Uhl GR (1993) Dopamine transporter mRNA expression is intense in rat midbrain neurons and modest outside midbrain. Mol Brain Res 18:181-186.

Cheramy A, Leviel V, Glowinski J (1981) Dendritic release of dopamine in the substantia nigra. Nature 289:537-542.

Ciliax BJ, Heilman C, Demchyshyn LL, Pristupa ZB, Ince E, Hersch SM, Niznik HB, Levey AI (1995) The dopamine transporter: immunocytochemical characterization and localization in brain. J Neurosci 15:1714-1723.

Cline EJ, Scheffel U, Boja JW, Mitchell WM, Carroll FI, Abraham P, Lewin AH, Kuhar MJ (1992) In vivo binding of [^{125}I]RTI-55 to dopamine transporters: pharmacology and regional distribution with autoradiography. Synapse 12:37-46.

Cuello AC, Iversen LL (1978) Interactions of dopamine with other neurotransmitters in the rat substantia nigra: a possible functional role of dendritic dopamine. In: Interactions Between Putative Neurotransmitters in the Brain (Garattini S, Pujol JF, Samanin R), pp 127-149. New York: Raven.

Doucet G, Descarries L, Garcia S (1986) Quantification of the dopamine innervation in adult rat neostriatum. Neuroscience 19:427-45.

Dray A (1981) Serotonin in the basal ganglia: functions and interactions with other neuronal pathways. J Physiol (Paris) 77:393-403.

Edwards RH (1992) The transport of neurotransmitters into synaptic vesicles. Curr Opin Neurobiol 2:586-594.

Edwards RH (1993) Neuronal degeneration and the transport of neurotransmitters. Ann Neurol 34:638-645.

Elverfors A, Nissbrandt H (1991) Reserpine-insensitive dopamine release in the substantia nigra? Brain Res 557:5-12.

Erickson JD, Eiden L, Hoffman B (1992) Expression cloning of a reserpine-sensitive vesicular monoamine transporter. Proc Natl Acad Sci, USA 89:10993-10997.

Freed C, Revay R, Vaughan RA, Kriek E, Grant S, Uhl GR, Kuhar MJ (1995) Dopamine transporter immunoreactivity in rat brain. J Comp Neurol 359:340-349.

Garris PA, Wightman RM (1994) Different kinetics govern dopaminergic transmission in the amygdala, prefrontal cortex, and striatum: an *in vivo* voltammetric study. J Neurosci 14:442-450.

Geffen LB, Jessell TM, Cuello AC, Iversen LL (1976) Release of dopamine from dendrites in rat substantia nigra. Nature 260:258-260.

Gerfen CR, Herkenham M, Thibault J (1987) The neostriatal mosaic: II. Patch- and matrix-directed mesostriatal dopaminergic and non-dopaminergic systems. J Neurosci 7:3915-3934.

Giros B, El Mestikawy S, Bertrand L, Caron G (1991) Cloning and functional characterization of a cocaine-sensitive dopamine transporter. FEBS Lett 295:149-154.

Giros B, Jaber M, Jones SR, Wightman RM, Caron MG (1996) Hyperlocomotion and indifference to cocaine and amphetamine in mice lacking the dopamine transporter. Nature 379:606-612.

Graybiel AM, Moratalla R (1989) Dopamine uptake sites in the striatum are distributed differentially in striosome and matrix compartments. Proc Natl Acad Sci USA 86:9020-9024.

Groves PM, Linder JC (1983) Dendro-dendritic synapses in substantia nigra: descriptions based on analysis of serial sections. Exp Brain Res 49:209-217.

Groves PM, Linder JC, Young SY (1994) 5-Hydroxydopamine-labeled dopaminergic axons: three-dimensional reconstructions of axons, synapses, and postsynaptic targets in rat neostriatum. Neuroscience 58:593-604.

Hattori T, McGeer PL, McGeer EG (1979) Dendro axonic neurotransmission. II. Morphological sites for the synthesis, binding and release of neurotransmitters in dopaminergic dendrites in the substantia nigra and cholinergic dendrites in the neostriatum. Brain Res 170:71-83.

Heeringa MJ, Abercrombie ED (1995) Biochemistry of somatodendritic dopamine release in substantia nigra: an *in vivo* comparison with striatal dopamine release. J Neurochem 65:192-200.

Henry J-P (1989) Radioligands of the vesicular monoamine transporter and their use as markers of monoamine storage vesicles. Biochem Pharm 38:2395-2404.

Herkenham M (1987) Mismatches between neurotransmitter and receptor localizations in brain: observations and implications. Neuroscience 23:1-38.

Hurd YL, Ungerstedt U (1989) Influence of a carrier transport process on in vivo release and metabolism of dopamine: dependence on extracellular Na^+. Life Sci 45:283-293.

Kaakkola S, Teräväinen H (1990) Animal models of parkinsonism. Pharmacol Toxicol 67:95-100.

Kelly E, Jenner P, Marsden CD (1985) Evidence that [^{3}H]dopamine is taken up and released from nondopaminergic nerve terminals in the rat substantia nigra in vitro. J Neurochem 45:137-144.

Kilty JE, Lorang D, Amara SG (1991) Cloning and expression of a cocaine-sensitive rat dopamine transporter. Science 254:578-579.

Korf J, Zieleman M, Westerink BHC (1976) Dopamine release in the substantia nigra? Nature 260:257-258.

Kuhar MJ, Ritz MC, Boja JW (1991) The dopamine hypothesis of the reinforcing properties of cocaine. Trends Neurosci 14:299-302.

Liu Y, Peter D, Roghani A, Schuldiner S, Privé GG, Eisenberg D, Brecha N, Edwards RH (1992) A cDNA that suppresses MPP^+ toxicity encodes a vesicular amine transporter. Cell 70:539-551.

Llinás R (1979) The role of calcium in neuronal function. In: The Neurosciences: Fourth Study Program (Schmitt FO, Worden FG), pp 555-571. Cambridge, MA: The M.I.T. Press.

Lorang D, Amara SG, Simerly RB (1994) Cell-type-specific expression of catecholamine transporters in the rat brain. J Neurosci 14:4903-4914.

Maley BE, Engle MG, Humphries S, Vascik DA, Howes KA, Newton BW, Elde RP (1990) Monoamine synaptic structure and localization in the central nervous system. J Electron Microscop Tech 15:20-33.

Manaye KF, Sonsalla PK, Barnett G, Heikkila RE, Woodward DJ, Smith WK, German DC (1989) 1-Methyl-4-(2'-methylphenyl)-1,2,3,6-tetrahydropyridine (2'-CH_3-MPTP)-induced degeneration of mesostriatal dopaminergic neurons in the mouse: biochemical and neuroanatomical studies. Brain Res 491:307-315.

Matteoli M, Reetz T, De Camilli P (1991) Small synaptic vesicles and large dense-core vesicles: secretory organelles involved in two modes of neuronal signalling. In: Volume Transmission in the Brain: Novel Mechanisms for Neural Transmission (Fuxe K, Agnati LF), pp 181-193. New York: Raven Press, Ltd.

Mercer L, del Fiacco M, Cuello AC (1979) The smooth endoplasmic reticulum as a possible storage site for dendritic dopamine in substantia nigra neurones. Experientia 35:101-103.

Nicholson C, Rice ME (1991) Diffusion of ions and transmitters in the brain cell microenvironment. In: Volume Transmission in the Brain: Novel Mechanisms for Neural Transmission (Fuxe K, Agnati LF), pp 279-294. New York: Raven Press, Ltd.

Nirenberg MJ, Liu Y, Peter D, Edwards RH, Pickel VM (1995) The vesicular monoamine transporter 2 is present in small synaptic vesicles and preferentially localizes to large dense core vesicles in rat solitary tract nuclei. Proc Natl Acad Sci, USA 92:8773-8777.

Nirenberg MJ, Chan J, Liu Y, Edwards RH, Pickel VM (1996a) Ultrastructural localization of the vesicular monoamine transporter-2 in midbrain dopaminergic neurons: potential sites for somatodendritic storage and release of dopamine. J Neurosci 16:4135-4145.

Nirenberg MJ, Vaughan RA, Uhl GR, Kuhar MJ, Pickel VM (1996b) The dopamine transporter is localized to dendritic and axonal plasma membranes of nigrostriatal dopaminergic neurons. J Neurosci 16:436-447.

Peter D, Jimenez J, Liu Y, Kim J, Edwards RH (1994) The chromaffin granule and synaptic vesicle amine transporters differ in substrate recognition and sensitivity to inhibitors. J Biol Chem 269:7231-7237.

Peter D, Liu Y, Sternini C, de Giorgio R, Brecha N, Edwards RH (1995) Differential expression of two vesicular monoamine transporters. J Neurosci 15:6179-6188.

Pickel VM, Milner TA (1987) Electron microscopy of central catecholamine systems. In: Psychopharmacology: The Third Generation of Progress (Meltzer HY), pp 49-59. New York: Raven Press.

Pickel VM, Johnson E, Carson M, Chan J (1992) Ultrastructure of spared dopamine terminals in caudate-putamen nuclei of adult rats neonatally treated with intranigral 6-hydroxydopamine. Brain Res Dev Brain Res 70:75-86.

Pickel VM, Sesack SR (1995) Electron microscopy of central dopamine systems. In: Psychopharmacology: The Fourth Generation of Progress (Bloom FE, Kupfer DJ), pp 257-268. New York: Raven Press, Ltd.

Revay R, Vaughan R, Grant S, Kuhar MJ (1996) Dopamine transporter immunohistochemistry in median eminence, amygdala, and other areas. Synapse 22:93-99.

Ritz MC, Lamb RJ, Goldberg SR, Kuhar MJ (1987) Cocaine receptors on dopamine transporters are related to self-administration of cocaine. Science 237:1219-1223.

Roth RH, Elsworth JD (1995) Biochemical pharmacology of midbrain dopamine neurons. In: Psychopharmacology: The Fourth Generation of Progress (Bloom FE, Kupfer DJ), pp 227-243. New York: Raven Press, Ltd.

Rudnick G, Clark J (1993) From synapse to vesicle: the reuptake and storage of biogenic amine neurotransmitters. Biochim Biophys Acta 1144:249-263.

Schmitt FO (1984) Molecular regulators of brain function: a new view. Neuroscience 13:991-1001.

Schuldiner S (1994) A molecular glimpse of vesicular monoamine transporters. J Neurochem 62:2067-2078.

Schuldiner S, Shirvan A, Stern BY, Steiner MS, Yelin R, Laskar O (1994) From bacterial antibiotic resistance to neurotransmitter uptake. A common theme of cell survival. Ann N Y Acad Sci 733:174-184.

Shimada S, Kitayama S, Lin CL, Patel A, Nanthakumar E, Gregor P, Kuhar M, Uhl G (1991) Cloning and expression of a cocaine-sensitive dopamine transporter complementary DNA [published erratum appears in Science 1992 Mar 6; 255(5049):1195]. Science 254:576-578.

Shimada S, Kitayama S, Walther D, Uhl G (1992) Dopamine transporter mRNA: dense expression in ventral midbrain neurons. Mol Brain Res 13:359-362.

Smith AD, Justice JB (1994) The effect of inhibition of synthesis, release, metabolism and uptake on the microdialysis extraction fraction of dopamine. J Neurosci Methods 54:75-82.

Sotelo C (1971) The fine structural localization of norepinephrine-^{3}H in the substantia nigra of the rat. An autoradiographic study. J Ultrastruct Res 36:824-841.

Stamford JA, Kruk ZL, Palij P, Millar J (1988) Diffusion and uptake of dopamine in rat caudate and nucleus accumbens compared using fast cyclic voltammetry. Brain Res 448:381-385.

Stern-Bach Y, Keen J, Bejerano M, Steiner-Mordoch S, Wallach M, Findlay J, Schuldiner S (1992) Homology of a vesicular amine transporter to a gene conferring resistance to 1-methyl-4-phenylpyridinium. Proc Natl Acad Sci, USA 89:9730-9733.

Sulzer D, Maidment NT, Rayport S (1993) Amphetamine and other weak bases act to promote reverse transport of dopamine in ventral midbrain neurons. J Neurochem 60:527-535.

Tennyson VM, Heikkila R, Mytilineou C, Côté L, Cohen G (1974) 5-Hydroxydopamine "tagged" neuronal boutons in rabbit neostriatum: interrelationship between vesicles and axonal membrane. Brain Res 82:341-348.

Tennyson VM, Mytilineou C, Heikkila R, Barrett RE, Cohen G, Côté L, Duffy PE, Marco L (1975) Dopamine-containing neurons of the substantia nigra and their terminals in the neostriatum. UCLA Forum in Medical Sciences 227-264.

Uhl GR, Hartig PR (1992) Transporter explosion: update on uptake. Trends Pharmacol 13:421-425.

Vander Borght TM, Sima AAF, Kilbourn MR, Desmond TJ, Kuhl DE, Frey KA (1995) [3H]Methoxytetrabenazine: a high specific activity ligand for estimating monoaminergic neuronal integrity. Neuroscience 68:955-962.

Wassef M, Berod A, Sotelo C (1981) Dopaminergic dendrites in pars reticulata of substantia nigra and their striatal input. Combined immunocytochemical localization of tyrosine hydroxylase and anterograde degeneration. Neuroscience 6:2125-2139.

Wilson CJ, Groves PM, Fifkova E (1977) Monoaminergic synapses, including dendro-dendritic synapses in the rat substantia nigra. Exp Brain Res 30:161-174.

Yelin R, Schuldiner S (1995) The pharmacological profile of the vesicular monoamine transporter resembles that of multidrug transporters. Febs Lett 377:201-207.

Neurotransmitter Transporters: Is Glycosylation Necessary for Function?

Amrat P. Patel

National Institutes of Health, National Institute on Drug Abuse, Intramural Research Program, Neuroscience Branch, Molecular Pharmacology Section, Baltimore, MD 21224.
(Present address: U.S. Food and Drug Administration, Center for Drug Evaluation and Research, Metro Park North 2, HFD-658, Rockville, MD 20855, U.S.A.)

1 Introduction

1.1 Glycosylation of Proteins:

A detailed description of the glycosylation of proteins is beyond the scope of this chapter. A brief outline of the process of glycosylation is described (Fig. 1). Emerging nascent polypeptide in the rough endoplasmic reticulum acquires a core of oligosaccharide from dolichol pyrophosphate catalyzed by glycosyltransferase (Hirschberg and Snider, 1987). This initial core of sugar residues (two N-acetylglucosamine(GlcNAc), nine mannose and three glucose residues) undergoes post-translational processing in endoplasmic reticulum and golgi to yield a final form of the oligosaccharide core. The specificity of sugar residues in the core is due to the presence of specific glycosyltransferases which are cell and tissue specific (Hubbard and Ivat, 1981; Kornfeld and Kornfeld, 1985).

1.2 Glycosylation sequence motif:

The consensus sequence for potential N-glycosylation includes the asparagine residue followed by any amino acid except proline, followed in turn by either serine or threonine (Asn-Xaa-Ser/Thr). Not all potential glycosylation sites are necessarily glycosylated. Furthermore, the glycosylation pattern of a protein may differ when it is expressed in foreign cell systems (see below). Glycosylation is affected by the presence of certain amino acids, disulfide bonds, protein folding regions and proximity to the amino or carboxy terminus (Opdenakker et al., 1993). The O-linked glycosylation sequence motif is not well defined. The hydroxyl group of the serine and threonine can accept GlcNAc as the starting sugar residue in an oligosaccharide. Mucins are example of O-linked glycoproteins, that are highly O-glycosylated.

The two major forms of glycosylations are N-linked and O-linked. A third form is an attachment of a glycosyl-phosphatidylinostiol (GPI) membrane anchor, the C-terminal amino acid residue is attached to a glycan moiety through an ethanolamine phosphate bridge; the glycan is attached to diacylglycerol through an inositol

NATO ASI Series, Vol. H 100
Neurotransmitter Release and Uptake
Edited by Şakire Pöğün

phosphate bridge (Fig.2). The N-linked glycosylation is a covalent modification catalyzed by specific glycosyltransferases at the side chain nitrogen atom of asparagine residues. Similarly, O-linked glycosylation is a covalent oligosaccharide modification of the hydroxyl group of serine, threonine and sometimes hydroxylysine.

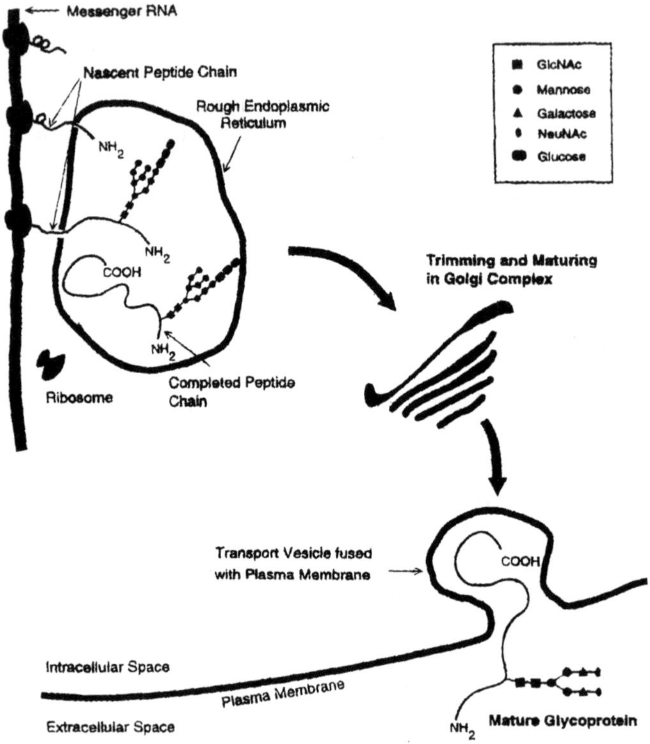

Fig 1. Biosynthesis of N-linked Oligosaccharides

1.3 Neurotransmitter Transporters: Glycosylation Differences

The putative N-glycosylation sites are localized in the large extracellular loop between transmembrane spanning domains 3 and 4. One of the potential N-glycosylation site is conserved in dopamine, norepinephrine and serotonin transporter. Sugars contribute between 15 to 50% of the molecular mass of the transporter and its role in the transporter stability and function is discussed below.

A. DopamineTransporter:

The biochemical properties of the dopamine transporters (DAT) have been studied with photoaffinity radioligands of two classes of compounds, 1,4-diphenyl piperazine (GBR series of compounds) and cocaine congener (RTI-82). Initial estimates of DAT molecular mass from rat and dog brain membranes labeled with GBR based compounds was > 65 kDa (Grigoriadis et al., 1989; Salle et al., 1989; Lew et al.,1991). Later studies labeled a 80 kDa DAT from rat brain membranes with [^{3}H] GBR12783, [125I]DEEP and [^{125}I]RTI-82 (Berger et al.,1991; Patel et al.,1991). The difference in the molecular mass of rat brain membrane DAT determined earlier is possibly due to differences in denaturing gel conditions and protein size markers source and quality. The rat striatal DAT treated with neuraminidase and N-glycanase suggested the presence of sialic acid residues and complex N-linked sugars, respectively (Grigoriadis et al., 1989; Salle et al., 1989; Lew et al., 1992; Patel et al., 1993). The rat brain DAT or rDAT expressed in COS-7 cells (monkey kidney epithelial cells) did not show presence of O-linked glycosylation. The rDAT expressed in COS-7 cells is insensitive to neuraminidase, suggesting either very low levels or absence of sialic acid residues in the sugar core (Patel et al., 1993). The molecular mass of N-deglycosylated DAT in dog, rat, human caudate membranes, and rDAT COS-7 cell membranes has a molecular mass of ≈ 50 kDa (Patel et al., 1993).

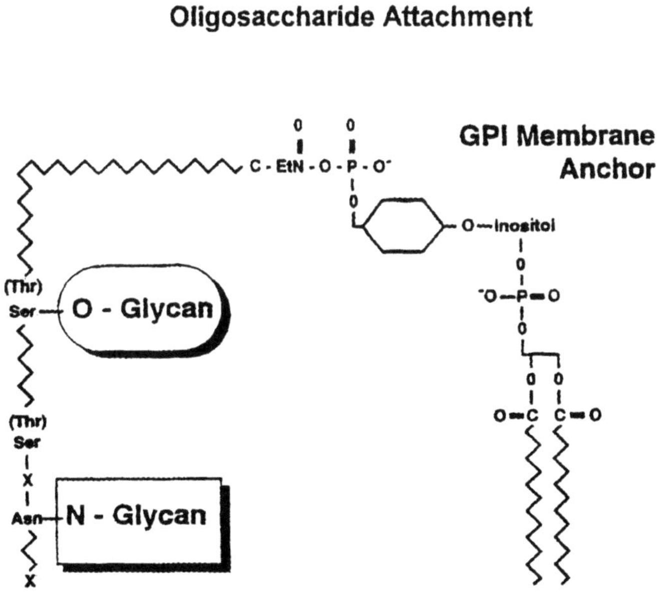

Fig 2. The three main forms of eukaryotic protein glycosylation

The glycosylation of dopamine transporter protein has been shown to be cell, tissue and species-specific and is regulated during development (Patel et al., 1993 and 1994). The molecular mass of rDAT expressed in COS-7 cells and LLC-PK1 cells (porcine kidney epithelial cells), is 110 kDa and 100 kDa, respectively; and human DAT stably expressed in CHO cells(Chinese hamster ovary cells) is 93 kDa (Patel et al., 1993; Patel, 1996). The rat DAT from nucleus accumbens and that in striatum differ in molecular mass (Lew et al., 1991). The difference in molecular mass is due to differences in core sugars and not in the sialic acid content (Lew et al.,1992). In contrast, similar difference in the molecular mass of human DAT from nucleus accumbens and caudate nucleus was not observed when labeled with [^{125}I]DEEP or [^{125}I]RTI-82 (Patel, 1996). Similarly, no apparent difference in the molecular mass was observed in human DAT from caudate nucleus and putamen (Patel et al., unpublished data; Niznik et al., 1991). Differences in glycosylation of DAT between species has also been observed. The molecular mass of photolabeled DAT from dog caudate is 78 kDa, the human caudate nucleus is 62 -74 kDa, and rat striatum is 80 kDa (Niznik et al., 1991; Patel et al., 1993). The rat DAT specific anti-peptide antibodies cross reacts with DAT from different tissues and species (Vaughan et al., 1996). Developmentally regulated glycosylation of rat DAT exists between postnatal days 4, 14 and 60 due to differences in the N-linked core sugars and not from differences in sialic acid content or differences in polypeptide (Patel et al., 1994).

B. GABA Transporter:

The first neurotransmitter transporter protein to be purified to homogeneity was GABA transporter from rat brain. On denaturing polyacrylamide gel, the transporter migrates as a protein of 80 kDa (glycosylated) and following enzymatic N-deglycosylation with a molecular mass of 60 kDa (Radian et al., 1986; Kanner, 1994). Peptide sequence information from the purified protein led to a successful cloning of the rat and human brain GABA (GAT1) transporters with predicted molecular mass of 67 kDa (Guastella et al., 1990; Nelson et al., 1990). The rat GABA transporter expressed in mouse Ltk- cells, L-cells and HeLa cells has a molecular mass of 70 kDa and 60 kDa in cells grown in the presence of tunicamycin, an inhibitor of N-linked glycosylation (Keynan et al., 1992). The GABA transporter function was abolished in mammalian cells expressing GAT1, when cells were grown in the presence of tunicamycin (Keynan et al., 1992).

C. Glutamate Transporter:

Like the GABA transporter, purification of the glutamate transporter to homogeneity from the rat brain has also been achieved. It has been identified as being of glial origin. It is a glycoprotein with sugars representing 14% of the molecular mass. In denaturing polyacrylamide gel, the glycosylated form has a molecular mass of 73 - 80 kDa with a pI of 6.2. The N-deglycosylated form has a molecular mass of 63 kDa (Danbolt et al.,1990; Danbolt et al., 1992; Danbolt, 1994). Conradt et al., (1995) identified two glycosylated glutamate transporters (GLAST-1); with molecular masses of 64 kDa and 70 kDa in the rat brain. These transporters have been separated by lentil and wheat germ agglutinin lectin columns, respectively. This indicates that the transporters are differentially glycosylated. The molecular mass of the rat GLAST-1 expressed in Xenopus oocytes has a molecular mass of 65 kDa which is reduced to 56 kDa following N-deglycosylation. It is suggested that the $\approx$100 kDa form is the homodimer of 65 kDa form (Conradt, et al., 1995).

D. Glycine Transporter:

The glycine transporter has also been purified to homogeneity from pig brain stems.

It has a molecular mass in denaturing polyacrylamide gels of 100 kDa (Nunez and Aragon, 1994). The transporter is not O-glycosylated and may contain low levels of sialic acid (Nunez and Aragon, 1994). Complete N-deglycosylation of the glycine transporter resulted in inhibition of transport activity. Similar to GAT1 expressing mammalian cells grown in the presence of tunicamycin. However, it is not clear whether this reflects failure of the transporter reaching the plasma membrane or rapid turnover of the protein, as shown for the serotonin and human norepinephrine transporters (Tate and Blakely, 1994; Melikian et al., 1994 and 1996; Nguyen and Amara, 1996).

Olivares et al., (1995), on Western blots identified three bands of GLYT1 from COS-7 cells expressing glycine transporter. A broad 80 - 100 kDa fully glycosylated band, a 57 kDa partially glycosylated band and a 47 kDa nonglycosylated band. The GLYT1 mutant, with all four potential N-glycosylation sites conservatively replaced with glutamine, also has a molecular mass of 47 kDa (nonglycosylated form) in denaturing polyacrylamide gel. All four potential N-glycosylation sites are glycosylated in COS-7 cells expressing GLYT1 (Olivares et al.,1995). However, whether all of the GLYT1 potential N-glycosylation sites are utilized in the native state or other cell system should not be assumed. Glycosylation is cell, tissue and species-specific and yields different glycosylation patterns. The differences in glycosylation may well be the result of a variable oligosaccharide core and or the utilization of a variable number of potential glycosylation sites (Jenkins 1995; Kusui et al., 1995).

E. Norepinephrine Transporter:

The structural information of the native norepinephrine transporter has not been available due to lack of suitable tools, such as photoaffinity ligands or specific antibodies. This information is now forthcoming from the use of human norepinephrine transporter antipeptide antibodies. The biochemical properties of hNET was studied in LLC-PK1 cells stably expressing hNET. The anti-hNET antibodies immunoprecipitated [^{35}S]-methionine labeled proteins with molecular masses in denaturing polyacrylamide gels of 54 and 80 kDa. Enzymatic N-deglycosylated hNET has a molecular mass of 46 kDa as does LLC-hNET cells grown in the presence of tunicamycin (Melikian et al., 1994). This suggests that the fully glycosylated form of the hNET transporter expressed in LLC-PK1 cells has a molecular mass of 80 kDa. The partially glycosylated form has a molecular mass of 54 kDa and the nonglycosylated form has a molecular mass of 46 kDa. However, HeLa cells expressing hNET, a 54 kDa band and a 46 kDa band was detected after enzymatic N-deglycosylation. This would suggest that the 54 kDa band in HeLa cells is the fully glycosylated form of hNET (Melikian et al. 1994).

The hNET transporter expressed in COS cell has molecular mass of 54 kDa and 46 kDa, glycosylated and non-glycosylated forms respectively, similar to that described in HeLa cells (Melikian, et al., 1996). Bruss et al.,(1995) also report similar molecular mass for the hNET transporter expressed in COS cells, 50 kDa and 58 kDa. However, it is not clear whether these represent the partially glycosylated and the fully glycosylated forms, or the nonglycosylated and the partially glycosylated forms of the hNET. This is in contrast to the differences in glycosylation of the rat dopamine transporters expressed in LLC-PK1 and COS cells.

However, Nguyen and Amara (1996) have shown presence of 60 kDa band and 90 kDa band by immunoprecipitation with hNET-antisera from hNET expressed in HeLa and Madin-Darby canine kidney (MDCK) cells. Interestingly, in these cell lines hNET appears to be glycosylated to similar extent of glycosylation. Although, the 60 kDa band is in agreement with that seen previously in HeLa and COS cells, the 90 kDa band had not been identified previously. It would appear that the fully glycosylated form of hNET has a molecular mass of 90 kDa, partially glycosylated form 60 - 54 kDa and nonglycosylated form 46 - 50 kDa. It is not clear why the 90 kDa band of hNET was not detected previously in HeLa or COS cells; differences in antisera sensitivity may be responsible. Like other transporter proteins, hNET is also cell-specifically glycosylated.

F. Serotonin Transporter:

The serotonin transporter labeled with different photoaffinity ligands has shown varied molecular mass on denaturing polyacrylamide gels. [^{3}H]2-nitroimipramine specifically photolabeled 30 - 35 kDa band in the human platelet and rat brain membranes (Wengole et al., 1985). Paroxetine and cyanoimipramine specifically photolabeled a 68 kDa glycoprotein in human platelet membranes. The smaller molecular mass of protein labeled with nitroimipramine could be the tricyclic binding domain of the serotonin transporter or a degradation product. There has been a great effort to purify the serotonin transporter. The serotonin transporter from human platelet membranes has been purified to homogeneity with a molecular mass in denaturing gels of 68 kDa (Launay et al., 1992). Recently, a 67 kDa serotonin transporter protein was purified from the human striatal brain to homogeneity. The purified transporter binds to tricyclic antidepressants, selective serotonin reuptake inhibitors and 5-HT, which suggests overlapping binding sites on a single protein (Rotondo et al., 1996).

The cell-specific differential glycosylation of the serotonin transporters from rat tissues, 76 kDa in brain, 80 kDa in lungs, and 94 kDa is reported (Qian et al., 1995). These results also show species, rat and human, related differences in post-translational processing of the serotonin transporter. HeLa cells, transfected with rat and human serotonin cDNAs, expresses 61 kDa glycosylated and a 56 kDa deglycosylated serotonin transporter (Qian et al., 1995). In Sf9 insect cells, the rat serotonin transporter has a molecular mass of 54 kDa (Tate and Blakely, 1994). The serotonin transporter, like the dopamine transporter glycosylation is cell, tissue and species-specific.

1.4 Glycosylation and Transporter Function:

In general, the glycosylation of proteins is cell, tissue and species-specific and transporter proteins are no exception. The role of sugars associated with proteins is varied, for example proper folding, stability and targeting/trafficking of the protein to membranes. In some instances, the absence of glycosylation results in compromised function, reduced stability and lower expression at cell surface of membrane associated proteins.

For the neurotransmitter transporters, sugars have been implicated to play a role in transporter function. The possible role of sialic acid residues in dopamine uptake was first shown in rat synaptosomes. Enzymatic removal of sialic acids resulted in the decrease in V_{max} without affecting K_m for [^{3}H] dopamine uptake (Zaleska and Erecinska, 1988). Human platelets treated with neuraminidase to release sialic acid residues decreased the binding of [^{3}H]5HT (Launay et al., 1992). Whereas, the enzymatic incorporation of N-acetylneuraminic in human platelets, in vitro, accelerated the uptake of [^{3}H]5HT compared to control platelets (Szabados et al., 1975). These studies show that the uptake and or binding of substrates to neurotransmitter transporters is modulated by the addition or removal of sialic acid residues. The importance of sialic acid residues may explain the ten fold difference in K_m for [^{3}H]DA uptake between rat synaptosomal (0.1μM) and dopamine transporters expressed in various mammalian cell lines (1μM). Sialic acid was not detected in rat dopamine transporter expressed in COS-7 cells; and this may account for the differences in K_m of dopamine uptake in COS-7 cells and other mammalian cells expressing the dopamine transporter to that observed in rat synaptosomes (Patel et al., 1993). Interestingly, human DAT expressed in baculovirus infected Sf9 insect cells has a 10 fold higher affinity for uptake of amphetamine and dopamine compared to hDAT transiently expressed in COS-7 cells (McConkey, et al., 1995). The K_m for dopamine uptake in striatal synaptosomes is similar to that observed with hDAT expressed in Sf9 cells, 100 nM and 300 nM, respectively (Lee, et al., 1995; McConkey, et al., 1995). In contrast, desialation of the rat striatal membranes did not affect the binding of [^{3}H]CFT to the dopamine transporter (Lew et al., 1992) or the functioning of the glycine and serotonin transporters (Nunez and Aragon, 1994; Melikian et al., 1994).

The difference in glycosylation, however, has not appreciably affected the functioning of the serotonin transporter protein, expressed in Sf9 insect cells compared to that expressed in rat brain (Tate and Blakely, 1994). In Sf9 cells, neither uptake of [^{3}H]5HT nor the binding of [^{125}I]RTI-55 to the serotonin transporters required complete glycosylation (Tate and Blakely, 1994). Similarly, the functional properties of hNET or GLYT1 transporters were not compromised by varying degrees of glycosylation (Melikian et al., 1994 and 1996; Nguyen and Amara, 1996; Nunez and Aragon, 1994; Conradt et al., 1995). The apparent decrease in the uptake of substrates can largely be attributed to a lower number of transporters at the cell surface. The affinity of substrate and inhibitors of norepinephrine transporters were not altered in partially- or non-glycosylated transporters (Melikian et al., 1996; Nguyen and Amara, 1996). Glycosylation does not play an important role in glutamate transporter function (B. Kanner, personal communication). The GLUT2 glucose transporter lacks N-glycosylation motif and glycosidase enzymes for N-and O-linked sugars, and sialic acid failed to affect the mobility of GLUT2 polypeptide in denaturing polyacrylamide gels (Brant et al., 1992). Thus, it appears that in nature, GLUT2 transporter functions normally in the absence of glycosylation.

The stability and trafficking/targeting of the glycine, norepinephrine and serotonin transporters are greatly dependent on the presence of sugars (Nunez and Aragon, 1994; Conradt et al., 1995; Melikian et al., 1994 and 1996; Nguyen and Amara, 1996; Tate and Blakely, 1994). Glycosylation is important for the GLUT1 glucose transporter protein expressed in CHO cells for intracellular targeting and protein stability (Asano et al.,1993). Similarly, N-glycosylation of the human erythrocyte band 3 anion transport protein is not necessary for anion transport function, but plays a role in correct folding of the protein (Groves and Tanner, 1994). The turnover of the unglycosylated proteins is faster than that of the glycosylated proteins. The faster turnover of the transporter protein may be due to the absence of oligosaccharide; N- and O-linked oligosaccharide structures may hinder the access of cytosolic proteases and thereby slowing the turnover of proteins.

Glycosylation of mammalian proteins in Sf9 cells results in limited or incomplete glycosylated protein. The murine gastrin-releasing peptide receptor (mGRP-R) expressed in Sf9 cells is smaller by 62% compared to that in native tissues (Kusui, et al., 1995). The nature of glycosylation of "foreign" proteins expressed in "host" cells differ in oligosaccharide composition and in utilization of potential glycosylation sites (Jenkins, 1995; Kusui, et al., 1995).

Thus, caution is advised in generalizing information concerning glycosylation of the transporter proteins obtained in host cell systems to that in native membranes. The glycosylation information is best obtained from native membranes or purified proteins.

2 Concluding Comments:

The glycosylation of the neurotransmitter transporters are cell, tissue and species-specific and vary during development and aging. Recent evidence shows that the role of oligosaccharides is important for the stability and targeting of the transporter to the plasma membrane and less for transporter function. Although, expression of the neurotransmitter transporters in host cell lines are useful models for studying transporter function, caution is advised in extrapolating information concerning the utilization of sites, and the extent and nature of glycosylation to the native transporter protein. The role of glycosylation and certain functional properties of proteins can be generalized, however, it is best evaluated on an individual basis.

Acknoledgements

The author is appreciative of Arlene A. Patel for her assistance in the preparation of this chapter and thanks Oxford GlycoSystem for providing Figs. 1 and 2.

References

Asano, T., Takata, K., Katagiri, H, Ishihara, H., Inukai, K., Anai, M., Hirano, H., Yazaki, Y., and Oka, Y. (1993). The role of N-glycosylation in the targeting and stability of GLUT1 glucose transporter. Febs Lett. 324:258-261.

Berger, P., Martenson, R., Laing, P., Thurcauf, A., DeCosta, B., Rice, K. C., and Paul, S. M. (1991). Photoaffinity labeling of the dopamine reuptake carrier protein with 3-azido[3H]GBR-12935. Mol. Pharmacol., 39: 429-435.

Brant, A.M., Gibbs, M.E., and Gould, G.W. (1992) Examination of the glycosidation state of five members of the human facilitative glucose transporter family. Biochem Soc. Trans. 20: 235S.

Bruss, M., Hammermann, R., Brimijoin, S., and Bonisch, H. (1995). Antipeptide antibodies confirm the topology of the human norepinephrine transporter. J. Biol. Chem. 270: 9197-9201.

Conradt, M., Storck, T., and Stoffel, W. (1995). Localization of N-glycosylation sites and functional role of the carbohydrate units of GLAST-1, a cloned rat brain L-glutamate/L-aspartate transporter. Eur. J. Biochem. 229: 682-687.

Danbolt, N.C., Pines, G., and Kanner, B.I. (1990). Purification and reconstitution of the sodium- and potassium-coupled glutamate transport glycoprotein from rat brain. Biochemistry 29: 6734-6740.

Danbolt, N.C., Storm-Mathisen, J., and Kanner, B.I. (1992). A [Na+- K+]coupled L-glutamate transporter purified from rat brain is located in glial cell processes. Neuroscience 51: 295-310.

Danbolt, N.C. (1994). The high affinity uptake system for excitatory amino acids in the brain. Prog. Neurobiol. 44: 377-396.

Grigoriadis, D. E., Wilson, A. A., Lew, R., Sharkey, J. S., and Kuhar, M. J. (1989). Dopamine transporter sites selectively labeled by a novel photoaffinity probe: [125I]DEEP. J. Neurosci. 9: 2664-2670.

Groves, J.D. and Tanner, M.J. (1994). Role of N-glycosylation in the expression of human band 3-mediated anion transport. Mol. Membr. Biol. 11: 31-38.

Guastella, J.., Nelson, N., Nelson, H., Czyzyk, L., Keynan, S., Miedel, M.C., Davidson, N.C., Lester, H.A., and Kanner, B.I. (1990). Cloning and expression of a rat brain GABA transporter. Science 249: 1303-1306.

Hirschberg, C.B. and Snider, M.D. (1987) Topography of glycosylation in the rough endoplasmic reticulum and golgi apparatus. Ann. Rev. Biochem. 56: 63-87.

Hubbard, S.C. and Ivatt, R.J. (1981) Synthesis and processing of asparagine-linked oligosaccharides. Ann. Rev. Biochem. 50: 555-83.

Jenkins, N. (1995). Monitoring and control of recombinant glycoprotein heterogeneity in animal cell cultures. Biochem. Soc. Trans., 23: 171-175.

Kanner, B.I. (1994). Sodium-coupled neurotransmitter transport: structure, function and regulation. J. Exp. Biol. 196: 237-249.

Keynan, S., Suh, Y.J., Kanner, B.I., and Rudnick, G. (1992). Expression of a cloned gamma-aminobutyric acid transporter in mammalian cells. Biochem. 31: 1974-1979.

Kornfeld, R. and Kornfeld, S. (1985) Assembly of asparagine-linked oligosaccharides. Ann. Rev. Biochem. 54: 631-664.

Kusui, T., Hellmich, M.R., Wang, L-H, Evans, R.L., Benya, R.V., Battey, J.F., and Jensen, R.T. (1995). Characterization of gastrin-releasing peptide receptor expressed in Sf9 insect cells by Baculovirus. Biochem. 34: 8061-8075.

Launay, J-M., Geoffroy, C., Mutel, V., Buckel, M., Cesura, A., Alouf, J.E., and Da Prada, M. (1992). One-step purification of the serotonin transporter located at the human platelet plasma membrane. J. Biol. Chem. 267: 11344-11351.

Lee, F.J.S., Pristupa, Z.B., and Niznik, H.B. (1995). The human dopamine transporter: Functional consequences of carboxyl tail truncation/substitution. Soc. Neurosci. 21:374.

Lew, R., Vaughan, R., Simantov, R., Wilson, A., and Kuhar, M.J. (1991). Dopamine transporters in the nucleus accumbens and the striatum have different apparent molecular weights. Synapse, 8: 152-153.

Lew, R., Patel, A., Vaughan, R. A., Wilson, A., and Kuhar, M.J. (1992). Microheterogeneity of dopamine transporters in rat striatum and nucleus accumbens. Brain Res. 584: 266-271.

McConkey, F., Pristupa, Z.B., and Niznik, H.B. (1995). Pharmacological characterization and functional regulation of the human dopamine transporter in Sf9 cells. Soc. Neurosci. 21: 375.

Melikian, H.E., McDonald, J.K., Gu, H., Rudnick, G., Moore, K.R., and Blakely, R.D. (1994). Human norepinephrine transporter, biosynthetic studies using a site-directed polyclonal antibody. J. Biol. Chem. 269: 12290-12297.

Melikian, H.E., Ramamoorthy, S., Tate, C.G., and Blakely, R.D. (1996). Inability to N- glycosylate the human norepinephrine transporter reduces protein stability, surface trafficking, and transport activity but not ligand recognition. Mol. Pharmacol., in press.

Nelson, H., Mandiyan, S., and Nelson, N. (1990). Cloning of the human brain GABA transporter. FEBS Lett. 269: 181-184.

Nguyen, T and Amara, S.G. (1996). N-Linked oligosaccharides are required for cell surface expression of the norepinephrine transporter but do not influence substrate or inhibitor recognition. J. Neurochem., 67: 645-655.

Niznik, H. B., Fogel, E. F., Fasso, F. F., and Seeman, P. (1991). The dopamine transporter is absent in Parkinsonian putamen and reduced in the caudate nucleus. J. neurochem., 56: 192-198.

Nunez, E. and Aragon, C. (1994). Structural analysis and functional role of the carbohydrate component of glycine transporter. J. Biol. Chem. 269: 16920- 16924.

Olivares, L., Aragon, C., Gimenez, C., and Zafra, F. (1995). The role of N-glycosylation in the targeting and activity of the GLYT1 glycine transporter. J. Biol. Chem. 270: 9437-9442.

Opdenakker, G., Rudd, P.M., Ponting, C.P., and Dwek, R.A. (1993). Concepts and principles of glycobiology. FASEB J. 7: 1330-1337.

Patel, A., Boja, J. W., Lever, J., Lew, R., Simantov, R., Carroll, F. I., Lewin, A. H., Phillip, A., Gao, Y., and Kuhar, M. J. (1991). A cocaine analog and a GBR analog label the same protein in rat striatal membranes. Brain Res., 576: 173-174.

Patel, A., Uhl, G., and Kuhar, M.J. (1993). Species differences in dopamine transporters: Postmortem changes and glycosylation differences. J. Neurochem. 61: 496-500.

Patel, A. P., Cerruti, C., Vaughan, R. A., and Kuhar, M.J. (1994). Developmentally regulated glycosylation of dopamine transporter. Developmental Brain Res., 83: 53-58.

Patel, A.P. (1996). Neurotransmitter transporter proteins: Posttranslational modifications. In Neurotransmitter Transporters: Structure, Function, and Regulation (Ed. M.E.A. Reith), Humana Press Inc., NJ. pp 241-262.

Qian, Y., Melikian, H.E., Rye, D.B., Levey, A.I., and Blakely, R.D. (1995). Identification and characterization of antidepressant-sensitive serotonin transporter proteins using site-specific antibodies. J. Neurosci. 15: 1261-1274.

Radian, R., Bendahan, A., and Kanner, B.I. (1986). Purification and identification of the functional sodium- and chloride-coupled g-aminobutyric acid transport glycoprotein from rat brain. J. Biol. Chem. 261: 15437-15441.

Rotondo, A., Giannaccini, G., Betti, L., Chiellini, G., Marazziti, D., Martin, C., Lucacchini, A., and Cassano, G.B. The serotonin transporter from human brain: Purification and partial characterization. Neurochem. Int. 28: 299-307, 1996.

Sallee, F. R., Fogel, E. L., Schwartz, E., Choi, S. M., Curran, D. P., and Niznik, H. B. (1989). Photoaffinity labeling of the mammalian dopamine transporter. FEBS Lett., 256: 219-224.

Szabados, L., Mester, L., Michal, F., and Born, G.V.R. (1975). Accelerated uptake of 5- Hydroxytryptamine by human blood platelets enriched in a sialic acid. Biochem. J. 148: 335-336.

Tate, C.G. and Blakely, R.D. (1994). The effect of N-linked glycosylation on activity of the Na+ and Cl- dependent serotonin transporter expressed using recombinant baculovirus in insect cells. J. Biol. Chem. 269: 26303-26310.

Vaughan, R.A., Brown, V.L., McCoy, M.T., and Kuhar, M.J. (1996). Species- and Brain Region-Specific Dopamine Transporters:Immunological and Glycosylation Chacteristics. J Neurochem, 66, 2146-2152.

Wennogle, L.P., Ashton, R.A., Schuster, D.I., Murphy, R.B., and Meyerson, L.R. (1985). 2-Nitroimipramine: a photoaffinity probe for the serotonin uptake/tricyclic binding site complex. EMBO J. 4: 971-977.

Zaleska, M. M. and Erecinska, M. (1987). Involvement of sialic acid in high-affinity uptake of dopamine by synaptosomes from rat brain. Neurosci. Lett., 82: 107-112.

Dopamine Transporter: Function and Imaging

Michael J. Kuhar, Ph.D.

Yerkes Primate Research Center, Emory University, Atlanta, GA 30322, U.S.A.

1 Introduction

The dopamine transporter was discovered as a unique transporter entity in the late 1960's (Glowinski and Iversen, 1966; Snyder and Coyle, 1969). The next two decades were spent largely dealing with the pharmacology of the transporter, but progress was made regarding its molecular nature and in the development of binding ligands. Late in 1980, it became clear that the transporter was an important target or "receptor" for cocaine (Ritz et al., 1987; Bergman et al., 1989) and irreversible photolabeling probes began a productive study of its molecular nature (Grigoriadis et al., 1989).

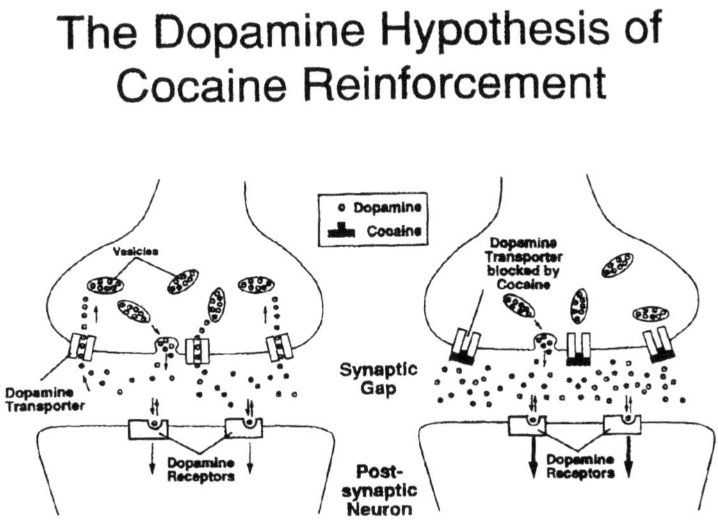

Fig 1. A dopaminergic synapse under normal conditions (left) and when cocaine is present (right). Released dopamine is removed from the synaptic cleft by reuptake via the dopamine transporter. Cocaine blocks this process. It is believed that the latter action in limbic areas somehow causes the reinforcing actions of cocaine. From Kuhar et al., 1991.

The transporter is an important entity of the dopaminergic nerve terminal. It is a major mechanism of inactivation of released dopamine, as it removes dopamine from the synaptic cleft and terminates the process of neurotransmission (Fig. 1). Drugs

NATO ASI Series, Vol. H 100
Neurotransmitter Release and Uptake
Edited by Şakire Pöğün

known to inhibit this reuptake or termination are known to have profound effects on behavior. The dopamine transporter is also believed to mediate neurotoxicity by such agents as 6-hydroxydopamine or MPTP. Drugs which block the transporter are known to block the toxicity induced by these agents. It has been proposed that the transporter is somehow involved in mediating the effects of environmental neurotoxins which ultimately result in Parkinson's Disease in certain exposed and/or vulnerable individuals. The transporter, as mentioned above, is also a target for psychostimulant drugs including cocaine and amphetamine, and is therefore a topic of great current interest (Fig. 1). It has been postulated that the transporter is a key mechanism in the locomotor, rewarding and reinforcing properties of psychostimulant drugs. An antidepressant, bupropion, has potent inhibitory effects at the dopamine transporter, leading to the suggestion that dopamine is involved in the antidepressant properties of this drug. On the other hand, a major metabolite of this compound is an inhibitor of norepinephrine transport. Given this background, it is not surprising that the dopamine transporter has been the focus of intense and recent studies. The dopamine transporter has also benefitted from recent progress in transporter studies in general (Amara and Kuhar, 1993; Uhl and Hartig, 1992; Giros and Caron, 1993).

2 Biochemical Characterization

Significant developments in the biochemical characterization of the dopamine transporter have been the design of radiolabeled photoaffinity ligands, the cloning of a cDNA for the transporter, and the development of specific high affinity antibodies. These successes in cloning are clearly one of the most important developments in transporter research.

Table 1. Characteristics of the Dopamine Transporter

Na, Cl dependent
60 - 80K Da molecular weight
N glycosylated
Carbohydrate contains sialic acids
Regional differences in M.W.
Ontogenetic changes in M.W.

Table 1 lists the characteristics of the dopamine transporter. In studies involving the photoaffinity labeling of the transporter protein, it became clear that the molecular weight of the protein was in the range of 60,000 to 80,000 Da (Grigoriadis et al., 1989; Sallee et al., 1989). Treatment with enzymes known to digest carbohydrates, revealed that the transporter is a glycoprotein containing N-linked carbohydrates which in turn contain sialic acid residues. The molecular weight of the transporter protein appears to be slightly different in different brain regions, with the nucleus accumbens having a slightly higher molecular weight than the transporter in the striatum (Lew et al., 1991). This appears to be due mainly to differences in the attached carbohydrates, rather than in differences in the polypeptide itself. Also, the apparent molecular weight of the transporter increases over ontogeny and aging, at least in the rat, and again this appears to be due mainly to changes in attached carbohydrates (Patel et al., 1994). All of these characteristics, which were deduced in biochemical studies were confirmed and their existence supported by the results of cDNA cloning studies. Studies of the turnover and half-life of the dopamine transporter suggest that it is a long-lived protein similar to receptors. For example, determination of the half-life utilizing an irreversible binding probe revealed a return of binding over with a half-life of about six days (Fleckenstein et al., 1995).

3 Molecular Cloning

In 1991, several groups, including our own at the National Institute on Drug Abuse, cloned cDNA's for the dopamine transporter (Kilty et al., 1991; Shimada et al., 1991; Giros et al., 1991; Usdin et al., 1991). The predicted polypeptide had a molecular weight of about 70,000 Da and 12 transmembrane regions with intracellular N-terminal and C-terminal tails. A large extracellular loop between transmembrane regions 3 and 4 possessed several consensus glycosylation sites. *In situ* hybridization studies using probes based on the cloned cDNA revealed intense and selective hybridization of the probe over known dopaminergic cell bodies. Also, production of antibodies based on the predicted amino acid sequence yielded antibodies and immunostaining of known and established dopaminergic cells (Freed et al., 1995; Revay et al., 1996). These findings indicated that the cDNA cloned from the libraries corresponded to mRNA's produced by the dopaminergic cells and to transporter proteins expressed by these cells. A schematic of the cloned transporter protein is shown in Fig. 2.

Studies involving the construction and study of both chimeras and specifically mutated transporter proteins suggest that key residues in the transporter structure are involved in the binding of dopamine and inhibitors to the transporter protein. These include acidic amino acid residues in transmembrane region 1 and serine residues in transmembrane regions 7 (Kitayama et al., 1992). Identification of the overall structure and a proposed 3-dimensional structure based on reasonable constraints involving distances based on this model have produced a reasonable model of the transporter protein. While progress up to this point in time has been very rapid, the next challenge of discovering the mechanism by which the substrate is transported and moved across the membrane is a difficult one.

The proposed structure of the protein with 12 transmembrane regions will require confirmation. However, a number of studies strongly support the general proposed structure. Electron microscopic immunohistochemistry using immunogold labeling of the transporter, where the antibody was directed against an amino terminal peptide predicted to be inside the cell, revealed intracellular staining within neurons of the brain (Nirenberg et al., 1996). Detailed peptide mapping studies are consistent with the large extracellular loop containing glycosylation sites. The success of the immunohistochemical studies in determining topology is striking and is likely to be continued and explored further.

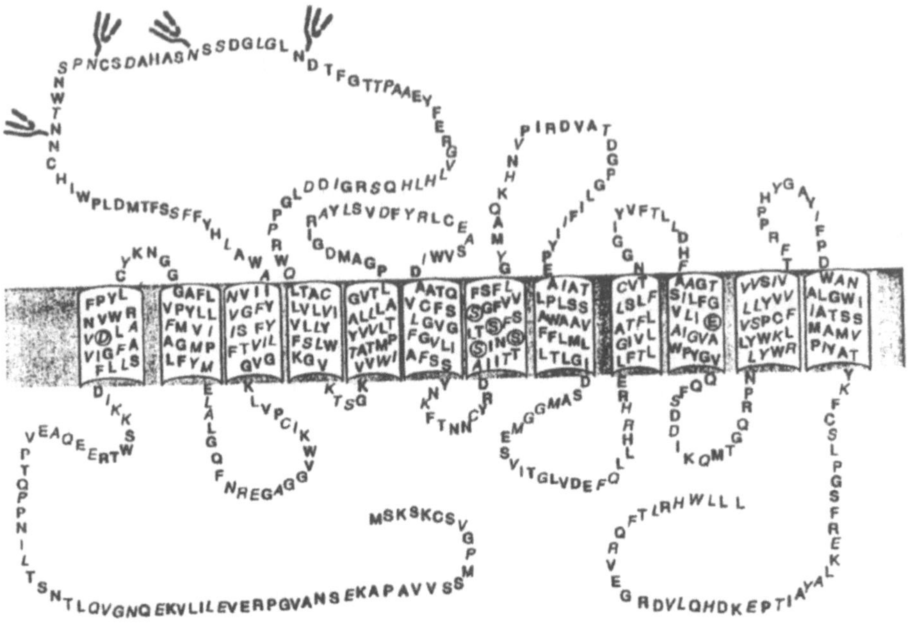

Fig 2. The proposed dopamine transporter. From Shimada et al., 1991.

4 Regulation of the Transporter

Cloning of the transporter revealed several consensus phosphorylation sites in the predicted polypeptide. Some studies involving expression of the cloned transporter suggest that protein kinase C phosphorylates the transporter such that the activity of the transporter is reduced (Kitayama et al., 1994). Determination of whether or not the transporter in the brain is phosphorylated and regulated by phosphorylation remains to be determined. Preliminary results from our group suggest that the native transporter is indeed phosphorylated. Regulation by phosphorylation and its consequences is a frontier in this area.

5 Imaging the Dopamine Transporter

Imaging the transporter can be carried out at both the cellular and clinical levels. Imaging at the cellular level involves immunocytochemistry and the use of antibodies to identify antigenic sites within cells. There have been clear advances in the production and use of such antibodies (Freed et al., 1995). Neuroanatomical mapping using *in situ* hybridization probes is also a means of imaging the transporter at the cellular level. Numerous laboratories have carried out successful studies by this approach. A surprising finding with immunohistochemistry is that transporters are not found or are present in reduced amounts at synaptic thickenings

(Nirenberg et al., 1996). If this proves to be the case, then some thoughts about transporter function may change. If there is little or no transporter at synapses, then transporter may capture dopamine that has diffused to the sides of the synapse. In that case, both diffusion *and* reuptake would be mechanisms of inactivation of dopamine. Also, lack of transporters at synaptic thickenings would eliminate the problem of keeping plasmallemal transporter out of vesicles which are presumably formed by membrane recycling at synaptic thickenings.

Yet another approach to imaging the transporter, an approach of clinical significance, is the use of external imaging techniques such as PET or SPECT. The imaging of radiolabeled probes inside the brain by external cameras requires the incorporation of specific radioactive atoms into transporter ligands. The technical aspects of this approach, and the atoms and their relative desirability and advantages have been explained in numerous publications. The procedure essentially relies upon the high binding affinity of the transporter for specific compounds or ligands following intravenous injection. Following injection, these compounds move into the brain where they bind to the transporter. This binding to the transporter, with its subsequent slow off rate, combined with the loss of transporter from the blood and extracellular space due to excretion, results in a preferential localization of binding ligands to the transporter protein. Subsequent imaging of the location and quantity of radioactivity inside the brain results in a measurable signal reflecting specific binding of ligand to the transporter protein (Boja et al., 1994).

A number of ligands have proven useful for such imaging. Some of the earliest ligands used include nomifensine and cocaine. These compounds, even though they have a low affinity for the transporter, resulted in successful imaging studies (Boja et al., 1994). However, most studies carried out today utilize higher affinity ligands which provide higher specific to nonspecific binding ratios and better signal to noise ratios, affording a practical advantage in clinical imaging studies. Following the identification of the phenyltropane analogs of cocaine as ligands with advantageous properties (Scheffel et al., 1989), there has been the development of many phenyltropane analogs as useful ligands for the transporter. Other chemical classes have been used as ligands as well (Boja et al., 1994). In any case, the field of dopamine transporter imaging has thrived over the past several years, and is reaching the point where it is reliable and predictable enough to be applied to routine clinical imaging, such as in the diagnosis of Parkinson's Disease, for example.

Because the dopamine transporter is a unique constituent of dopaminergic neurons and nerve terminals, it is a suitable marker for the presence of these nerve terminals. Hence, as mentioned above, brain imaging can be used to determine the relative presence of dopaminergic nerve terminals. In Parkinson's Disease, where there is a large loss of neurons, and in experimental models of this disease, it has been shown that dopamine transporter imaging is a useful predictor of the presence or absence of these nerve terminals (See Fig. 3). The use of transporter imaging to show that dopamine systems appear to be reduced in Lesch-Nyhan's Syndrome (Wong et al., 1996) shows the usefulness of such approaches for determining the involvement of specific neurons in these diseases.

Aside from determining the relative presence or absence of dopaminergic nerve terminals, brain imaging techniques have additional uses. For example, it has been proposed to use low affinity ligands in conjunction with competition for the naturally occurring endogenous ligand, dopamine, to develop a procedure for identifying the concentration of dopamine in the synaptic cleft. While it may be impractical to identify specific concentrations, relative changes in the quantities of released dopamine would be discernable. Preliminary studies suggest that this is feasible. Yet another potential use of imaging, again, perhaps using low affinity ligands, would be the determination of doses of therapeutic drugs that occupy significant doses of the transporter. By carrying out *in vivo* competition studies, it would be possible to determine which doses of new therapeutic drugs produce substantial binding of such drugs at the transporter site. Behavioral studies suggest, for example, that large numbers of the transporters need to be occupied before significant behaviors can be observed (Cline et al., 1992).

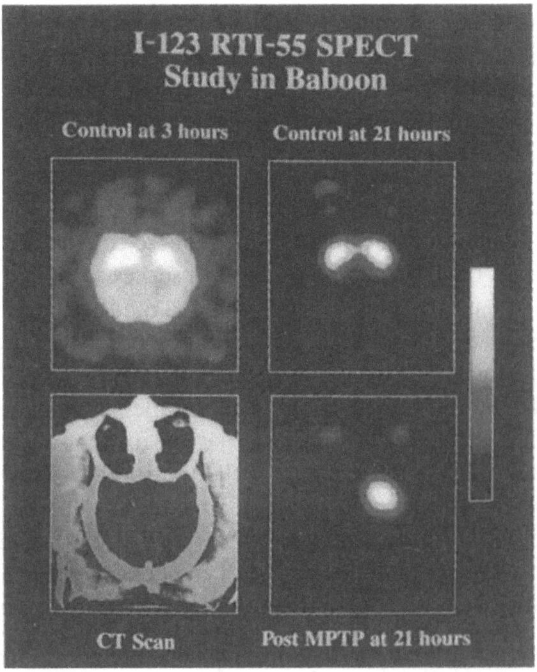

Fig 3. Imaging the Dopamine Transporter. [^{123}I]-RTI-55 labels the dopamine transporter *in vivo* and allows external imaging by single photon emission computed tomography (SPECT). The two areas of accumulation of [^{123}I]-RTI-55 in the top two frames are the basal ganglia. The CT scan in the lower left is used to identify the anatomical structures associated with the SPECT images. In the bottom right, MPTP, a neurotoxin that destroys dopaminergic nerve terminals was injected unilaterally to destroy dopaminergic neurons on the left side; there is a consequent loss of transporter binding on the left. From Shaya et al., 1992.

At this stage, it is safe to say that imaging the dopamine transporter can be reliably carried out. The frontier that remains is the development of a procedure will that will allow absolute quantification of the number of transporter binding sites in the brain.

6 Conclusion

It is clear that tremendous progress has been made in understanding the dopamine transporter over the past five to ten years. Yet, important frontiers and questions remain. While we understand the biochemical nature of the transporter due to studies with probes and cloning, we still do not have a fundamental understanding of how the transporter operates. While we are beginning to understand its three dimensional structure, we have no idea about the mechanisms involved in the translocation of dopamine which is driven by sodium and chloride. Perhaps even more fundamental, we only have a proposed three dimensional structure which will require confirmation and clarification by molecular and biophysical techniques. The identification of consensus phosphorylation sites is very suggestive that phosphorylation is involved in regulation of the transporter. Studies in expressed transporter suggest that this may be the case, but confirmation and elucidation with the native transporter is needed. Finally, while brain imaging techniques have been employed with the transporter, there is still room for fundamental improvement in being able to determine the properties and numbers of transporters by external imaging approaches.

References

Amara S, Kuhar MJ (1993) Neurotransmitter transporters: recent progress. Annu Rev Neurosci 16:73-93.

Bergman J, Madras BK, Johnson SE, Spealman RD (1989) Effects of cocaine and related drugs in nonhuman primates III self-administration by squirrel monkeys. JPET 251:150-155.

Boja JW, Vaughan R, Patel A, Shaya EK, Kuhar MJ (1994) The Dopamine Transporter, in: *Dopamine Receptors and Transporters*. Hyman Niznik, ed. Marcel Dekker, Inc., New York, pp.611-644.

Cline EJ, Scheffel U, Boja JW, Carroll FI, Katz JL, Kuhar MJ (1992) Behavioral effects of novel cocaine analogs: a comparison with *in vivo* receptor binding potency. JPET 260(3):1174-1179.

Fleckenstein AE, Carroll FI, Pogun S, Kuhar MJ (1995) Recovery of dopamine transporter binding and function following administration of the irreversible inhibitor RTI-76. Society for Neurosci 21(1):376.

Freed C, Revay R, Vaughan RA, Kreik E, Grant S, Uhl GR, Kuhar MJ (1995) Dopamine transporter immunoreactivity in rat brain. J Comp Neurol 359:340-349.

Giros B, Caron MG (1993) Molecular characterization of the dopamine transporter. TiPS 14:43-49.

Giros B, Mestikawy SE, Bertrand L, Caron MG (1991) Cloning and functional characterization of a cocaine-sensitive dopamine transporter. FEBS Lett 295(1,2,3):149-154.

Glowinski J, Iversen LL (1966) Regional studies of catecholamines in the rat brain--I. The disposition of [^{3}H] norepinephrine, [^{3}H]dopamine and [^{3}H]dopa in various regions of the brain. J Neurochem 13:655-669.

Grigoriadis DE, Wilson AA, Lew R, Sharkey JS, Kuhar MJ (1989) Dopamine transport sites selectively labeled by a novel photoaffinity probe: ^{125}I-DEEP. J Neurosci 9:2664-2670.

Kilty J, Lorang D, Amara SG (1991) Cloning and expression of a cocaine-sensitive rat dopamine transporter. Science 254:578-579.

Kitayama S, Dohi T, Uhl G (1994) Phorbol esters alter functions of the expressed dopamine transporter. Eur J Pharmacol 268:115-119.

Kitayama S, Shimada S, Xu H, Markham L, Donavan DM, Uhl GR (1992) Dopamine transporter site-directed mutations differentially alter substrate transport and cocaine binding. Proc Natl Acad Sci 89:7782-7785.

Kuhar MJ, Ritz MC, Boja JW (1991) The dopamine hypothesis of the reinforcing properties of cocaine. Trends in Neurosci 14:299-302.

Lew R, Vaughan R, Simantov R, Wilson A, Kuhar MJ (1991) Dopamine transporters in the nucleus accumbens and the striatum have different apparent molecular weights. Synapse 8:152-153.

Nirenberg MJ, Vaughan RA, Uhl GR, Kuhar MJ, Pickel VM (1996) The dopamine transporter is localized to dendritic and axonal plasma membranes of nigrostriatal dopaminergic neurons. J Neurosci 16(2):436-447.

Patel AP, Cerruti C, Vaughan RA, Kuhar MJ (1994) Developmentally regulated glycosylation of dopamine transporters. Develop Brain Res 83:53-58.

Revay R, Vaughan R, Grant S, Kuhar MJ (1996) Dopamine transporter immunohistochemistry in median eminence, amygdala, and other areas of the rat brain. Synapse 22:93-99.

Ritz MC, Lamb RJ, Goldberg SR, Kuhar MJ (1987) Cocaine receptors on dopamine transporters are related to self-administration of cocaine. Science 237:1219-1223.

Sallee FR, Fogel EL, Schwartz E, Choi SM, Curran DP, Niznik HB (1989) Photoaffinity labeling of the mammalian dopamine transporter. FEBS Lett 256:219-224.

Scheffel U, Boja JW, Kuhar MJ (1989) Cocaine receptors: *in vivo* labeling with ^{3}H-(-)cocaine, ^{3}H-WIIN 35,065-2, and ^{3}H-WIN 35,428. Synapse 4:390-392.

Shaya EK, Scheffel U, Dannals RF, Ricaurte GA, Carroll FI, Wagner HN Jr., Kuhar MJ, Wong DF (1992) *In vivo* imaging of dopamine reuptake sites in the primate brain using single photon emission computer tomography (SPECT) and Iodine-123 labeled RTI-55. Synapse 10:169-172.

Shimada S, Kitayama S, Lin C-L, Patel A, Nathankumar E, Gregor P, Kuhar M, Uhl G (1991) Cloning and expression of a cocaine-sensitive dopamine transporter complementary DNA. Science 254:576-578.

Snyder SH and Coyle JT (1969) Regional differences in H3-norepinephrine and H3-dopamine uptake into rat brain homogenates. JPET 165:78-86.

Uhl GR, Hartig PR (1992) Transporter explosion: update on uptake. TiPS 13(12):421-425.

Usdin TB, Mezey E, Chen C, Brownstein MJ, Hoffman BJ (1991) Cloning of the cocaine-sensitive bovine dopamine transporter. Proc Natl Acad Sci 88:11168-11171.

Wong DF, Harris JC, Naidu S, Yokoi F, Marenco S, Dannals RF, Ravert HT, Yaster M, Evans A, Rousset O, Bryan RN, Gjedde A, Kuhar MJ, Breese GR (1996) Dopamine transporters are markedly reduced in Lesch-Nyhan disease *in vivo*. PNAS (In press).

Turnover of Rat Dopamine Transporter Protein in rDAT-LLC-PK1 Cells

Amrat P. Patel[1,4] F.Ivy Carroll[2] and Michael J. Kuhar[3]

[1]National Institutes of Health, National Institute on Drug Abuse, Intramural Research Program, Neuroscience Branch, Baltimore, MD 21224.
[2]Research Triangle Institute, 3040 Cornwallis Road, Research Triangle Park, NC 27709.
[3]Yerkes Primate Center, Emory University, 954 Gatewood NE, Atlanta, GA 30322.
[4]Present address: U.S. Food and Drug Administration, Center for Drug Evaluation and Research, MPN 2, HFD-658, Rockville, MD 20855, U.S.A.

Abstract

We have determined the turnover of the dopamine transporter protein in a porcine kidney epithelial cell line, LLC-PK1, that stably expresses a rat dopamine transporter (rDAT-LLC-PK1). A cocaine analog, RTI-76 (3β-(*p*-chlorophenyl)tropan-2β-carboxylic acid-*p*-isothiocyanatophenylethyl ester), was used to block dopamine transporter binding irreversibly in intact cells, and the rate of reappearance of transporter binding was measured. RTI-76 inhibited [^{125}I]RTI-55 binding to the dopamine transporter rapidly (within 10 min.) with an IC_{50} of 23.8 nM. After exposure to RTI-76, dopamine transporter binding reappeared in the cells over time with monoexponential kinetics with a half-life of 23 ± 1.6 h. Cell lines stably expressing DAT and irreversible binding ligands will be useful to study perturbations of transporter activity and their consequent effects on the transporter protein half-life.

1 Introduction

The dopamine transporter (DAT) is a member of the sodium and chloride dependent neurotransmitter transporter family which is responsible for terminating synaptic transmission at a variety of synapses. Many experiments characterizing the transporter protein have been carried out. DAT is a glycoprotein with an apparent molecular weight of about 80,000 Da (Patel et al., 1991; Kuhar, 1993). The protein is postulated to have 12 transmembrane segments, and domains binding dopamine and psychostimulants have been suggested (Kitayama et al., 1992; Giros and Caron, 1993). While much has been learned about the structure of DAT, little is known about its dynamics. In this communication, we report on efforts to measure the half-life of the transporter protein in porcine kidney cells that stably express the rat dopamine transporter (rDAT LLC-PK1; Gu et al., 1994). RTI-76, a potent irreversible blocker of ligand binding at DAT (Carroll et al., 1992), is used. This approach will be useful in studying changes in metabolism and turnover of the transporter protein.

NATO ASI Series, Vol. H 100
Neurotransmitter Release and Uptake
Edited by Şakire Pöğün

2 Methods and Materials

2.1 Materials:

[^{125}I]RTI-55 (2200 Ci/mmol) was purchased from New England Nuclear (Boston, MA). RTI-76 was synthesized as described previously (Carroll et al., 1992). Cocaine.HCl was obtained from National Institute on Drug Abuse/ Addiction Research Center (Baltimore, MD). Reagents for buffers were from Sigma Chemical Co. (St.Louis, MO.), tissue culture media and Nunc 24 well culture plates and flasks were from Gibco-BRL (Rockville, MD).

2.2 Expression of rat DAT cDNA in LLC-PK1 cells:

rDAT-LLC-PK1 cells were kindly provided by Dr. Gary Rudnick (Yale University). The cells were maintained in DMEM culture media containing 2 mM glutamine, 10% fetal calf serum, 100 U/ml penicillin and 100 μg/ml streptomycin in humidified atmosphere of 95% air and 5% CO_2 as described by Gu, et.al. (1994).

2.3 Treatment of rDAT-LLC-PK1 cells with RTI-76:

Cells in a log phase (50 - 70% confluent) were treated with 100 or 300 nM of RTI-76 for 10 minutes at 37°C, rinsed with HANK'S buffer and returned to the incubator. At each time point untreated (control) and RTI-76 treated cells were assayed. The turnover rate was determined from six separate experiments (in triplicate) with recovery time intervals ranging from 3 to 48 h The dopamine transporter binding activity remaining at each time point was determined as a fraction of that in the untreated cells. The [^{125}I]RTI-55 specific binding was carried out as described below. The binding at 48 h in the RTI-76 treated cells was > 98% of the control untreated cells.

2.4 [^{125}I]RTI-55 Binding to DAT in intact cells:

Cells were grown in 24 well culture plates and incubated with [^{125}I]RTI-55 (0.1 nM) in binding buffer (phosphate buffered saline supplemented with 0.1 mM $CaCl_2$ and 1 mM $MgCl_2$, pH 7.4 at 21° C) in presence or absence of 10 μM (-)cocaine for 50 min at 21° C (Boja et al., 1991). At the end of the incubation, cells were quickly washed with 3x1ml cold binding buffer and lysed with 0.2 ml of 0.2 M NaOH containing 0.1% SDS. The bound radioactivity was counted in LKB 5600 Gamma Counter. Same samples were then used for the assay of protein with bovine serum albumin as the standard (Patel and Linden, 1986).

3 Results and Discussion:

Mammalian cell systems allow study of the cloned neurotransmitter transporter genes. A porcine epithelial kidney cell line, LLC-PK1, has been transfected with rDATcDNA (Shimada, et.al. 1991) and stably expresses the transporter protein (Gu et.al. 1994). The pharmacological properties of the dopamine transporter in these cells are similar to those for the brain dopamine transporter (Gu et.al., 1994).

We have utilized RTI-76, a cocaine analog with an isothiocyanate moiety that blocks the dopamine transporter in a wash resistant fashion (Carroll et al., 1992; figure 1). RTI-76 interacts with the dopamine transporter with high potency (IC_{50} ≈24 nM) as determined by inhibition of [^{125}I]RTI-55 binding (figure 1). RTI-76 reacts very rapidly, since greater than 95% of [^{125}I]RTI-55 binding sites are blocked after exposure of cells with 100 nM RTI-76 for 10, 30 or 60 min at 21°C or 37 °C (not shown) and more than 98% of the dopamine transporters were blocked with 0.5 μM of RTI-76 (not shown).

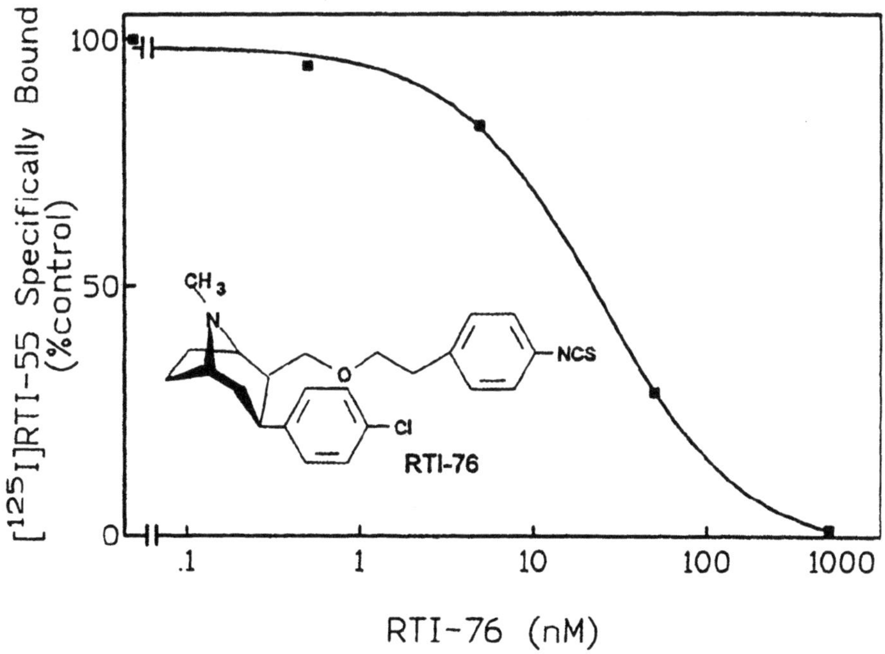

Fig 1 A representative dose response curve of RTI-76 on the specific binding of [^{125}I]RTI-55 to rDAT-LLC-PK1 cells. Inset shows chemical structure of RTI-76 (3*β*-(*p*-Chlorophenyl)tropan-2*β*-carboxylic acid-*p*-isothiocyanatophenylethyl ester). The rDAT-LLC-PK1 cells were grown in Dulbecco's modified eagle medium containing 10% fetal bovine serum with 2 mM L-glutamine, at 37°C in a humidified incubator with 5% CO_2. Cells (70% confluent) in 24 well plates were rinsed with PBS (phosphate buffered saline, pH 7.4), then incubated with RTI-76. DAT binding was determined with [^{125}I]RTI-55 in the presence or absence of 10 μM (-) cocaine in PBS for 50 min at 21°C. At the end of the incubation, the cells were washed with cold PBS, lysed with 0.2 N NaOH and 0.1 % SDS and radioactivity in the cell lysate measured in a gamma counter. RTI-76 inhibited binding of [^{125}I]RTI-55 to the dopamine transporter with an IC_{50} of 24 nM.

Turnover of the dopamine transporter protein in rDAT-LLC-PK1 cell was determined by inhibiting the transporter with 0.1 or 0.3 μM RTI-76 and analyzing the reappearance of the transporter binding at various times after RTI-76 treatment as described under methods. Figure 2, shows result of a representative experiment. The half-life of the transporter determined from six separate experiments over the time interval ranging from 3 to 48 h after treatment resulted in $t_{1/2}$ of 23 ± 1.6 h (mean ± SEM) hours. While the use of this general approach to measure half-life is not uncommon, it assumes that the reappearance of [^{125}I]RTI-55 binding is due to synthesis of new transporter protein rather than the loss of RTI-76 from the existing DAT. This seems reasonable since a similar half-life of DAT was found when cells were treated with the protein synthesis inhibitor cyclohexamide (not shown). Dissociation of RTI-76 from DAT seems unlikely since much longer half-lives are measured *in vivo* in rat brain (see below). Moreover, cell counts and staining with trypan blue did not reveal a toxicity associated with RTI-76. Also, this method gives results similar to those from other experiments. The half-life of norepinephrine transporters stably expressed in LLC-PK1 cells was found to be ≈24 h (Melikian et.al., 1994) which is in excellent agreement with that of the DAT in the same cell line studied here.

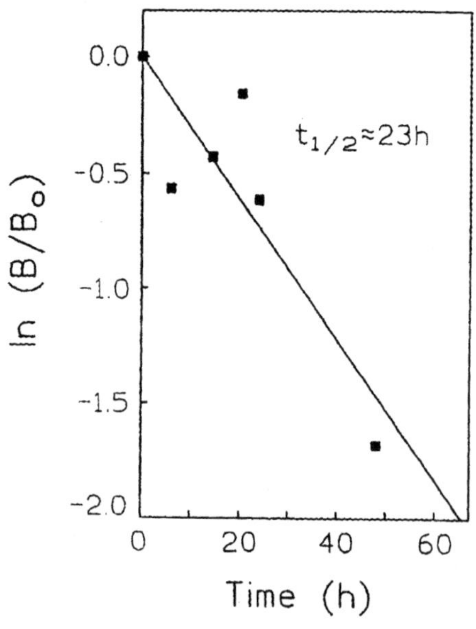

Fig 2. A representative measurement of DAT halflife in rDAT-LLC-PK1 cells. Cells in a log phase were treated with 100 nM of RTI-76 for 10 minutes at 37°C, rinsed with HANK'S buffer and returned to the incubator. At each time point, cells that were untreated as well as the RTI-76 treated cells were assayed. The dopamine transporter binding activity remaining at each time point was determined as a fraction of that in the untreated cells. Measurement of [^{125}I]RTI-55 specific binding was carried out as described in legend to figure 1. The turnover rate determined from six experiments with time intervals ranging from 3 to 48 h gave a $t_{1/2}$ of 23 ± 1.6 (mean ± SEM) hours.

The binding data shown here were determined with a single concentration of ^{125}I-RTI-55. The changes found are most likely due to B_{max} changes as RTI-76 irreversibly blocks binding and produces B_{max} changes after *in vivo* injection (Fleckenstein 1995). However, it seems that subsequent application of this method should include B_{max} measurements.

However, the half-life of the dopamine transporter *in vivo* in the rat striatum, determined after direct intrastriatal injection of RTI-76, was found to be about 6 days (Fleckenstein, et.al., 1995). This difference between our results and those of Fleckenstein et al. (1995) could be due to a number of factors. The striatum contains dopaminergic nerve terminals rather than cell bodies, and regulatory factors in nerve terminals could be different from those in perikarya. Also, the DAT protein in nigrostriatal neurons is likely to be subject to different processing influences than in the rDAT-LLC-PK1 cells. It may be argued that because the dopamine transporter is not constitutively expressed in LLC-PK1 cells, internal regulatory events would not affect transporter in the same way as in dopaminergic neurons. For example, in COS7 cells, another nonneuronal cell type that expresses the dopamine transporter, it has been shown that the apparent molecular weight of DAT is greater than in brain because of more extensive glycosylation (Patel et al., 1993). The molecular weight of DAT in the rDAT-LLC-PK1 cells is also greater than in brain (Patel et.al.,1995). Nevertheless, the cell lines stably expressing DAT may be useful to study factors that regulate the transporter protein half-life.

References

Boja, J., Patel, A., Carroll, F. I., Rahman, M.A., Philip, A., Lewin, A.H., Kopajtic, T.A. and Kuhar, M.J. (1991) [^{125}I]RTI-55: a potent ligand for dopamine transporters. Eur. J. Pharmacol. 194, 133-134.

Carroll, F.I., Gao, Y., Abraham, P., Lewin, A.H., Lew, R., Patel, A., Boja, J.W. and Kuhar, M.J. (1992) Probes for the cocaine receptor. Potentially irreversible ligands for the dopamine transporter. J. Med. Chem. 35, 1813-1817.

Fleckenstein, A.E., Caroll, F.I., Pogun, S. and Kuhar, M.J. (1995) Recovery of dopamine transporter binding and function following administration of the irreversible inhibitor RTI-76. Soc. Neurosci. 21, 376.

Giros, B. and Caron, M.G. (1993) Molecular characterization of the dopamine transporter. Trends in Pharmacol. Sci. 14, 43-49.

Gu, H., Wall, S.C. and Rudnick, G. (1994) Stable expression of biogenic amine transporters reveals differences in inhibitor sensitivity, kinetic, and ion dependence. J. Biol. Chem. 269, 7124-7130.

Kitayama, S., Shimada, S., Xu, H., Markham, L., Donovan, D.M. and Uhl, G.R. (1992) Dopamine transporter site-directed mutations differentially alter substrate transport and cocaine binding. Proc. Natl. Acad. Sci. 89, 7782-7785.

Kuhar, M.J. (1993) The Otto Krayer Award Lecture: Neurotransmitter transporters as drug targets: Recent research with a focus on the dopamine transporter. The Pharmacologist 35(1), 28-33.

Melikian, H.E., McDonald, J.K., Gu, H., Rudnick, G., Moore, K.R. and Blakely, R.D. (1994) Human norepinepherine transporter, biosynthetic studies using a site-directed polyclonal antibody. J. Biol. Chem. 269, 12290-12297.

Patel, A. and Linden, J. (1988) Photoaffinity labeling of adenosine receptors. In: Adenosine Receptors, D.M.F. Cooper and D. Londos (eds.). Alan R. Liss, Inc. New York, Volume 11, pp 27-41.

Patel, A., Boja, J.W., Lever, J., Lew, R., Simantov, R., Carroll, F.I., Lewin, A.., Philip, A., Gao,Y. and Kuhar, M.J. (1992) A cocaine analog and a GBR analog label the same protein in rat striatal membranes. Brain Res. 576, 173-174.

Patel, A., Uhl, G.R. and Kuhar, M.J. (1993) Species differences in dopamine transporters: Postmortem changes and glycosylation differences. J. Neurochem. 61, 496-500.

Patel, A.P., Martel, J-C., Vandenberg, D.J., Uhl, G.R. and Kuhar, M.J. (1995) Cell lines expressing human and rat dopamine transporter cDNAs: Different ligands yield different radiolabeling patterns. Soc. Neurosci. 21, 782.

Shimada, S., Kitayama, S., Lin, C-L., Patel, A., Nathankumar, E., Gregor, P., Kuhar, M. and Uhl, G. (1991) Cloning and expression of a cocaine-sensitive dopamine transporter complementary DNA. Science 254, 576-578.

Neurotransporters at the Juncture of Drug Action: Role in Programmed Cell Death, and Toxicity of Abused MDMA

Rabi Simantov

Department of Molecular Genetics, Weizmann Institute of Science, Rehovot 76100

Abstract

Neurotransmitter transporters (neurotransporters) play a central role in the transmission of signals between neurons, and in delivery of various compounds into neurons. During the last few year several neurotransporters were cloned, and this opened a new era of studies aimed to unravel the involvement of these membrane-bound glycoproteins in normal neurophysiological activities, as well as in certain pathological conditions. Herein we address the involvement of dopamine and serotonin neurotransporters in programmed cell death. The role of dopamine transporters as mediators and modulators of dopamine neurotoxicity, and the usefulness of antisense nucleic acid oligonucleotides in blocking the expression of dopamine transporters, and thus protecting the cells from dopamine and similar toxic agents, will be addressed. The study has implications regarding the mechanism of dopaminergic nerve degeneration in Parkinson's disease and in aging. As to serotonin transporters, human serotonergic cells were used to verify the toxic properties of the abused drug MDMA ('ecstasy'). We show that MDMA-induced apoptosis in cultured serotonergic but not dopaminergic human cells. The neurochemical and molecular mechanisms involved in this programmed cell death are summarized.

1 Introduction: Transporters as Delivery Systems

Recent studies indicate that transporters in general, and those in charge of amine neurotransmitter uptake, in particular, should be viewed not only as uptake sites, but also as delivery systems (1-3, and references included). Table 1.1 shows a list of compounds interacting with neurotransporters; these may be divided into three classes. The first group includes both neurotransmitters and closely related neurotransmitter derivatives, such as 6-hydroxydopamine (6-OHDA), and 5,7-dihydroxytryptamine. 6-OHDA, as well as other dopamine oxidation products, were suspected to be produced in vivo under certain conditions (4), possibly playing, therefore, a role in neurodegeneration. Being toxic to the neurons that take them up, such compounds became excellent experimental tool. The second group of agents that interact with neurotransporters, and may enter neurons with the aid of these proteins, are synthetic chemicals, often used in the industry or agriculture. In this group one can include, e.g., pyridine derivatives, as suggested by Snyder and D'Amato (5). Neurodegeneration induced by 4-methyl-1,2,5,6-tetra-hydroxypyridine

NATO ASI Series, Vol. H 100
Neurotransmitter Release and Uptake
Edited by Şakire Pöğün

(MPTP), first oxidized to MPP^+, has been well documented and reviewed (6). The third group of agents interacting with neurotransporters are abused drugs of the amphetamine and methamphetamine family. These interact primarily with dopamine and serotonin transporters, apparently the best example of transporters involved in delivering the three types of compounds listed in Table 1.1.

Table 1.1. Neurotransporters as drug-delivery systems

Neurotransmitters and related analogues (e.g. 6-OHDA)
Synthetic chemicals (e.g. pyridines)
Abused drugs (e.g. methamphetamines)

Compounds interacting with amine neurotransporters include psychoactive drugs and psychostimulants (amphetamines, methamphetamines, cocaine, reserpine, benztropine), tricyclic antidepressants and selective serotonin uptake inhibitors (desipramine, imipramine, amitriptyline, nortriptyline, maprotyline, fluoxetine, paroxetine, sertraline, citalopram, zimelidine), and others (mazindol, tetrabenazine, 6-OHDA, MPP+, DSP-4, GBR 12935, and bupropion). See Ref. 1,2,5,6 for a more complete list.

2 Apoptotic Death of Dopaminergic Neurons

Dopamine is one of the major classic neurotransmitters in the mammalian brain, and dopaminergic neurons are involved with several normal physiological activities, such as cognition, behaviour, motor function, and neuroendocrine systems. During the last three decades it was established that dopaminergic cells die in large numbers in Parkinson's disease (7,8), and in aging (9-12). Studies conducted in many laboratories indicate that multiple factors are involved in degeneration of dopaminergic neurons, among them iron (12), toxic oxygen radicals (14-16 and references included), and others (17-22). One of the interesting developments regarding the degeneration of dopaminergic neurons is the finding that the metabolism of dopamine plays a major role in cell death; dopamine itself became therefore a suspected neurotoxic agent. This suggests that an increased level, or enhanced overflow of dopamine (e.g. upon continuous stimulation), can induce cell death (see Ref. 23). Several biochemical pathways and cellular compartments (e.g. membranes, mitochondria) are clearly involved in degeneration of dopaminergic cells, but the details of this neurotoxicity are yet to be elucidated.

Studies with fibroblast and other cells have indicated that cell growth, proliferation and apoptosis are interconnected, and often involve similar growth regulators and biochemical pathways. It has been reported that dopamine has a developmental role in the nervous system, and dopaminergic neurons regulate the development of target neurons. In analogy to growth factors, cytokines and hormones that are involved in apoptosis, dopamine may also be involved in programmed cell death. Indeed, it was shown that dopamine induces apoptosis of cultured chick sympathetic neurons (24). Additional support to the notion that dopaminergic neurons can die in an apoptotic fashion was obtained upon excitatory stimulation of the brain during development

(25). More recently, we have found that dopamine induces apoptosis of human neuronal cells (26), and glutathione plays a central role in protecting the cells from this process (27). These recent studies are briefly reviewed below.

2.1 Dopaminergic Synapse and the Dopamine Transporter DAT

The key protein controlling dopamine levels in the synaptic cleft is the sodium-dependent dopamine transporter DAT, a membrane-bound glycoprotein, which is also a cocaine receptor (see Ref. 1-3, and Fig. 2.1.). Blocking the activity of this transporter in vivo increases extracellular dopamine levels, thus inducing a whole range of behavioral effects, hyperactivity and enhanced locomotion. Isolation of a cDNA coding the dopamine transporter (28-32) opened a variety of possibilities in studying dopamine metabolism in the brain. The report (12) that the dopamine transporter mRNA content in human substantia nigra decreases upon aging by about 95% established the notion that presynaptic dopaminergic elements suffer a severe loss in aged subjects (9-11). Thus DAT can serve as a useful experimental tool to follow degeneration of dopaminergic neurons.

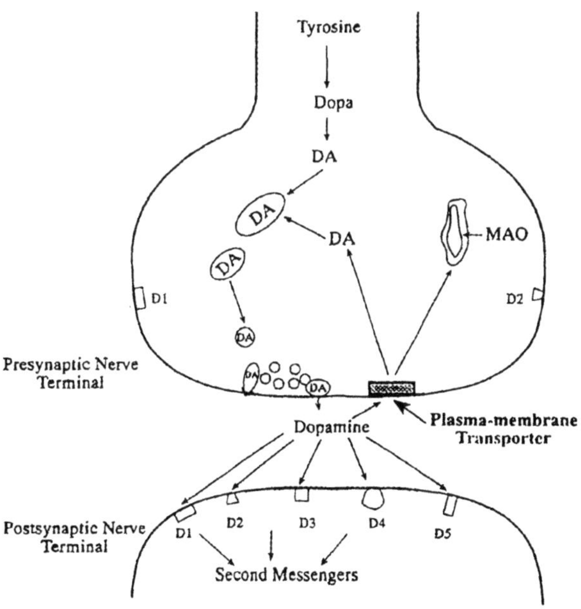

Fig 2.1. A schematic presentation of a dopaminergic synapse in the brain

2.2 Primary Mesencephalic Cultures; the Advantage of a Homogeneous Neuronal Cell Line

Brain mesencephalic primary cell cultures were used for years to investigate dopaminergic cell growth and function (e.g. 33). Such neuronal cultures, free of glial

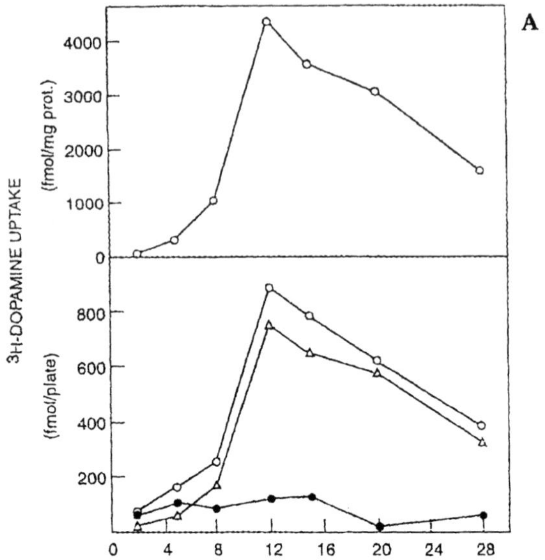

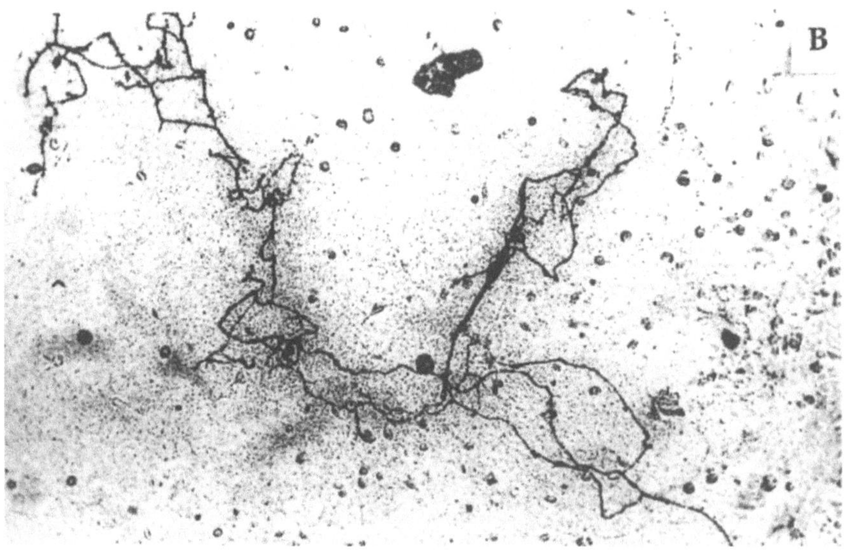

Fig 2.2. **A**. ^{3}H-Dopamine uptake in primary rat mesencephalic cultures. **B**. Autoradiography of dopaminergic neurons in mesencephalic rat brain cultures, labeled with ^{3}H-dopamine.

cells, were prepared from the ventral midbrain of E15 rat embryos, and used to study the uptake of dopamine and the expression of DAT (Fig. 2.2. A).

^{3}H-dopamine uptake in such cultures was increased within the first 12-15 days in vitro (DIV), and then remained stable for a few days, before gradually decreasing. This was also confirmed by binding experiments with a selective dopamine transporter blocker. The dopaminergic cells were also visualized by autoradiography of 18-DIV cultures (Fig. 2.2. B), and the release of ^{3}H-dopamine by these cells was stimulated by potassium chloride. It is clear, however, that the percentage of dopaminergic cells in such primary cultures was low (1-5%), and they were not suitable for studying biochemical and molecular processes, such as apoptosis. We have searched, therefore, for a more suitable system were this issue may be addressed. The human neuroblastoma NMB cell line (34) was found to possess a very efficient DAT system.

The V_{max} of ^{3}H-dopamine uptake by NMB cells was more than 3 times higher than rat striatal synaptosoms, and the drug-selectivity of the transporter was similar to that of previously reported DATs (26). As a homogeneous cell population, these cells were useful to verify the neurotoxic activity of dopamine, and the approaches available to restrain this activity.

It was shown that dopamine induced a time- and dose-dependent programmed cell death in NMB cells, and this process was inhibited by the protein synthesis blocker cycloheximide (26). The dopamine receptor antagonist haloperidol (10 mM) did not block dopamine neurotoxicity, and the dopamine receptor agonist 2-bromo-ergocryptine had no toxic activity in NMB cells (Fig. 2. 3), suggesting that dopamine receptors were not involved in the neurotoxic activity of dopamine in these cells.

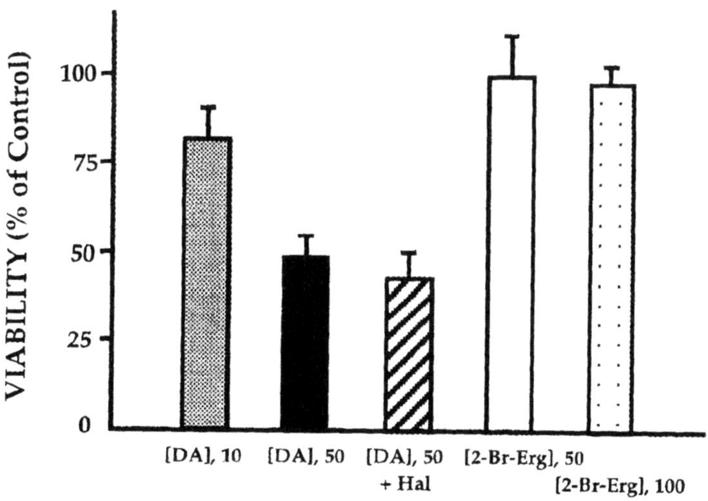

Fig 2.3. Effect of haloperidol and 2-bromo-ergocryptine on NMB cells (Concentrations in μM.

Using the antisense knockout approach, it was found that an 18-mer antisense oligonucleotide of DAT (OLIGO-3) inhibited dopamine uptake and dopamine-induced cytotoxicity (26). A 'mismatch' control oligonucleotide, with identical nucleotides content as OLIGO-3 but in a different sequence, did not block ^{3}H-dopamine uptake (Fig. 2. 4). This indicates that inhibition of DAT expression might be useful in blocking the neurotoxicity of agents entering the cells via this neurotransporter.

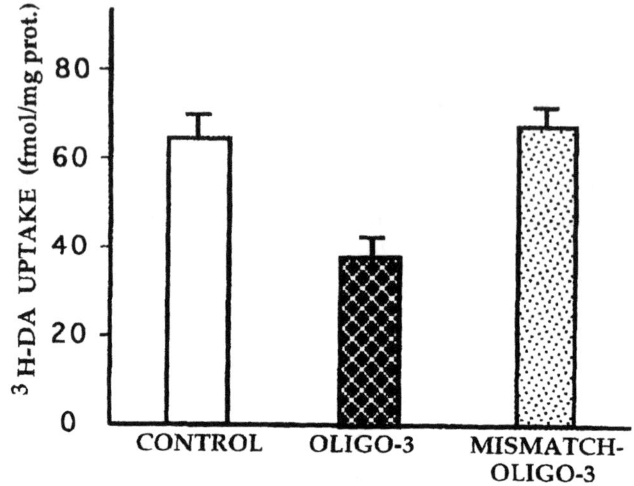

Fig 2.4. Effect of mismatch OLIGO-3 on ^{3}H-dopamine uptake

2.3 Free Radicals, Antioxidants and Dopamine Neurotoxicity

Hydrogen peroxide and free radicals are toxic to various cell types, and oxidative stress play a key role in the pathogenesis of many diseases, in the nervous system and in other organs (35). The pivotal role of endogenous antioxidants, such as vitamins C and E, glutathione, superoxide dismutase and catalase in protecting mammalian cells from free radicals has been under an extensive research during the last decade. Fig. 2. 5. illustrates some of the biochemical pathways involved in these activities.

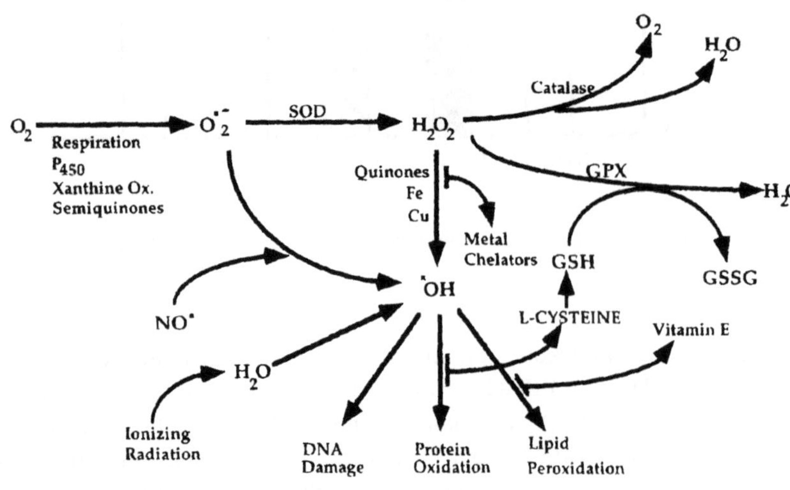

Fig 2.5. A schematic presentation of oxidative and antioxidative processes in neuronal cells

In the brain, enzymatic as well as non-enzymatic oxidation of dopamine, producing free radicals and semiquinones (Table 2.1), participate in dopamine-induced cell death. Indeed, excess lipid peroxidation was observed in dopaminergic brain regions of Parkinsonian patients (36). The important finding that the level of reduced glutathione (GSH) is lower in key brain regions of Parkinsonian patients, raised the possibility that certain endogenous antioxidants play a role in modulating dopamine neurotoxicity (37-39).

This issue has been recently addressed in experiments with NMB cells (27). It appeared that antioxidants such as glutathione protect these neurons from apoptosis, induced by dopamine or 6-OHDA. The glutathione precursor amino acid L-cysteine, and an inhibitor of glutathione synthesis have been used to verify glutathione selectivity, and to overcome the problem of its slow penetration into mammalian cells. The experiments with NMB cells have indicated that vitamin E and vitamin C did not have a similar protective effect (27). Interestingly, high concentrations of vitamin C enhanced dopamine neurotoxicity, being in line with previous reports indicating that this antioxidant can interact with Fe^{+3}, and produce Fe^{+2} and free radicals.

Table 2.1. Dopamine as a source for free radicals: Enzymatic and non-enzymatic mechanisms

1. Enzymatic Oxidation by MAO	
$DA + O_2 + H_2O \xrightarrow{MAO} 3{,}4$ dihydroxyphenyl-acetaldehyde$+ NH_3 + H_2O_2$	
2. Auto-oxidation	
$DA + O_2 \rightarrow SQ\bullet + O_2^- + H^+$	(SQ: Semiquinone)
$DA + O_2^- \rightarrow 2H^+ \rightarrow SQ\bullet + H_2O_2$	
3. Fe+2 containing Enzyme cofactors	Cytchrome C Cytchrome P450 Others
$DA \rightarrow$ 2 − OHDA 5 − OHDA 6 − OHDA $\rightarrow H_2O_2$ + 6 − OHDA quinone	
4. Hydroxylation by Hydroxyl Radical (OH•)	

3 Interactions of Serotonin Transporters with Methamphetamines: Cytotoxicity of MDMA (Ecstasy)

During the last decade MDMA (3,4-methylenedioxy-methamphetamine, also called 'ecstasy') became a popular recreational abused drug. Potent stimulatory, mood-altering and hallucinogenic effects result from usage of this drug. Recently it became apparent, however, that MDMA is also cytotoxic, primarily to serotonergic neurons in the brain. These neurons play a central role in a variety of physiological and behavioural activities, such as food intake, sleep, pain perception, cognition, depression, and anxiety. Indeed, MDMA abuse can induce long-term neuropsychiatric abnormal behaviours, such as panic and psychosis (40-44). Unravelling the biochemical systems that are responsible for 'ecstasy'-induced death of serotonergic neurons has, therefore, important physiological and medical implications. This may clarify whether the long-term effects of the drug are related to its cytotoxic activity.

Neuroanatomical studies with rodents and monkeys indicate that 'ecstasy' causes degeneration of neuronal processes or cell bodies. This issue, for obvious reasons, is less clear with humans. A system useful in investigating 'ecstasy' toxicity in human cells is not thus far available. We have recently found that 'ecstasy' is cytotoxic to a human serotonergic cell line. Treatment with the drug induces morphological and biochemical changes characteristic to apoptosis, as described below.

3.1 Amphetamines: Behavioural, Neurochemical, and Addictive Properties

Amphetamines are mostly known for their psychostimulant activities. They induce various behavioural and psychological effects such as the following: decrease fatigue and hunger, increase locomotion, aggressiveness and sexual activity, modified learning and sleep pattern, and a general feeling of 'physical strength and energy' (reviewed in 45). These effects are transient, and often followed by fatigue, hallucination, tremors, sweating, depression, psychosis, compulsive behaviour, and antisocial acts. Furthermore, in vivo studies with experimental animals, as well as clinical observations, indicate that amphetamines are addictive; they are reinforcing, and induce self-administration and dependence, as widely described in the literature (45,46).

Neurochemical and neurpharmacological experiments established the notion that amphetamines interact with dopaminergic and serotonergic neuronal pathways. Briefly, amphetamines block the uptake of the two neurotransmitters, enhance their release, and also inhibit monoamine oxidase activity. With regard to the dopaminergic system, amphetamines work as indirect ('false') dopamine agonists, with the characteristic increase in extrapyramidal motor activity and stereotype behaviours. These properties of amphetamines have been studied extensively (45,46), and, thus, will not be addressed in detail here.

Despite the established view that certain important in vivo effects of amphetamines (particularly on mood, cognition and abnormal food intake) are associated with altered function of serotonergic cells, the interaction of amphetamines with the serotonergic system is less understood. These aspects of amphetamine activity are especially relevant to the mode of action of MDMA, as discussed below.

3.2 MDMA- An Abused Neurotoxic Methamphetamine Analogue

During the last decade MDMA has become one of the most popular abused drugs in western countries. MDMA, a chemically unstable methamphetamine analogue (with a methylenedioxy moiety bound to the aromatic ring), is quickly metabolized in vivo to methamphetamines and dopamine analogues (47,48). This could reflect the two types of behavioural effects of the drug: **a.** a psychedelic, LSD-like hallucinogenic activity, caused by activation of serotonin receptors, possibly due to the stimulation of serotonin release (see below), and **b.** an amphetamine-like psychostimulant, similar to other indirect dopamine agonists.

One of the important issues of MDMA abuse is its toxicity to serotonergic neurons (49-53). This supports the current notion that MDMA is not just a transiently-active psychostimulant, but has long lasting behavioural and physiological effects. As indicated above, panic, suicidal and psychotic behaviours were reported after MDMA abuse (40-44). The first anatomical studies in rodents indicated that MDMA destroyed serotonergic axonal terminals (50-53). From the mechanistic point of view, it is interesting that not all axons were equally sensitive to the drug; e.g., fine serotonergic terminals in the dorsal neocortex were more vulnerable than beaded serotonin projections of the amygdala (54). The significance of this selectivity, with regard to the long-term effects of the drug, has yet to be analysed.

Based on a variety of experiments, it was determined that serotonin transporters play a key role in the mechanism of MDMA cytotoxicity (55-58). Briefly, MDMA is a potent blocker of Na^+-dependent imipramine binding to the serotonin uptake site (the transporter). Also, MDMA stimulates the efflux of serotonin via the serotonin transporter. In line with these observations, selective inhibitors of the serotonin transporter also inhibit MDMA neurotoxic activity. This indicates that this plasma-membrane transporter is a key target protein for MDMA in serotonergic neurons. Yet, recent experiments have shown that certain amphetamine derivatives that interact with the serotonin transporter, and induce serotonin release, are not neurotoxic (59). Thus, MDMA mode of action, and its interaction with the serotonin transporter, are not fully clear. This is particularly true in human serotonergic cells.

3.3 Using Cultured Human Serotonergic Cells to Study the Role of the Serotonin Transporter in MDMA-Induced Programmed Cell Death

Analyzing MDMA mode of action in human cells is important from several aspects. First, the metabolism of amphetamines in primates (monkeys) is different from their metabolism in rodents (60). Secondly, the neurotoxic activity of MDMA in rodents was mainly observed at the level of neuronal processes, whereas in monkeys the cell bodies were also destroyed (51-54). This may be related to the finding that at the level of the whole animal, primates are more sensitive than rodents to the toxic activity of this drug. Overall, rodents show transient anatomical abnormalities upon treatment with the drug, whereas the changes in monkeys are long-lasting.

MDMA was found to be cytotoxic to the serotonergic human placenta JAR cells; cell counting indicated a time- and dose-dependent decrease in the number of viable cells, and a dose-dependent increase in the amount of the cytosolic enzyme lactate dehydrogenase released into the culture medium (61). Under the same experimental conditions, MDMA was not toxic to the human neuronal NMB cells, expressing dopamine- but not serotonin- transporters (26). MDMA cytotoxicity to JAR cells was partially blocked by the serotonin transporter blocker imipramine, but not by cocaine, thus further indicating the key role of the serotonin transporter .

Flow cytometry (FACS) experiments indicated that MDMA altered the cell cycle profile of JAR cells. The inhibitor of protein synthesis, cycloheximide, blocked the cytotoxic effect of the drug, and the alteration in the cell cycle profile. Under the same conditions MDMA had no significant effect on the cell cycle profile of NMB cells (61). DNA damage by MDMA was indicated by intranucleosomal DNA fragmentation, which appeared as a characteristic DNA ladder. Again, no such effect was observed in MDMA-treated NMB cells. Lastly, we investigated whether MDMA cytotoxicity in JAR cells involved the production of the endogenous toxic agent nitric oxide (NO). The selective inhibitor of the enzyme nitric oxide synthase, L-NAME, inhibited MDMA toxicity in a dose-dependent way; L-NAME was more effective than D-NAME. Furthermore, the nitric oxide donor, sodium nitroprusside, induced cell death in a dose-dependent fashion, as well as DNA fragmentation (61).

These experiments demonstrate that MDMA induces an apparent programmed death of human serotonergic cells, and nitric oxide plays a key role in this process. Other mechanisms involved in this process, as well as the approaches available to block MDMA toxicity need further analysis.

Acknowledgment:

The studies conducted in this laboratory were supported by The Israel Antidrug Authority, Israel Ministry of Health, and Rosen Sidney Foundation, Leo & Julia Forchheimer Center for Molecular Biology, and Nella & Leon Benoziyo Center for Neurosciences, Weizmann Institute of Science. I wish also to thank my colleagues H. Blinder, M. Gabbay, S. Porat, T. Ratovitski, G. Tafet and M. Tauber.

References

1. Amara S. G. and Kuhar M. J. *Annu. Rev. Neurosci.* **16:** 73, 1993.
2. Rudnick G. and Clark J. *Biochim. Biophys. Acta* **1144:** 249, 1993.
3. Simantov R. *Biochem. Pharmacol.* **50:** 435, 1995.
4. Slivka A. and Cohen G. *J. Biol. Chem.* **260:** 15466, 1986.
5. Snyder S. H. and D'Amato R. J. *Neurology* **36:** 42, 1986
6. Tipton K. F. and Singer T.P. *J. Neurochem.* **61:** 1191, 1993.
7. Hornykiewicz O. *Fed. Proc.* **32:** 183, 1973.
8. Calne D. and Langston J. W. *Lancet* **ii:** 1457, 1983.
9. Carlsson A. and Winblad B. *J. Neural Transm.* **38:** 271, 1976.
10. McGeer P. L., McGeer E. G. and SuzukiJ. S. *Arch. Neurolog.* **34:** 33, 1977.
11. De Keyser J., Ebinger G. and Vauquelin G. Ann. Neurol. **27:** 157, 1990.
12. Bannon M. J., Poosch M. S., Xia Y., Goebel D. G., Cassin B. and Kapatos G. *Proc. Natl. Acad. Sci. USA* **89:** 7095, 1992.
13. Youdim M. B. H., Ben-Shachar D. and Riederer P. *Movement Disorders* **8:** 1, 1993
14. Graham D. G. *Mol. Pharmacol.* **14:** 633, 1978.

15. Olanow C. W. *Trends Neurosci.* **16:** 439, 1993.
16. Coyle J. T. and Puttfarcken P. *Science* **262:** 689, 1993.
17. Hirsch E. C., Graybiel A. M. and Agid Y. *Nature* **334:** 345, 1988.
18. Youdim M. B. H. and Riederer P. *J. Neural. Transm.* **91:** 181, 1993.
19. Meldrum B. S. and Garthwaite J. *Trends Pharmacol. Sci.* **11:** 379, 1990.
20. Perry T. L., Godin D. V. and Hansen S. *Neurosci. Lett.* **33:** 305, 1982.
21. Kish S. J., Morito C. L. H. and Hornykiewicz O. *Neurochem. Pathol.* **4:** 23, 1986.
22. Schapira A. H. V., Cooper J. M., Dexter D., Jenner P., Clark J. B. and Marsden C.D. *J. Neurochem.* **54:** 823, 1989.
23. Zigmond M. J., Hastings T. G. and Abercrombie E. D. *Ann. N. Y. Acad. Sci.* **648:** 71, 1992.
24. Ziv I., Melamed E., Nardi N., Luria D., Achiron A., Offen D. and Barzilai A. *Neurosci. Lett.* **170:** 136, 1994.
25. Macaya A., Munell F, Gubits R. M. and Burke R. E. *Proc. Natl. Acad. Sci. USA* **91:** 8117, 1994.
26. Simantov R., Blinder H., Ratovitski T., Tauber M., Gabbay M. and Porat S. *Neuroscience* **74:** 39, 1996.
27. Gabbay M., Tauber M., Porat S. and Simantov R. *Neuropharmacology* in press.
28. Shimada S., Kitayama S., Lin C-L., Patel A., Nanthakumar E., Gregor P., Kuhar M. J. and Uhl G. *Science* **254:** 576, 1991.
29. Kilty J.E., Lorang D. and Amara S. G. *Science* **254:** 578, 1991.
30. Giros B., El Mestikawy S.,Bertrand L. and Caron M. G. *FEBS Lett.* **295:** 149, 1991.
31. Usdin T. B., Mezey E., Brownstein M. B. and Hoffman B. J. *Proc. Natl. Acad. Sci. USA* **88:** 11168, 1991.
32. Giros B., El Mestikawy S., Godinot N., Zheng K., Han H., Yang-Feng T. and Caron M. G. *Mol. Pharmcol.* **42:** 283, 1992.
33. Prochiantz A., Di Porzio U., Kato A., Berger B. and Glowinski J. *Proc. Natl. Acad. Sci. USA* **76:** 5387, 1979.
34. Brodeur G. M., Sekhon G. S and Goldstein M. N. *Cancer* **40:** 2256, 1977.
35. Guttridge J. M. C. *Free Rad. Res. Commun.* **19:** 141, 1993.
36. Dexter D. T., Carter J. C, Wells F. R., Javoy-Agid F., Agid Y., Lees A., Jenner P., and Marsden C. D. *J. Neurochem.* **52:** 381, 1989.
37. Riederer P., Sofic E., Rausch W. D., Schmidt B., Reynolds G. P., Jellinger K. and Youdim M. B. H. *J. Neurochem.* **52:** 515, 1989.
38. Jenner P., Dexter D. T., Sian J., Schapira A. H. V. and Marsden C. D. *Ann. Neurol.* **32:** S82, 1992.
39. Sofic E., Lange K. W., Jellinger K. and Riederer P. *Neurosci. Lett.* **142:** 128, 1992.
40. Schifano F. *Lancet* **338:** 1335, 1991.
41. Benazzi, F. and Mazzoli M. *Lancet* **338:** 1520, 1991.
42. McCann U. D. and Ricaurte G. A. *J. Clin. Psychopharmacol.* **11:** 302, 1991.
43. Creighton F. J., Black D. L. and Hyde C. E. *Brit. J. Psychiatry* **159:** 713, 1991.

44. McGuire P. and Fahy T. *Br. J. Med.* **302:** 697, 1991.
45. Iversen L. L., Iversen, S. D. and Snyder, S. H. *Handbook of Psychopharmacology* **11:** Plenum Press, New York, 1980.
46. Seiden L. S., Sabol K. E. and Ricaurte G. A. *Annu. Rev. Pharmacol. Toxicol.* **32:** 639, 1993.
47. McCann U. D. and Ricaurte G. A. *Brain Res.* **545:** 279, 1991.
48. Steele T. D., Brewster W. K., Johnson M. P., Nichols D. E. and Yim G. K. *Pharmacol. Biochem. Behav.* **38:** 345, 1990.
49. Commins D. L., Vosmer G., Virus R. M., Woolverton W. L., Schuster C. R., and Seiden L. S. *J. Pharmacol. Exp. Ther.* **242:** 338, 1987.
50. Ricaurte G. A., DeLanney L. E., Irwin I., Langston J. W. *Brain Res.* **446:** 165, 1988.
51. Ricaurte G. A, Forno L. A Wilson M. A., DeLanney L. E., Irwin I., Molliver M. E., Langston J. W. *JAMA* **260:** 51, 1988.
52. O'Hearn L., Battaglia G., De Souza E. B., Kuhar M. J., Molliver M. E. *J. Neurosci.* **8:** 2788, 1988.
53. Insel T. R., Battaglia G., Johannessen J. N., Marra S. and DeSouza E. B. *J. Pharmacol. Exp. Ther.* **249:** 713, 1989.
54. Wilson M. A., Ricuarte G. A., and Molliver M. E. *Neuroscience* **28:** 121, 1989.
55. Ricaurte G. A., Bryan L., Strauss L., Seiden L., Schuster C. R. *Science* **229:** 986, 1985.
56. Rudnick G., Wall S. C. *Proc. Natl. Acad. Sci. USA* **89:** 1817, 1992.
57. Battaglia G., Brooks B. P., Kulsakdinun C., DeSouza E. B. *Eur. J. Pharmacol.* **149:** 159, 1988.
58. Steele T. D., McCann U. D., Ricaurte G.A. *Addiction* **89:** 539, 1994.
59. Rudnick G., Wall S. C. *Mol. Pharmacol.* **43:** 271, 1993.
60. Caldwell J., Dring L. G. and Williams R. T. *Biochem. J.* **129:** 11, 1976.
61. Simantov R. and Tauber M. Submitted for publication.

Voltammetric Studies on the Kinetics and Mechanism of Catecholamine Transporters

J. B. Justice, Jr.[1], M. D. Bailey[1], E. L. Barker[2], and R. D. Blakely[2]

[1] Department of Chemistry, Emory University, Atlanta, GA 30322 USA
[2] Department of Pharmacology, Vanderbilt University, Nashville, TN USA

1 Introduction

The plasma membrane transporters for the catecholamine neurotransmitters norepinephrine (NE) and dopamine (DA) are integral membrane proteins whose primary function appears to be the removal of respective transmitters from the extracellular fluid, terminating action. The role of NE in depression and other physiological processes has made its transport a subject of intense study (Bunney et al., 1965; Willner, 1985). The clear role that dopamine transporter plays as a target for the abused drugs cocaine and amphetamine has made it heavily studied. The recent developments in the cloning and expression of catecholamine transporters have created new opportunities to better understand the kinetics and mechanism of these transporters. In addition to the advances in the molecular biology of these transporters, a recent innovation in electrochemistry allows detailed observation of the kinetics of these transporters for a very wide range of substrates, without the need for radioactive labeling.

This chapter describes voltammetric work on the kinetics of the wild type human norepinephrine transporter (hNET) that lays the groundwork for studies to test specific hypotheses regarding interaction of substrates with these residues. It also describes preliminary results on effects of mutation of asp 75 to glu 75 in hNET on uptake of dopamine and dihydroxybenzylamine measured by liquid chromatography with electrochemical detection.

2 Structure-Activity Studies in Catecholamine Transporters

The human NE and DA transporters have now been cloned (Pacholczyk et al., 1991; Giros et al., 1992) and their amino acid sequences determined. These transporters show a high degree of sequence identity or similarity, and are believed to belong to a large family of neurotransmitter transporters whose endogenous substrates include biogenic amines and amino acids (Nelson, 1994). Recombinant DNA technology has been used to express transporters in cultured mammalian cells where they do not occur naturally (Pacholczyk 1991; Giros et al., 1992; Blakely et al., 1993; Buck, 1994).

NATO ASI Series, Vol. H 100
Neurotransmitter Release and Uptake
Edited by Şakire Pöğün

Analysis of the catecholamine transporter mechanism has evolved in part from the results of mutations in the β-adrenergic receptor (βAR). In the βAR, an aspartate (Asp113) in transmembrane region 3 and two serines (Ser 204 and 207) in TM 5 have been suggested to interact with the amino group and hydroxyl groups, respectively, of the catecholamines (Strader et al., 1988, 1989a,b, 1991).

Building on the work in the βAR of Strader et al., and earlier work implicating a negatively charged amino acid in transporters interacting with tricyclic antidepressants (Depaulis et al 1978; Maxwell et al., 1969), Kitayama et al., (1992) showed that in the DAT, aspartate 79 in TM1 was critical for transport. This amino acid is conserved in DAT, NET and SERT. Substitution of aspartate 79 with glutamate reduced V_{max} to 23.7% of that for wild type DAT. Substitution with alanine reduced activity to less than 5% of wild type. These results clearly implicate asp 79 in transport but do not prove that a direct interaction with the ligand occurs. The effect could be due to structural alterations which raise one or more energy barriers to conformational changes needed for complete cycle of states of the transporter. However, further support for interaction of an aspartate with the charged amino group of the substrate is provided by mutation of asp 33 to glu 33 in TM1 of vesicular monoamine transporter VMAT2, which reduced, but did not eliminate transport activity, while elimination of the negative charge did (Merickel et al, 1995).

As in the βAR, mutation of two serines has been shown to affect DAT activity. In the DAT, mutation of both serines 356 and 359 in TM7 to alanines reduces activity to 35% of wild type activity, while substitution by glycine reduced activity to 12% of the wild type (Kitayama et al., 1992). Because both serines were simultaneously mutated, it was not possible to examine specific interactions of meta and para hydroxyl groups in the ligand with individual serine residues, as was done for the βAR (Strader et al., 1989a). The accumulating evidence described above indicates that aspartate 79 in TM1 in DAT is critical to transport, and that serines 356 and 359 in TM7 also play an important role.

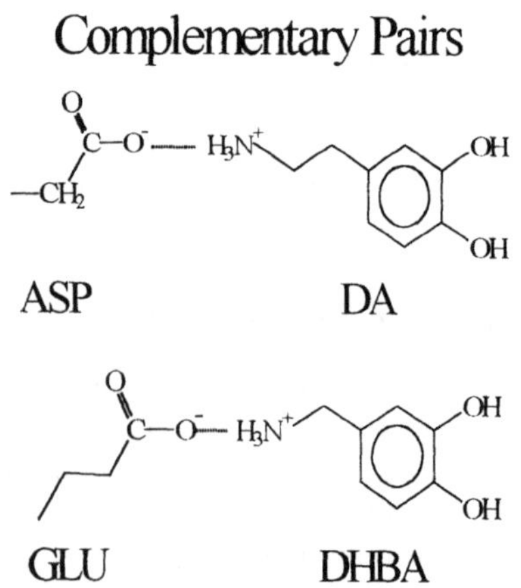

Fig 1. Complimentary pairs of amino acid residues at position 79 in dopamine transporter and substrate structures.

The hypothesis under investigation is that the amino group of catecholamines interacts directly with aspartate 75 of TM1 of NET, as has been suggested by studies on the βAR and the DAT. If this interaction exists between transporter and ligand, then lengthening the carboxylic side chain via site directed mutagenesis to glutamate should allow transport of dihydroxybenzylamine, DHBA, a dopamine analog whose side chain is one carbon shorter than dopamine. DHBA is transported by wild type NET at about 17% of DA (Burnette et al., 1996. A higher ratio of DHBA to DA transport in the glu 75 mutant would support the hypothesis that this residue directly interacts with the amine group of the substrate.

The modified cell line used in the present work (Gu et al., 1994; Melikian, 1994) contains functional NE transporters in the cell membrane, but lacks the mechanisms found in the NE nerve terminal for the synthesis, storage, exocytotic release and metabolism of NE. The absence of these complicating processes, in combination with the rapid voltammetric measurement of neurotransmitter concentration over the complete time course of an experiment, makes it possible to investigate catecholamine transporter kinetics more thoroughly than previously possible. The results demonstrate the utility of the voltammetric methodology to examine catecholamine transporters expressed in stable and transiently transfected cell lines.

2.1 Test of Asp 75/DA Interaction Using Glu75/DHBA Substitution

This experiment is designed to test the hypothesis that the positively charged amino group of catecholamine substrates interacts with the negatively charged carboxylic acid functionality of the aspartate 75 residue of transmembrane region 1 of hNET. As shown in the cartoon of Figure 1, lengthening the amino acid side chain by one CH_2 unit while also shortening the substrate amine side chain by one CH_2 maintains approximately the same distance between peptide backbone and catechol hydroxyl groups.

The experiment compares the relative rates of uptake of DA and DHBA in asp 75 (wild type) and glu 75 versions of DAT. COS were transiently transfected to express wild type or glu 75 mutant DAT. Within each cell culture, DA uptake was taken as control and compared to DHBA uptake.

The accumulation of DA and DHBA in the transiently transfected COS cells was measured by HPLC with electrochemical detection. Mutation of aspartate 75 to glutamate 75 in TM1 of NET and in the homologous position in rSERT has been described (Melikian et al., 1996). For uptake assays, transfected and nontransfected cells were grown in 24 well plates, washed, and 200 μL of 5 μM DA or DHBA added. After five minutes, the medium was removed and the cells washed three times. The cells were lysed in 200 μL buffer. The cell lysate was centrifuged, filtered and the supernatant frozen at -80 degrees C for later analysis by HPLC with electrochemical detection.

In preliminary work (figure 2) measuring initial rate of uptake of 5 μM substrate, DHBA had 15% the rate of DA in the wild type transporter, 0.79 +/- 0.13 vs. 5.29 +/-0.46 amol/min/cell. In the glu 75 mutant, both DHBA and DA had lower rates of uptake, 0.42 +/- 0.06 and 0.62 +/-0.03 amol/min/cell, respectively. In the mutant, the DHBA/DA ratio is four times what it was in the wild type transporter, with the

DHBA rate now at 67% of the rate of DA. This is consistent with an interaction of the carboxylic acid residue at position 75 with the amine group of the substrate. However, another way to look at the data is to conclude that the mutation had a large effect on DA transport, but little effect on DHBA transport. Were the asp75/amine

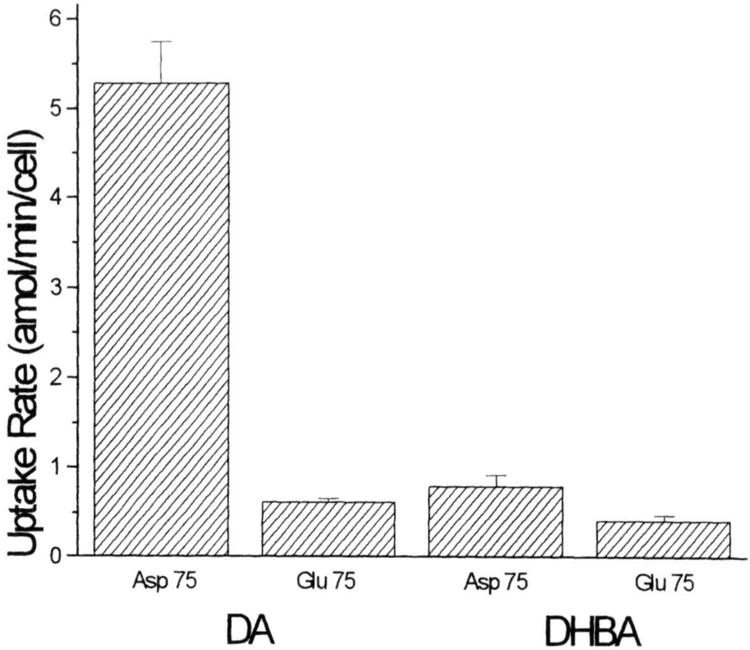

Fig 2. Rates of uptake of DA (n=6) and DHBA (n=3) in wild type and glu 75 mutants of NET

interaction the only significant interaction in transport, one might expect a reversal of relative rates of DA and DHBA in the glu mutant. To the extent that other interactions play a role in transport, the efficacy of the DHBA/asp combination will be diminished. Two types of interactions can be defined: those involving additional substrate protein interactions and those that do not involve the substrate. The former category includes binding and unbinding of the substrates on each face of the membrane. The latter category includes the conformational change that occurs as the empty carrier returns to the outer face. This step is often a slow step in transport and is a required step in the zero trans experiment. The larger glutamate residue may affect this conformational change in addition to its effect in the loaded carrier.

The conformational change involved with the return of the unloaded carrier can be bypassed by preloading the cells with appropriate substrates. Meta and para tyramine are electroactive analogs of dopamine that show trans acceleration of dopamine uptake and induce efflux of dopamine, as shown below. Meta and para hydroxybenzylamine can be used to preload the cells as equivalent analogs of DHBA. Other possible substrates for preloading are the phenethylamines. Beta-phenethylamine is a nonelectroactive analog of DA while d- and l- alpha phenethylamines are equivalent analogs of DHBA. Beta phenethylamine induces rapid and significant efflux of DA in wild type NET. The initial rate of DA efflux induced by 10 μM beta phenethylamine is 45% of the initial rate of 1 μM DA uptake. Alpha phenethylamines produce considerably less DA efflux (initial rate <5% of rate of DA uptake). Thus it would be interesting to compare DA uptake in cells preloaded with tyramine or beta-phenethylamine with DHBA uptake in cells preloaded with a hydroxybenzylamine or alpha-phenethylamine. The transport cycle would bypass the conformation change which occurs in the return of the empty carrier with its potentially rate limiting interaction of the larger glutamate residue in the empty carrier. The relative rates of uptake in the preloaded cells would provide a better evaluation of the possible direct interaction of residue 75 with the substrate. Such experiments are currently in progress.

3 Rotating Disk Electrode (RDE) Voltammetric Measurement of Transport

The measurement of accumulation of substrate by HPLC or by the more widely used use of radioactive substrates has advantages, but provides only a single point in time for each experiment. Recently, a novel application of a well established electrochemical method, rotating disk electrode voltammetry (RDE), was introduced which allows measurement of fast initial rates and also provides a record of the complete time course of uptake (Schenk et al., 1990). The method has been applied to the study of the dopamine transporter in tissue homogenates and synaptosomal preparations (McElvain et al., 1992; Meiergerd et al., 1994; Meiergerd et al., 1994; Meiergerd et al., 1994). In those studies, measurements of the initial time course of DA loss from the medium were made, rather than measurement of accumulation into tissue at the end of an incubation period. The initial decrease in DA concentration was shown to be linear with respect to time, a characteristic consistent with unidirectional transport. Such experiments can closely approximate zero trans conditions (Stein, 1986) (initial absence of substrate opposite (trans) the side of the measurement). The advantages of zero trans conditions are that it is well defined mathematically and the kinetic data complement the data obtained under equilibrium exchange conditions (Stein, 1986).

The remainder of this chapter describes the application of RDE voltammetry to catecholamine transporters expressed in cell lines. The present results are from wild type transporters, demonstrating the kind of data that can be obtained for use in mutation studies such as the mutation of asp 75 in hNET described above.

3.1 Methods

The methods have been described previously in Burnette et al., (1996). An abbreviated version is provided here. The cells used are LLC-NET cells, a line of porcine kidney cells stabley expressing the human NE transporter which were obtained from Dr. Gary Rudnick, Department of Pharmacology, Yale University (Gu et al., 1994).

3.1.1 Rotating Disk Electrode Voltammetry

RDE voltammetric measurements are made with an AFMSRX Analytical Rotator System (Pine Instrument Co.) equipped with a 3mm glassy carbon electrode, Pt auxiliary electrode, and Ag/AgCl reference electrode. An LC-4 Amperometric Detector (Bioanalytical Systems) is used as a potentiostat, and the output current amplified (Model 427, Keithley Instruments). All experiments are performed at a electrode rotation rate of 4000 rpm and an applied potential of +.45 V vs. Ag/AgCl. The temperature is maintained at 37^0C. Voltammetric measurements are acquired at a frequency of 128 Hz, averaged, and recorded at a frequency of 8 Hz on a 486 PC through a DT 2801A interface board (Data Translation) controlled by Origin data acquisition software (MicroCal Software).

3.1.2 Transport Assays

Cells are washed with 20 ml KRH buffer and harvested by scraping in two 2.5 ml volumes of room temperature Krebs/Ringer/HEPES (KRH) buffer, which were combined. The harvested cells are transferred to a single glass test tube and centrifuged at 2500 rpm (1000 × g) for 2 min. The pellet is divided for use in four replicate transport assays. At the end of each experiment the cell suspension is used for cell counting with a hemocytometer.

3.1.3 Kinetics of Transport

The pellet is re-suspended in 1.5 ml room temperature KRH buffer and transferred to a 1.5 ml centrifuge tube. The centrifuge tube is placed in a water bath at 37^0 C, and an O_2 stream is directed over the surface of the suspension. After five minutes, a 300 μl aliquot is placed in the electrochemical cell, while the rest of the suspension remains in the water bath. With the rotating glassy carbon electrode just below the surface of the solution, the potential is applied and the current monitored for 90 seconds prior to baseline acquisition. Following a 60 second period of baseline signal acquisition, 6 μl of DA or NE in 120 mM NaCl, 4.7 mM KCl, 2.2 mM $CaCl_2$, 1.2 mM $MgSO_4$ is added to the suspension to bring the solution to the desired initial concentration. After the addition, data are recorded for an additional 4 minutes.

3.1.4 Stimulated Efflux of DA

The pellet is resuspended in 2.5 ml KRH buffer saturated with a 95% O_2 / 5% CO_2 mixture and placed in a 12 × 75 mm test tube. The suspension is incubated in a 37^0 C water bath under O_2 for five minutes, after which a 400 μl aliquot is placed in the electrochemical cell under an O_2 stream. The RDE is introduced, rotation begun, and the potential applied. The current is monitored for 90 seconds, followed by a 60 second period of baseline signal acquisition. Six μL of DA is added to bring the concentration to 1 μM, and the decrease in DA then recorded for 6 minutes. The final 60 seconds are used to establish a baseline for measurement of subsequent DA efflux. To induce efflux, 8 μl of unbuffered electrolyte as a control, or 1, 10 or 100 μM of selected substrate (which does not oxidize at the potential used, +.45 V vs. Ag/AgCl) is added to the suspension.

3.1.5 Data Analysis

The baseline current preceding addition of catecholamine is subtracted and a linear regression is performed on seconds 1-11 of the clearance profile. A calibration factor for conversion of current to concentration is obtained by extrapolation of the regression line back to time of maximum rate of current increase. The initial rate of substrate clearance from the extracellular medium, v_0, is calculated using the slope and the calibration factor. Initial rates are normalized to cell count. To obtain kinetic parameter estimates for transport,the data are fit by nonlinear regression to the Michaelis-Menten function: $v_0 = (V_{max} \times [DA]) / (K_M + [DA])$. Initial rates of efflux of preloaded DA, caused by the addition of second substrate are estimated by linear regression over the 20 seconds following addition of second substrate.

3.2 RDE Measurement of Catecholamine Transporter Kinetics

3.2.1 Kinetics of NE and DA Transport

Figure 3 shows the rapid rise in current that occurs following addition of NE to a 300 μL suspension of LLC-NET cells. The time from addition of substrate to the maximum in the signal is less than one second. This indicates that rapid convective mixing occurs and also that diffusion through the medium to the cells is not limiting the delivery of substrate to the cells. Over the time course of the measurement, the initial rapid uptake rate declines to near zero, as the NE concentration in the medium decreases to a nonzero steady state level determined by the stoichiometry of coupled ions and the total intracellular volume (Stein, 1986). Although in this work the kinetic analysis is limited to the first few seconds of uptake, the complete time course of uptake obtained in this type of experiment can be expected to provide additional information about transporter function. As uptake progresses, the substrate accumulates intracellularly, slowing influx and increasing efflux. Eventually a steady state condition of equal flux in both directions is reached.

Figure 4 shows a fit of the Michaelis-Menten equation to initial rates of NE uptake. K_M for NE was estimated to be 1.30 +/- 0.21 μM (mean+/-sem), and Vmax was 200 +/-14 amol/min/cell. The K_M estimated for the transport of DA found here is larger than previously reported estimates, in which initial rates were measured as the accumulation of radiolabelled DA over periods ranging from 2 to 20 minutes (Giros et al., 1994; Buck, 1994; Gu et al., 1994; Melikian et al., 1994). Reported values of K_M are 200+/-100 nM (Melikian et al., 1994) and 670 +/-40 nM (Giros et al., 1994) for inward DA transport estimated by accumulation of radiolabel by the NE transporter expressed in cultured HeLa and COS-7 cells, respectively. Using the RDE method, Meiergerd et al. (1994) found a K_m of 1.2 +/- 0.1 μM for DA at the dopamine transporter in striatal homogenates, similar to that found in the present work on the NE transporter.

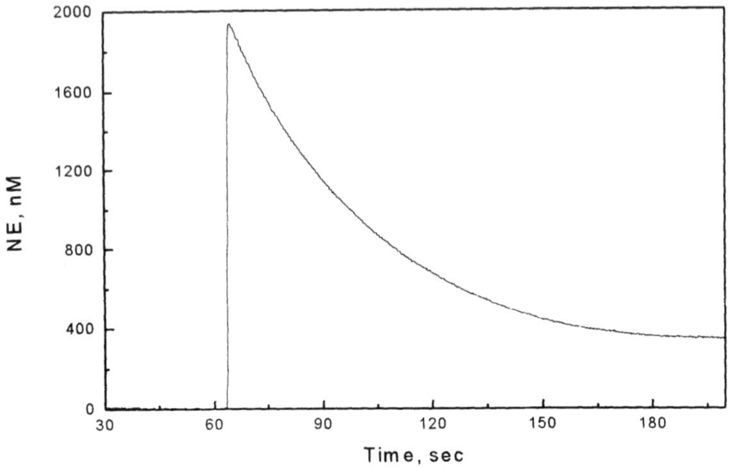

Fig 3. Uptake of NE. Concentration vs. time profile after baseline subtraction for addition of 2000 nM NE.

The V_{max} reported here for DA flux is about 3 times larger than a value of 60 amol/min/cell estimated by accumulation of tritiated DA in LLC-NET cells (Blakely, unpublished). Several factors may contribute to the difference. In the studies using radioactive uptake, the cells were a confluent monolayer on the surface of a tissue culture plate. This configuration reduces the cell surface area and therefore the number of transporters of each cell which are exposed to the medium. Also, the rapid convective mixing used in the present method reduces any effects of diffusion limited mass transport to the cell surface. Finally, the initial rate during the first few

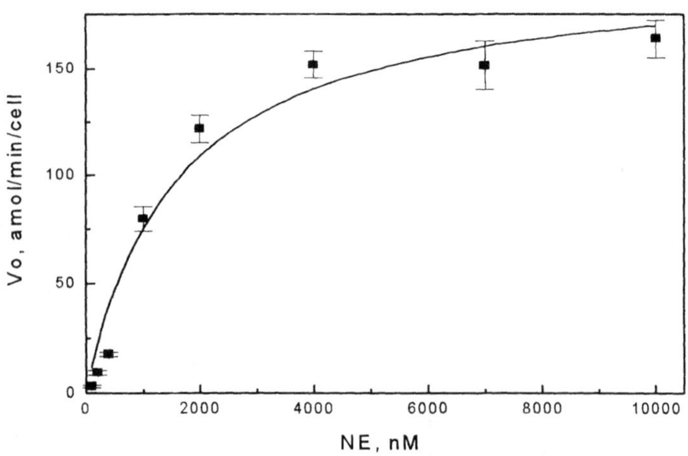

Fig 4. Kinetics of NE transport in LLC-NET cells. Points are the mean +/- S.E.M. initial rates of 4 replicates at each concentration. The solid line was generated by fitting the Michaelis-Menten function to the data: K_M = 1.64 +/- 0.46 μM, V_{max} = 197 +/- 17 amol/min/cell. From Burnette et al., (1996), with permission.

seconds of uptake may be higher than an initial rate estimated over longer periods of time, where rates may be slowed by intracellular amine accumulation. The experiment was repeated for NE over a range of concentrations. For NE estimates for K_M and V_{max} are 1.64 +/- 0.46 μM and 197 +/- 17 amol/min/cell (mean+/-sem), respectively. Reported values of K_M obtained by accumulation of radiolabelled NE by cloned NET expressed in cultured cells are 400 nM (Melikian, et al., 1994) , 457 nM (Pacholczyk, et al., 1991) and 2.6 μM (Giros, et al., 1994).

3.2.2 Stimulated Efflux of NE and DA

Clearance of 1μM DA followed by p-tyramine-induced efflux (100μM) is shown in Figure 5. Following addition of 1 μM DA, uptake proceeds until the system is at steady state with the concentration in the medium less than 200 nM. Addition of a second substrate induces immediate efflux. Following the initial rise, efflux

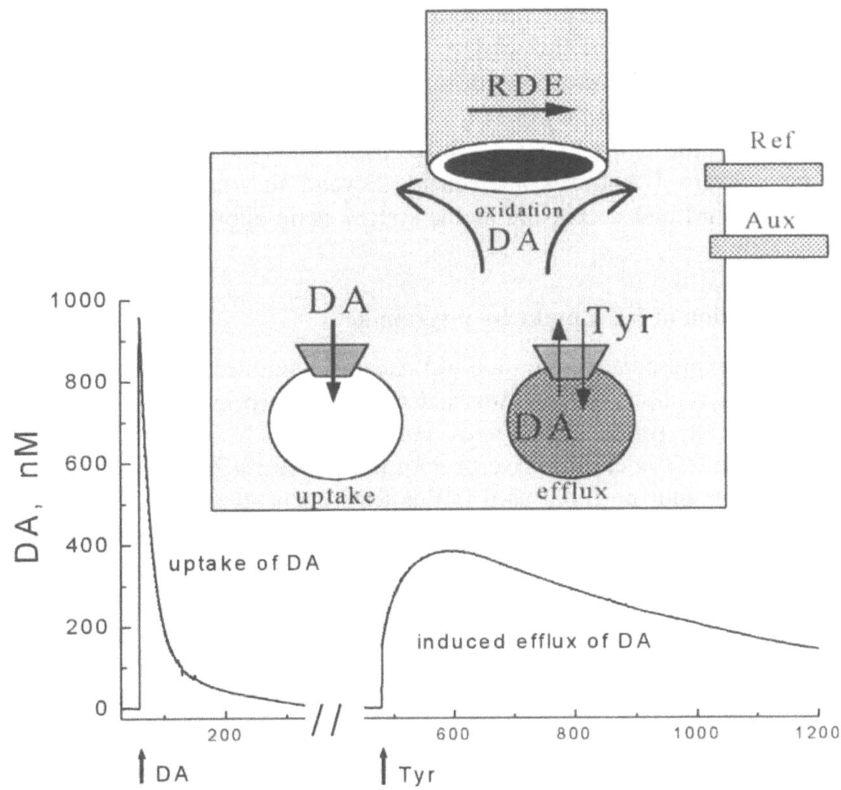

Fig 5. Uptake of DA followed by tyramine induced efflux of DA.Upper part of figure illustrates measurement of DA in the medium by RDE during the two parts of the experiment. Decline in added DA concentration is first obseved as DA is taken up into cells. Addition of tyramine induces DA efflux in second part of experiment.

continues until the concentration of DA in the medium has increased and intracellular DA decreased to the point that the net rate of transport changes sign and net uptake occurs.

The transport of tyramine was not directly observed in this experiment, but its ability to stimulate efflux of DA provides additional evidence that it is exchanged with DA through the transporter, in agreement with earlier studies of d-amphetamine and p-tyramine as substrates of the catecholamine transporters (Fischer et al., 1979; Brodie et al., 1969; Deves et al., 1979).

The efflux data can be understood as simple competition for the transporter on both sides of the membrane. Consider the stimulated efflux of DA by p-tyramine as an example. During the early rising phase of DA concentration following the addition of p-tyramine, net DA flux is outward, due to the high p-tyramine causing inward transport of p-tyramine, reducing the influx of DA. Initially there is little p-tyramine inside the cell to compete with the high inside concentration of DA for outward transport. DA can therefore bind to the available transporter, resulting in increased efflux of DA. As p-tyramine accumulates intracellularly, its competition increases, reducing outward flux of DA. At the same time, DA increases in the medium and p-tyramine decreases, shifting the competition outside more in favor of DA, increasing the inward flux of DA. The result is that the DA concentration in the medium goes through a maximum at the point when the increasing inward and decreasing outward DA fluxes are equal. Beyond the maximum, the net DA transport is inward and decreasing as the system again approaches the new steady state.

3.3.3 Acceleration of NE Uptake by p-tyramine

Figure 6 shows representative NE (400 nM) clearance profiles when either buffer, as a control, or p-tyramine (400 nM) was added to a cell suspension 2 minutes prior to the addition of NE (Inset: initial rates, mean +/- sem, N = 4 for both groups). Although p-tyramine is electroactive, it does not produce a measureable current at the concentration and potential used in this experiment (0.45 V vs. Ag/AgCl). It was therefore possible to observe the change in the initial rate of unidirectional NE transport uncomplicated by any current due to p-tyramine oxidation at the electrode. Preloading with p-tyramine caused the initial rate of NE clearance to increase by 60 percent, from 23.6 +/- 0.7 to 37.6 +/- 0.4 amol/min/cell. This result illustrates trans acceleration of NE transport, a characteristic which distinguishes transporters from channels. In models of carrier mediated transport, it is often postulated that a loaded transporter changes conformation between sides of the membrane faster than an unloaded one. In this view, the increased rate of NE uptake by p-tyramine loaded cells is thought to arise from the transporter returning faster loaded with p-tyramine which then dissociates instead of returning more slowly empty to a conformation ready for inward transport.

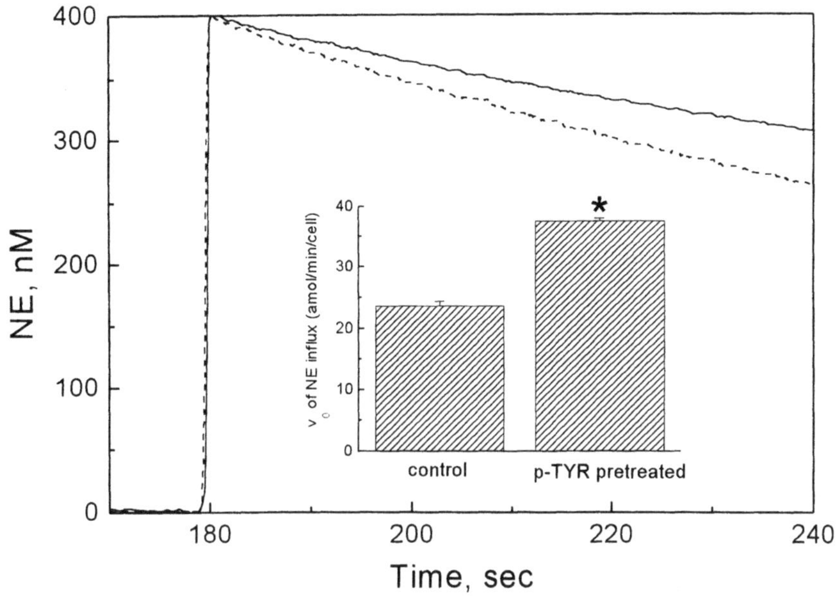

Fig 6. Trans acceleration of NE uptake by preloaded p-tyramine. Clearance of 400 nM NE where either buffer alone (solid line) or 400 nM p-tyramine, (p-TYR, (dotted line) was added two minutes prior to NE. Inset: Initial rates of transport. Control rate is 23.6 +/-0.7 anol/min/cell; NE + p-TYR rate is 37.6 +/- 0.8 amol/min/cell, p<0.001. From Burnette et al., (1996), with permission.

4. References

Blakely, R. D., Moore, K. M., Qian, Y., Tails of serotonin and norepinephrine transporters: deletion and chimeras retain function, *Molecular Biology and Function of Carrier Proteins.*; Rockefeller University Press: New York, 1993; 283-300.

Brodie, B. B., Cho, A. K., Stefano, F. J. E., Gessa, G. L., 1969, On mechanisms of norepienphrine release by amphetamine and tyramine and tolerance to their effect. *Adv. Biochem. Psychopharm.*,*1*, 219-38.

Buck, K. J., Amara, S. G., 1994, Chimeric dopamine-norepinephrine transporters delineate structural domains influencing selectivity for catecholamines and 1-methyl-4- phenylpyridinium, *Proc. Natl. Acad. Sci.*, 91, 12584-88.

Bunney, W. E., Jr., Davis, J. M., 1965 Norepinephrine in depressive reactions, *Arch Gen. Psychiatry.*, 13, 509-22.

Burnette, W. B., Bailey, M. D., Kukoyi, S., Blakely, R. D., Trowbridge, C.G., and Justice, J. B. Jr., Investigation of human norepinephrine transporter kinetics using rotating disk electrode voltammetry, *Anal. Chem.*, 1996, in press.

DePaulis, T., Kelder, D., and Ross, S. B.,1978, On the topology of the norepinephrine transport carrier in rat hyptthalmus: the site of action of tricyclic uptake inhibitors, *Mol. Pharm.*, 14, 596-606.

Deves, R., Krupka, R. M., 1979, A general kinetic analysis of transport:Test of the carrier model based on predicted relations among experimental parameters. *Biochim. Biophys. Acta.*, 556, 533-47.

Fischer, J. F., and Cho, A. K., 1979, Chemical release of dopamine from striatal homogenates: evidence for an exchange diffusion model. J. Pharmacol. Exp. Er., 208, 203-09.

Giros, B., El Mestikawy, S., Godinot, N., Zheng, K., Han, H., Yang-Feng, T., Caron, M. G.,1992, Cloning, pharmacological characterization, and chromosome assignment of the human dopamine transporter, *Mol. Pharmacol.*, *42,* 383-90.

Giros, B., Wang, Y.-M., Suter, S., McLeskey, S. B., Pifl, C., Caron, M. G., 1994, Delineation of discrete domains for substrate, cocaine and tricyclic antidepressant interactions using chimeric dopamine-norephinephrine transporters, *J. Biol. Chem.*, *269,* 15985-88.

Gu, H., Wall, S. C., Rudnick, G.,. 1994, Stable expression of biogenic amine transporters reveals differences in inhibitor sensitivity, kinetics and ion dependence, *J. Biol. Chem*, *269,* 7124-30.

Kitayama, S., Shimada, S., Xu, H., Markham, L., Donovan, D.M., and Uhl, G.R.,1992, Dopamine transporter site-directed mutations differentially alter substrate transport and cocaine binding, *Proc. Natl. Acad. Sci. USA*, 89, 7782-7785.

Maxwell, R. A., Keenan, P. D., Chaplin, E., Roth., B., Eckhardt, S. B., 1969, Molecular features affecting the potency of tricyclic antidepressants and structurally-related compounds as inhibitors of the uptake of tritiated norepinephrine by rabbit aortic strips. *J. Pharmacol. Exp. Ther.* 166, 320-329.

McElvain, J. S., Schenk, J. O., 1992, A multisubstrate mechanism of striatal dopamine uptake and its inhibition by cocaine, *Biochem. Pharmacol.*, *43,* 2189-99.

Meiergerd, S. M., Hooks, S. M., Schenk, J. O., 1994, The striatal transporter for dopamine in the rat may be kinetically up- regulated following three weeks of withdrawal from cocaine self-administration, *J. Neurochem.*, *63,*1277-81.

Meiergerd, S. M., McElvain, J. S., Schenk, J. O., 1994, The effects of cocaine and repeated cocaine followed by withdrawal: alterations of dopaminergic transporter turnover with no changes in kinetics of substrate recognition, *Biochem. Pharmacol.47,* 1627-34.

Meiergerd, S. M., Schenk, J. O., 1994, Kinetic evaluation of the commanality between the site(s) of action of cocaine and some other structurally similar and dissimilar inhibitors of the striatal transporter for dopamine, *J. Neurochem.*, *63,* 1683-92.

Meiergerd, S. M., Schenk, J. O., 1994, Striatal transporter for dopamine: Catechol structure-activity studies and susceptibility to chemical modification, *J. Neurochem. 62,*
998-1008.

Melikian, H.E., McDonald, J.K., Gu, H., Rudnick, G., Moore, K.R., Blakely, R.D., 1994, Human norepinephrine transporter: Biosynthetic studies using a site-b directed polyclonal antibody, *J. Biol. Chem.*, 269, 12290-12297.

Melikian, H. E., Moore, K. R., Qian, Y., Kimmel, H. L, Taylor,S.B., Gereau, R. W., Levey, A., and Blakely, R. D., 1993, Structure and function of plasma membrane serotonin transporters, *Soc. Neurosci . Abs.* 206.1, p494.

Merickel, A., Rosandich, P., Peters, D., and Edwards, R. H., 1995, Identification of residues involved in substrate recognition by a vesicular monoamine transporter, *J. Biol. Chem.*, 270, 25798-25804.

Nelson, N., and Lill, H., 1994, Porters and neurotrasmitter transporters, *J. Exp. Biol.*, *196*, 213-228.

Pacholczyk, T., Blakely, R. D., Amara, S. G., 1991, Expression cloning of a cocaine and antidepressant-sensitive human noradrenaline transporter, *Nature.*, *350,* 350-53.

Schenk, J. O., Patterson, T. A., McElvain, J. S., 1990, Rotating disk voltammetric measurements in neurobiology and neuropharmacology, *Trends Anal. Chem.*, *9,* 325- 30.

Stein, W. D., 1986, *Transport and Diffusion Across Cell Membranes*; Academic Press.

Strader, C.D. Sigal, I.S, Candelore, M.R., Rands, E., Hill,W.S and Dixon, R.A.F., 1988, Conserved aspartic acid residues 79 and 113 of the β-adrenergic receptor have different roles in receptor function, *J. Biol. Chem.*, 263, 10267-10271.

Strader, C.D., Candelore, M.R., Rands, E., Sigal, I.S, and Dixon, R.A.F., 1989a, Identification of two serine residues involved in agonist activation of the β-adrenergic receptor, *J. Biol. Chem., 264, 13572-13578.*

Strader, C.D. Sigal, I.S, and Dixon, R.A.F., 1989b, Structural basis of β-adrenergic receptor function, *FASEB Journal.*, 3, 1825-1832.

Strader, C.D., Gaffney, T., Sugg, E.E., Candelore, M.R., Keys, R., Patchett, A.A., and Dixon, R.A.F., 1991, Allele-specific activation of genetically engineered receptors, *J. Biol. Chem.*, 266, 5-8.

Willner, P., 1985, *Depression: a Psychobiological Synthesis*, Wiley-Interscience.

Coupling of Glutamate Uptake and Degradation in Transmitter Clearance: Anatomical Evidence

Amin Derouiche
Max-Planck-Institute for Brain Research, Deutschordenstr. 46,
D-60528 Frankfurt/M, Germany

Keywords. glutamine synthetase, glutamate transporter, GLAST

1. INTRODUCTION

One reason for investigating glutamate uptake in the CNS has been the requirement for rather low extracellular glutamate levels (ca. 1µM) and their tight homeostatic control. Without it, raised extracellular glutamate would lead to uncontrolled neuronal excitation and, ultimately, to neuronal death. The inactivation of glutamate seems to rely both on its uptake and its subsequent enzymatic degradation because - as hypothesized below- uptake is dependent on efficient intracellular removal of glutamate. Regarding the proteins mediating these processes, four high affinity glutamate transporters have been described (Danbolt, 1994; Fairman et al., 1995) and three of them localized in the rat CNS (Rothstein et al., 1994).

However, a defined mechanism of enzymatic degradation has not been established for glutamate released at synapses inspite of the presence of several glutamate degrading enzymes in the CNS. The situation is complicated by the subdivision of total brain glutamate into metabolic (60-70%) and transmitter (30-40%; Berl and Clarke, 1978) pools, which also do not interchange. The small, highly active pool is localized in astrocytes, and the larger in neurones (Balazs et al., 1972). This article focusses mainly on this issue of degradation of transmitter glutamate.

In anatomical terms, the requirements for a transmitter degrading enzyme would be its presence in the immediate synaptic micro-environment, and at least some degree of topographic correlation with glutamatergic terminal fields. Such an enzyme must also be present in cells with a transporter mediating high affinity glutamate uptake. Amongst several glutamate metabolizing enzymes present in the brain, three have been considered in principle for metabolism of transmitter glutamate (comp. Kugler, 1993); aspartate aminotransferase, glutamate dehydrogenase and glutamine synthetase.

A role of glutamine synthetase (EC 6.3.1.2.; GS) in degradation of transmitter glutamate was indicated by GS - immunoreactivity (ir) in astrocytes (Martinez-Hernandez et al., 1977). The reaction catalyzed by GS is the amidation of glutamate using *free* ammonia, thereby forming glutamine (Fig. 1). In addition, GS is the only enzyme in the CNS to metabolize free ammonia, a potent neurotoxin.

NATO ASI Series, Vol. H 100
Neurotransmitter Release and Uptake
Edited by Şakire Pöğün

Fig. 1
Glutamate degradation
by glutamine synthetase

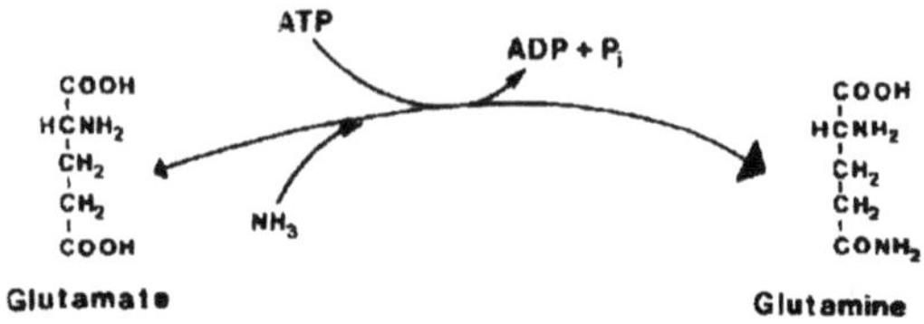

It is demonstrated here that the topographic requirements mentioned for a transmitter degrading enzyme are fulfilled in the case of GS, and that GS localization in the retina correlates with that of a particular glutamate transporter (L-Glutamate/ L-Aspartate Transporter, GLAST).

2.a THE COINCIDENCE OF GLUTAMINE SYNTHETASE IR WITH GLUTAMATERGIC TERMINAL FIELDS

For the investigation of a possible regional variation of GS-ir reflecting the extent of glutamatergic terminal fields one would have to carry out quantitative immunocytochemistry. In order to avoid the inherent technical

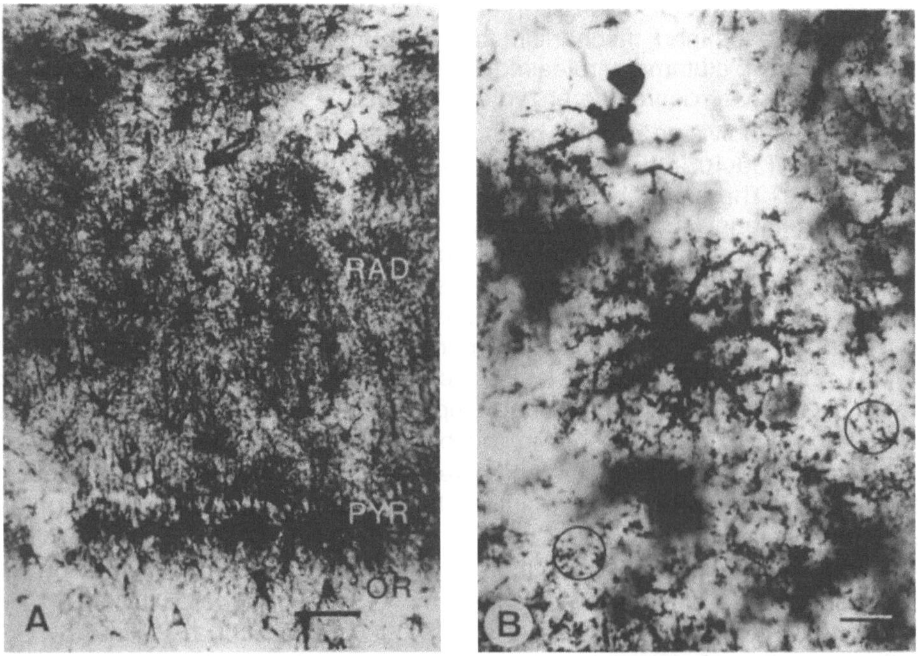

Fig. 2 GS-immunoreactive astrocytes in CA1 of rat hippocampus. (A) Astrocytes are densely spaced. At the lighter periphery of their territories, processes are less dense. (B) The "background" in (A) is seen to consist of the fine astrocytic processes (circles) in the neuropil. Note their "spiny" extensions. Stratum radiatum (RAD), stratum oriens (OR) and pyramidal cell layer (PYR), bars 50µm (A), 10µm (B)

difficulties, CNS regions with *laminated* glutamatergic terminal fields were chosen which differ markedly from adjacent laminae in their density of glutamatergic terminals. Thus, differences in GS immunoreactivity within the distance of several tens of micrometers in the same tissue section should be free of technical artefacts.

Rat brain and retina were fixed and prepared (as described by Derouiche and Frotscher, 1991, Derouiche and Rauen, 1995) and imunocytochemistry was carried out on free floating Vibratome sections, using diaminobenzidine as chromogen. A polyclonal rabbit anti-GS antiserum (Hallermayer and Hamprecht, 1984) was applied, which was the kind gift of Dr. B. Hamprecht . Sections for ultrastructural examination were silver-intensified (Derouiche and Frotscher, 1991), dehydrated, flat-embedded in araldite and further processed for light- and electron microscopic examination.

Since the first immunocytochemical demonstrations of GS (Martinez-Hernandez et al., 1977) it has been suggested that this enzyme may use transmitter glutamate as its substrate because immunoreaction was higher in forebrain regions such as stiatum and cortex than in midbrain or medulla.

In the present investigation, GS-ir was detected in all CNS regions, however, at differential intensity. The forebrain region displaying strongest ir was the hippocampus, which was camparable in staining to the olfactory bulb and the retina. Fields of differential staining within a given region were either merging into one another or clearly separated by lines.

Rat brain sections stained with the anti-GS antiserum reveal exclusively astrocytes and never neurones. The density of labelled cells suggests that all astrocytes are GS -immunoreactive (Fig. 2A), which has been confirmed in rat hippocampus by double-immunofluorescence with GFAP (not shown). The single astrocyte is labelled with its entire process tree and even its most distal, extremely fine processes which would not be revealed by GFAP-immunostaining (Fig. 2B). These most peripheral processes, which constitute approx. 75% of the cell surface and approx. 30% of all plasma membranes in cortex (Wolff, 1970), are therefore, in focus for a possible role of astrocytes in neurotransmission. The following regional comparisons are based on the homogeneous "background" in GS-immunocytochemistry, which is due to the abundance of these finest processes at the limit of light microscopic resolution. Some exemplary regions with a clearly laminated pattern of GS-ir are presented in the following, and the correspondence with glutamatergic terminal fields is explained.

The hippocampus displayed strongest labelling in the molecular layer of the fascia dentata with the outer two-thirds being markedly darker than the inner band (Fig. 3 A,B). These two layers have a clear line of separation. A similar line is apparent in ^{3}[H]-glutamate binding studies (Monaghan et al., 1983) and in glutamate-immunocytochemistry (Storm-Mathisen et al., 1983). Thus, the laminated pattern of GS-ir corresponds to the density distribution of glutamatergic terminals. The outer lamina receives a dense glutamatergic input from the entorhinal cortex (Steward, 1976; Witter et al., 1988). The

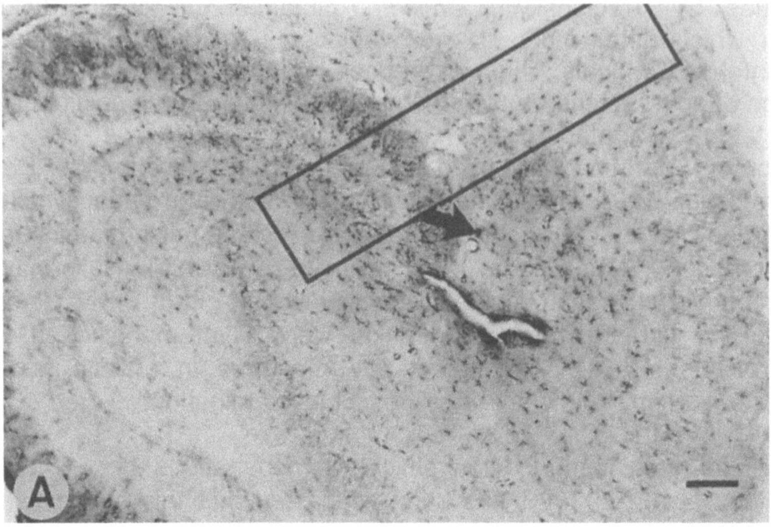

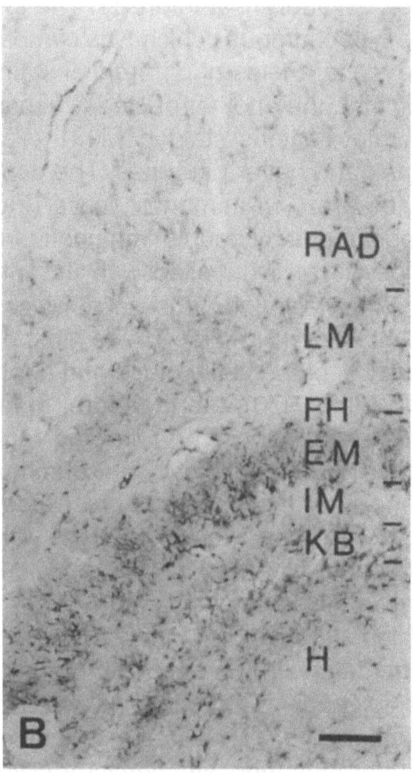

Fig. 3 A, B

A higher intensity of glutamine synthetase - ir is visible in the fascia dentata and the stratum lacunosum moleculare (A).

(B) Magnification of the boxed area in (A). The exterior portion of the molecular layer (EM) displays markedly stronger ir than its inner portion (IM). Similarly, the stratum lacunosum-moleculare (LM) of CA1 is darker than the adjacent stratum radiatum (RAD).
Hippocampal fissure (FH), granule cell layer (KB), hilus (H); bars 200μm (A), 100μm (B)

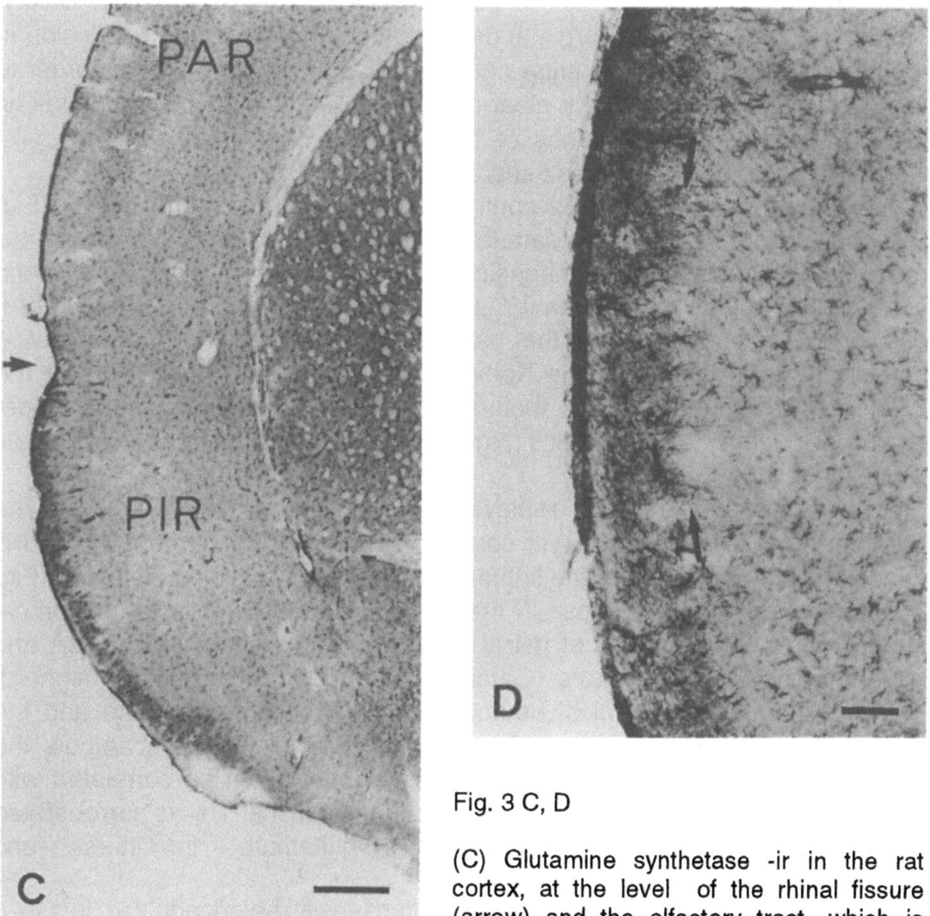

Fig. 3 C, D

(C) Glutamine synthetase -ir in the rat cortex, at the level of the rhinal fissure (arrow) and the olfactory tract, which is immunonegative. A dark band of heavy labelling is prominent in the most superficial layer of the prepiriform cortex (PIR). It extends from the olfactory tract to the rhinal fissure. Its clear demarcation to underlying strata is obvious even at higher magnification (D). The parietal cortex (PAR) in (C) displays a softer pattern. Both bands correspond to glutamatergic terminal fields (see text). Bars 500µm (C), 50µm (D) (reproduced from Derouiche et al., 1996, with permission from Cambridge Univ. Press)

inner lamina contains glutamatergic terminals from associational and commissural fibres and, in addition, non-glutamatergic synapses from septal (cholinergic) and intrinsic (GABAergic) afferents, which could explain the lower intensity of GS-ir in the inner portion of the the molecular layer.

In the prepiriform cortex, there is a conspicuous, dark band in the uppermost lamina (Fig. 3 C,D), which is only few cell layers in width and sharply set off against the deeper and lighter laminae. This dark band, corresponding to lamina IA of the prepiriform cortex extends from the olfactory tract to the rhinal fissure (Fig. 3C). The terminal field of the olfactory projection is lamina IA, but not lamina IB (Price, 1973; Westrum and Bakay, 1986). This is revealed by terminal degeneration after olfactory bulb lesion in the prepiriform cortex which shows as a small, subpial band with a border to the underlying strata similarly clear-cut as in GS-ir (Westrum and Bakay, 1986). The projection neurones of the olfactory bulb are the mitral cells, which are glutamatergic (Halász and Shepherd, 1983).

The parietal cortex, which is continuous with the prepiriform cortex (Fig. 3C) only displays a soft pattern of lamination visible only at low magnification. At higher magnification, the darker, superficial band corresponds to the lower lamina IV/ upper lamina V. This softer band of GS-ir can be seen particularly in the parietal cortex containing the (somato-) sensory afferents (for example König and Klippel, 1963). It appears at the level of the terminals of the thalamic projection (lower lamina IV/ upper lamina V), which also contains highest density of glutamate-ir terminals (Conti et al., 1987).

GS-ir in the olfactory bulb displays also a laminar pattern (Fig. 4A). The band of the outer plexiform layer contains strongest immunoreactivity and is thus clearly separated from the lighter mitral cell and granule cell layers (Fig. 4B). The predominant synapse in the outer plexiform layer is the reciprocal synapse between dendrites of mitral cells and granule cells with GABA and glutamate as the transmitters (Garcia et al., 1995; Halász and Shepherd, 1983). Although the chemical neuroanatomy of the other laminae and the glomeruli is less well described than for example in the hippocampus, the intensity of GS-ir in the outer plexiform layer may again be correlated with the high density of glutamatergic terminals in this lamina. This is corroborated by a very similar, laminated distribution of glutamate - ir (Ottersen and Storm-Mathisen, 1984).

The lamination of the cerebellar cortex can easily be recognized in GS-ir sections (Fig. 4C). The molecular layer is stained much darker than the granule cell layer, the light Purkinje cells being completely surrounded by immunoreactive Bergmann glia (Fig. 4D). The trilaminar pattern seen in GS-ir corresponds to the differential densities of glutamatergic synapses as shown by glutamate - ir (Ottersen and Storm-Mathisen, 1984). The lamina with highest GS-ir and highest density of glutamatergic synapses is the molecular

---

Fig. 4 A, B

The olfactory bulb with glutamine synthetase -ir, left: rostral. The laminated pattern is recognized at low power view in (A). Strongest immunoreaction is present in the external plexiform layer, it is weakest in the olfactory nerve fibre layer (below).
(B) shows the layers of the olfactory bulb, as indicated by arrowheads in (A). The dark label in the external plexiform layer (EPL) is clearly separated from the mitral cell layer (M) and the glomeruli (GLOM). GR, granule cell layer; ON, olfactory nerve; bars 500µm (A), 125µm (B)

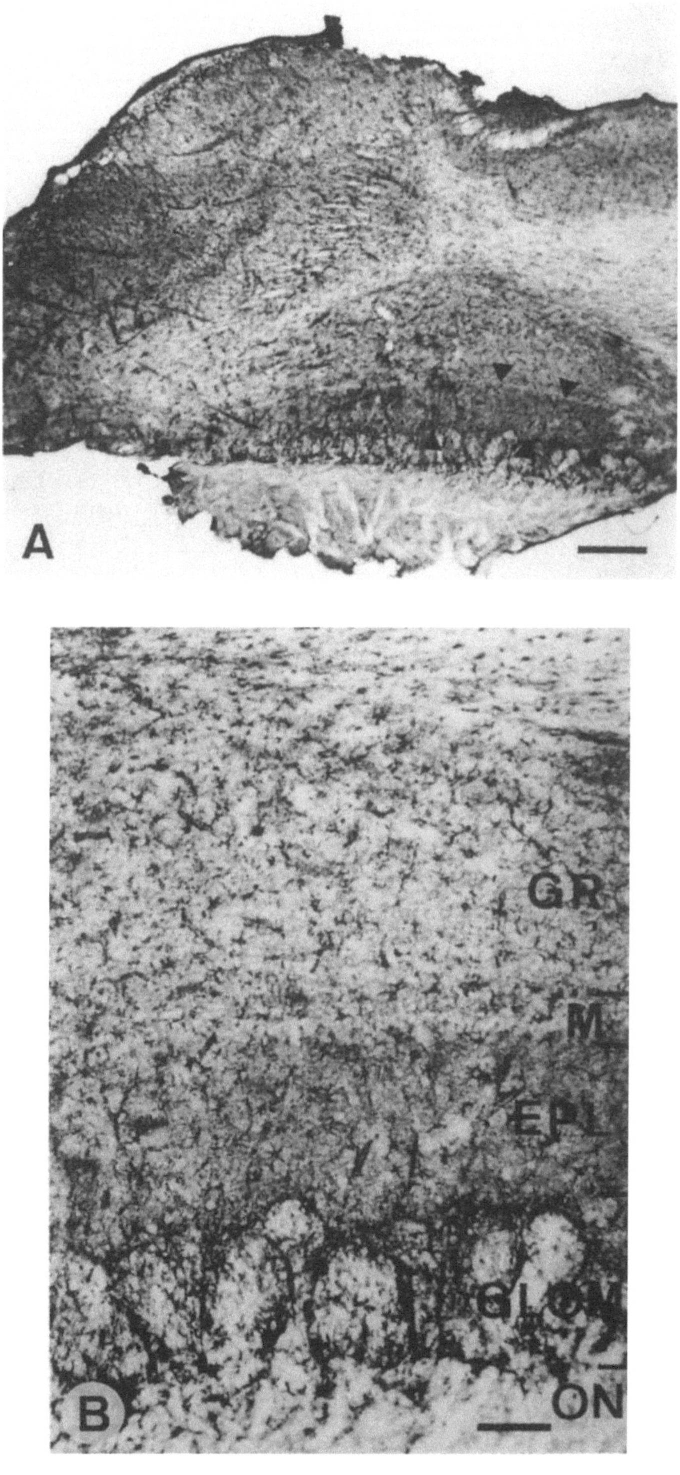
A
B
GR
M
EPL
GLOM
ON

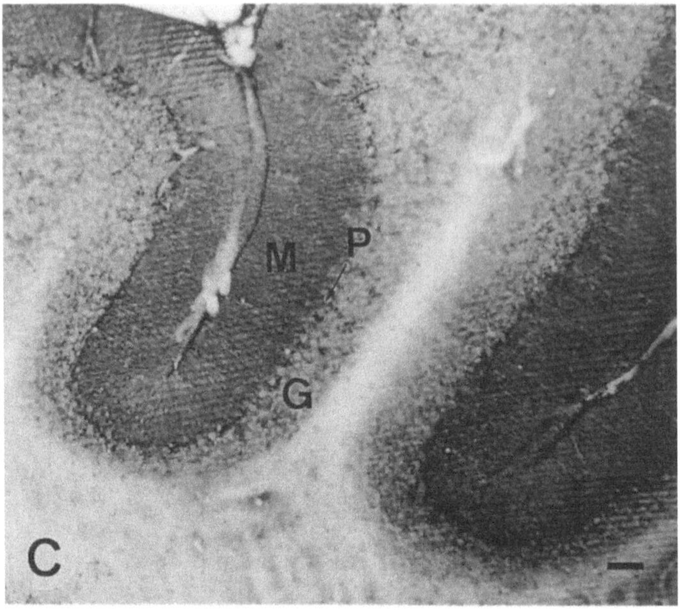
M
P
G
C

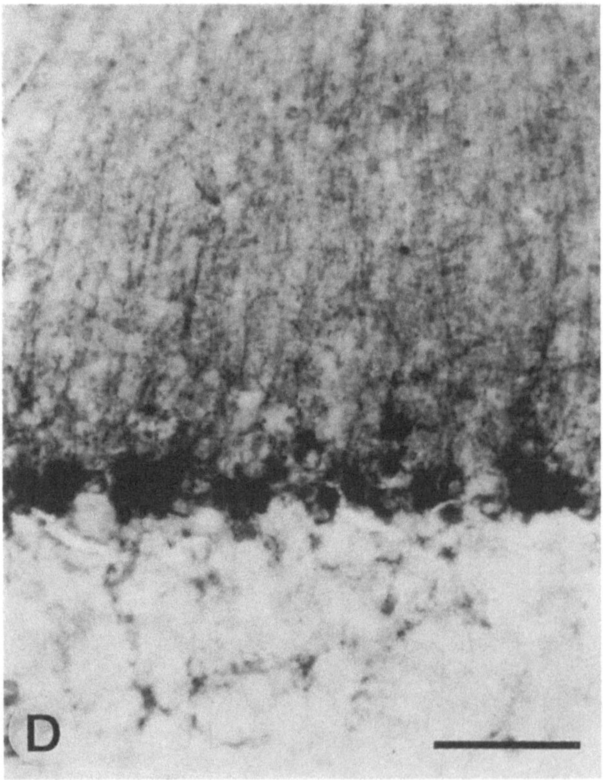
D

layer with its approx. 200 000 parallel-fibre synapses per Purkinje cell (Palay and Chan-Palay, 1974). The granule cells, from which the parallel fibres originate are known to be glutamatergic (Barbour et al., 1994).

2.b GLUTAMINE SYNTHETASE - IR IN GLIAL PROCESSES AROUND IDENTIFIED GLUTAMATERGIC SYNAPSES

The regions chosen for ultrastructural investigation of *identified* glutamatergic synapses where the outer portion of the fascia dentata in the dentate gyrus, and the rod sherule in the retina. Previous studies have shown that synapses in these regions use glutamate as transmitter: The perforant path supplies by far the most synapses in the outer molecular layer of the hippocampal fascia dentata (Steward, 1976; Witter et al., 1988), and it has been demonstrated coherently by glutamate receptor autoradiography and glutamate immunocytochemistry that the perforant path projection is glutamatergic (Maragos et al., 1988; Monaghan et al., 1983; Storm-Mathisen et al., 1983). Excitatory, i.e. asymmetric boutons in this region are almost exclusively perforant path terminals, and they can easily be distinguished from synapses of diffuse projections based on ultrastructural criteria.

The retinal rod spherule is well-characterized by localization, shape and its synaptic ribbons. It has been established that this synapse is glutamatergic (Massey, 1990; Massey and Redburn, 1987; Wässle and Boycott, 1991) although the position of the actual synaptic site within the spherule and the mechanism of synaptic transmission remain to be clarified.

The ultrastructural localization of GS-ir was exclusively in glial profiles, thus confirming the light microscopical findings. The method of silver intensification revealed GS-ir even in finest glial processes, which often consist of mere membrane appositions and are less than 50nm thick (Fig. 5A). The frequent association of these fine processes with identified glutamatergic synapses was a key finding: Two asymmetric, putatively glutamatergic synapses on dendritic spines in the fascia dentata are illustrated (Fig. 5A,B). The GS-ir glial process often approaches the synaptic cleft laterally or surrounds the spine and/ or the presynaptic element completely. Tiny protrusions from the fine glial profile into the synaptic cleft are consistently observed (Figs. 5 A,B). A similar association of GS-ir astrocyte processes with identified glutamatergic synapses has been shown in the case of the hippocampal

---

Fig. 4 C, D

(C) The layers of the cerebellar cortex are clearly revealed by glutamine synthetase - ir, the molecular layer (M) being labelled most intensely. The Purkinje cells (P), although not stained, are visible because of their tight glial coverage. (D) The cell types staining with this antiserum are the Bergmann - glia in the molecular layer, and sparse astrocytes in the granule cell layer (G). Bars 50µm

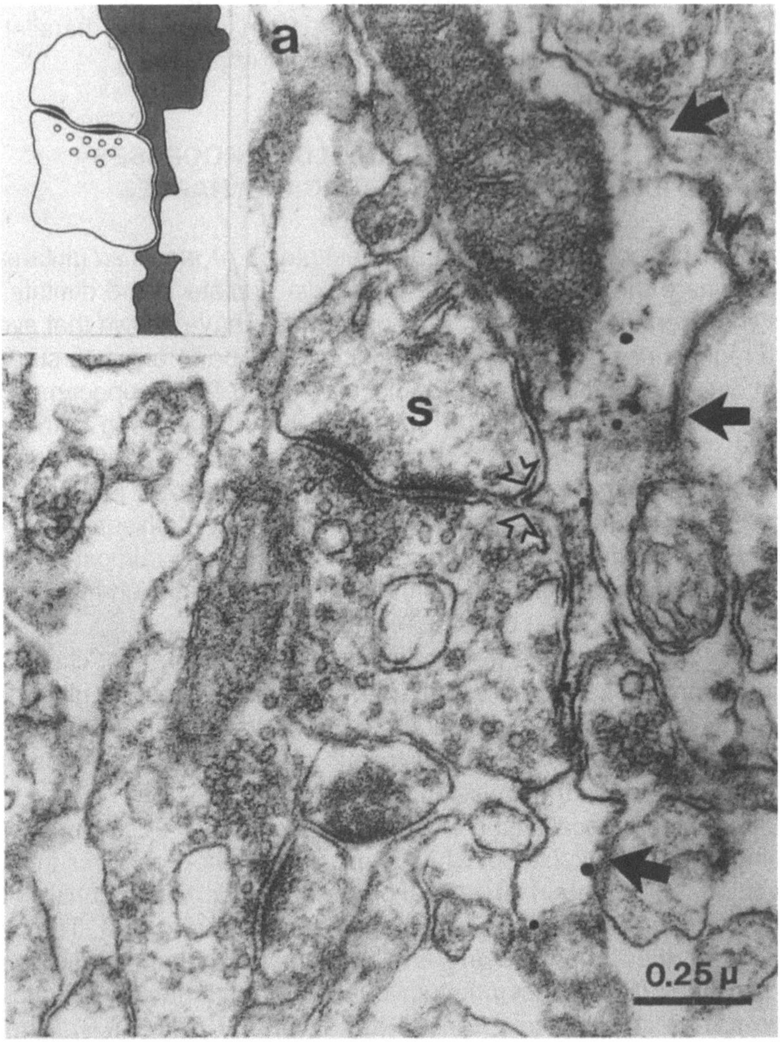

mossy fibre terminals in CA3, and the associational and commissural terminals in the statum radiatum of CA1 (Derouiche and Frotscher, 1991).

In the retina, the predominant member of the GS-ir astrocytic cell family is the Müller cell (Fig. 6B), but retinal astrocytes are also GS-ir, which can be detected, however, only ultrastructurally (Derouiche and Rauen, 1995). The terminals of rods in the outer plexiform layer are fully surrounded by GS-ir Müller cell cytoplasm (Fig. 7B), which can be seen already at light microscopical resolution (Fig. 6B). Ultrastructurally, the localization of GS-ir in fine, organelle-free glial processes is confirmed (Fig. 7B). The rod spherule is fully surrounded by GR-ir glia thus leading to an almost complete sequestration of the glutamtergic spherule from the surrounding neuropil.

Fig. 5

(A , opposite)

An astrocyte process (depicted schematically in inset) containing immunoreactivity for glutamine synthetase (silver grains, dark arrows) is tightly apposed to an asymmetrical spine (s) synapse. Situation in the outer molecular layer of the hippocampal fascia dentata. Open arrows point to a glial protrusion into the synaptic cleft.

(B)

An asymmetrical spine synapse taken from the same region as in (A). Here, the spine (s) is partly and the presynaptic element completely ensheathed by a glutamine synthetase - ir astrocyte process (dark arrows). Glial protrusions seal the synaptic cleft from both sides (open arrows).

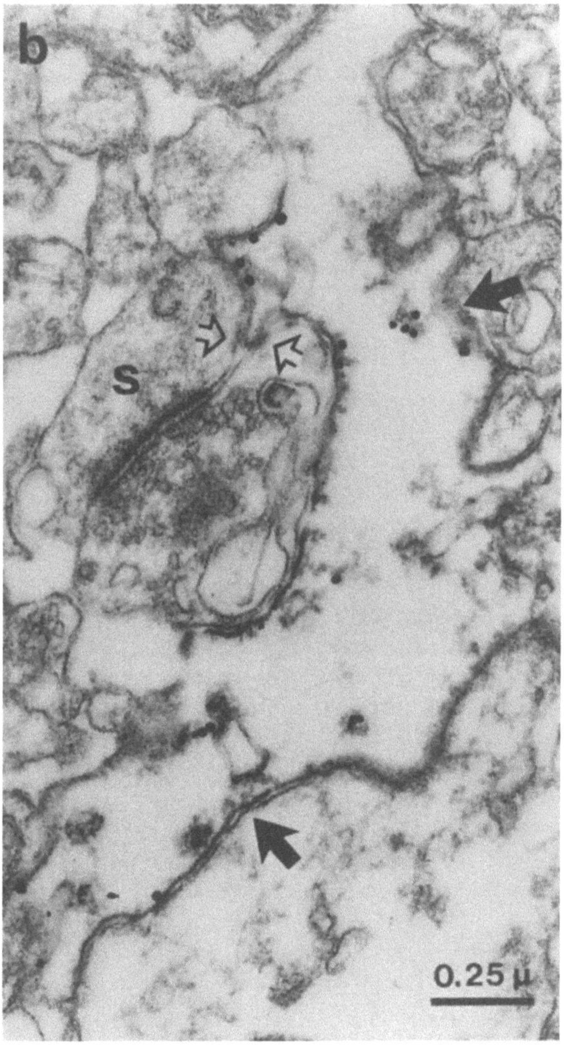

2.c COINCIDENCE OF GLUTAMINE SYNTHETASE - AND GLAST LOCALIZATION IN THE RETINA

In an attempt to establish which of the high affinity glutamate transporters may be involved in clearance of synaptically released glutamate immunostaining with a rabbit anti-GLAST antiserum was carried out. This antiserum was provided by Dr. W. Stoffel and has been characterized previously (Derouiche and Rauen, 1995). The rat retina was chosen for this part of the study because it contains the conspicuous, glutamatergic rod spherules and is clearly laminated, particularly with regard to glutamatergic circuitry (Massey, 1990; Massey and Redburn, 1987; Wässle and Boycott, 1991). The protocol for immunostaining was essentially the same as that for anti-GS immunocytochemistry.

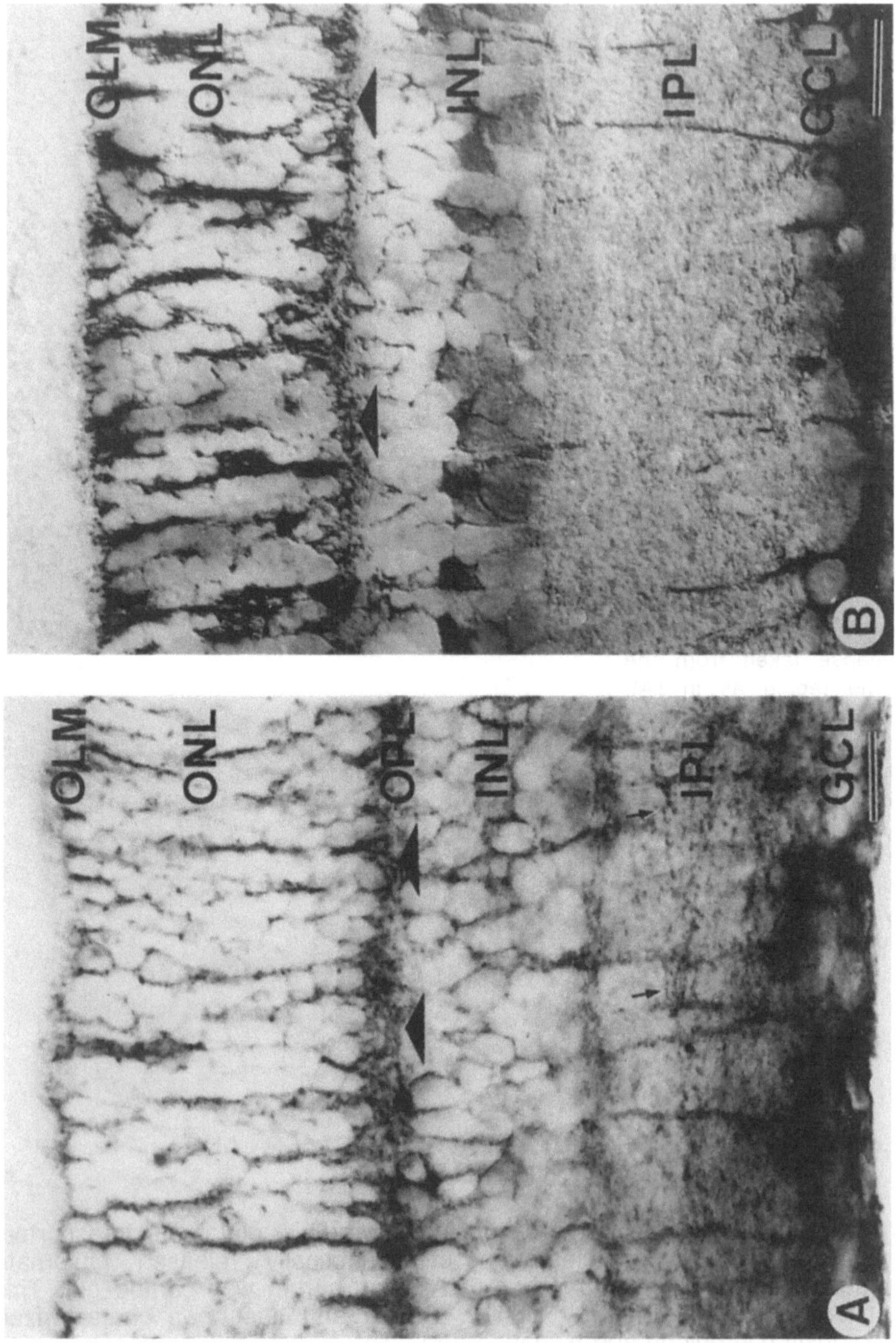
B
OLM
ONL
INL
IPL
GCL
A
OLM
ONL
OPL
INL
IPL
GCL

As can be seen in the light micrographs (Fig. 6A,B) the staining pattern was very similar to that of GS-ir: GLAST-ir was present in Müller cells and revealed a dark band in the outer plexiform layer. Ultrastructural investigation of this lamina showed GLAST-ir in Müller cell processes engulfing the rod spherules (Fig. 7A). These glial processes are as close to the rod spherule synapse as is possible without penetrating into the spherule. Also, the degree of insula-tion of the spherule appears similar to that seen in GS-ir specimens. Since Müller cells belong to the astrocyte family, the observed localization of GLAST-ir is in line with previous studies ascribing GLAST mRNA or immunoreactivity in the brain to astrocytes and Bergmann glia (Torp et al., 1994, Rothstein et al., 1994).

It may be of functional importance in the context of transmitter uptake that the retinal localization of GLAST-ir is very similar to that of GS-ir. The retinal cells stained by the two antisera are even identical, both label retinal astrocytes and pigment epithelial cells in addition to Müller cells (Derouiche and Rauen, 1995).

3. FUNCTIONAL CONSIDERATIONS

The anatomical data presented demonstrate a) that the distribution of GS-ir coincides with glutamatergic terminal fields at the regional level, b) that GS-ir is present in glial processes in the immediate vicinity of identified glutamatergic synapses, and C) that, in the retina, both GS-ir and GLAST-ir can be found in the same cell type and next to the glutamatergic rod photoreceptor synapse. Three questions arise regarding functional implications of the observations shown.

What functional data support the idea that GS degrades glutamate from the transmitter pool?

Müller cells are almost the only cells immunonegative in anti-glutamate staining of the rabbit retina (Pow and Robinson, 1994), which is easily explained by their GS content. If, however, GS enzyme activity is irreversibly inhibited (by methionine sulfoximine) prior to fixation of the retina, only Müller cells are glutamate-ir (Pow and Robinson, 1994). This speaks in favour of neurone-derived glutamate as the substrate of GS.

Fig. 6

GLAST - ir (A) and glutamine synthetase - ir (B) in rat retina. Both antisera stain Müller glial cells with their perpendicular processes. Many neurones in the inner and outer nuclear layer (INL and ONL) are ensheathed by immunoreactive Müller cell processes, but are not themselves stained. The intense labelling at the level of the outer plexiform layer (OPL, arrowheads) reveals immunonegative, globular structures which may correspond to rod spherules. OLM, outer limiting membrane; IPL, inner plexiform layer; GCL, ganglion cell layer; bars 20µm

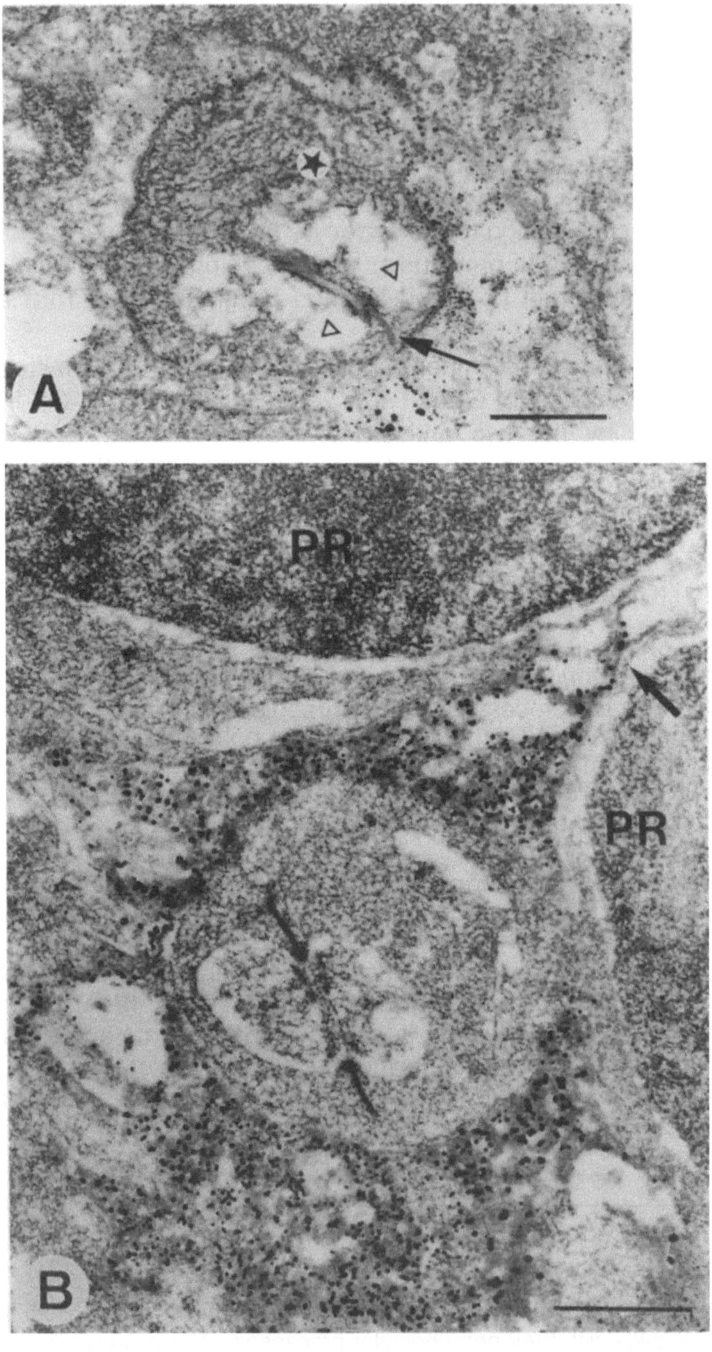
A
PR
PR
B

GS inhibitors, such as methionine sulfoximine or ferrous chloride, furthermore, seem to potentiate the neurotoxic effect of excitatory amino acids and have thus been experimentally applied as convulsants (Kolleger et al., 1991; Tiffany-Castiglioni et al., 1990). The seizure-prone gerbil may suffer from its partial GS defect (Laming et al., 1989), and GS plays a role in various forms of experimental hypoxia and cerebral ischemia (Oliver et al., 1990; Petito et al., 1992; Schor, 1988; Sher and Hu, 1990; Tanaka et al., 1992).

Other glutamate metabolizing enzymes in the CNS (reviewed by Kugler, 1993) are less well suited for metabolism of transmitter glutamate. Aspartate aminotransferase, for example is localized predominantly in neurones. Glutamate dehydrogenase, which occurs in glia - but also in photorecetors (Gebhard, 1992) - is not localized in the fine, organelle-free processes but only in mitochondria. Because of its reactivity for glutamate degradation, which is approximately 40 times less than that for glutamate synthesis (Chee et al., 1979), this enzyme is not likely to be involved in transmitter metabolism.

In the light of four glutamate transporters known, what points to a role of GLAST in transmitter clearance?

Glutamate uptake can be mediated by at least four distinct high affinity transporters (Danbolt, 1994; Fairman et al., 1995) or by low affinity, sodium-independent uptake (Logan and Snyder, 1972; Pin et al., 1984). However, high affinity, sodium dependent glutamate uptake appears to be the only significant influx route for glutamate (Barbour et al., 1993), but it has remained unknown which if any particular of the transporters is responsible for taking up synaptically released glutamate.

This question can be further analyzed by comparing the presently demonstrated localization of GLAST-ir in the retina with the anatomical distribution of $[^3H]$glutamate uptake, and with that of other high affinity glutamate transporters. In the rat retina, two more of these transporters have been localized immunocytochemically, namely Excitatory amino acid carrier-1 (EAAC1; Schultz and Stell, 1996) and Glutamate transporter 1 (GLT-1, Rauen and Kanner, 1994). EAAC1 is found in horizontal cells and a subtype of amacrine cells (Schultz and Stell, 1996), whereas GLT-1 is present in cone photoreceptors and cone bipolar cells (Rauen and Kanner, 1994). Only the localization of GLAST-ir in Müller cells corresponds to the observed

---

Fig. 7

Müller cell processes ensheathing rod sherules in the outer plexiform layer of the rat retina. In (A) , these processes contain silver grains revealing GLAST - immunoreactivity. A similar situation is shown in (B) from an anti-glutamine synthetase stained specimen. The rod spherules are fully embedded in these glial processes. The synaptic ribbons (arrow in (A)) and the light horizontal cell processes (triangles in (A)) are clearly visible. PR, nucleus of photoreceptor cell; bars 0.5µm

accumulation of [^{3}H]glutamate in these cells (Ehinger and Falck, 1971). This finding is also consistent with the restriction of endogenous glutamate-ir to Müller cells in the rabbit retina (when GS is inhibited; Pow and Robinson, 1994). There is no such correspondence with functional uptake in the case of GLT-1 - or EAAC1-ir.

Also, GLAST-ir displays a general and direct relation to rod spherules, which are known to display highest glutamatergic activity in the *absence* of stimulation; GLT-1-ir, in contrast, is present only in a restricted set of neuronal cells (Rauen and Kanner, 1994). Horizontal cells, although EAAC-1-ir, do not display detectable accumulation of [^{3}H]glutamate under *in vivo* conditions (Ehinger and Falck, 1971). In addition, it remains to be seen whether EAAC1-ir is present also in the horizontal cell tips participating in the rod photoreceptor synapse. These comparisons suggest that the glutamate clearance in the retina is essentially controlled by GLAST because the decisive requirements for such a role are met: Close and systematic relation to sites of glutamatergic neurotransmission, and colocalization with glutamate uptake.

Glutamate transporters not related to transmitter clearance may be specialized, as for example probably GLT-1 (see above). They may also subserve various other physiological functions such as intercellular signalling (Attwell et al., 1993), cytoskeletal changes (Shao et al., 1994), glutathione synthesis (Reichelt et al., 1994), or supply of precursor for GABA synthesis (Schultz and Stell, 1996).

Are GLAST and GS linked through their substrate?

The localization of GLAST- and GS-ir in the same cells and around identified glutamatergic synapses suggests that the substrate of GS is glutamate taken up from the extracellular space by GLAST.

Such a linkage of glutamate uptake with efficient intracellular glutamate metabolism appears to represent even a prerequisite for glutamate clearance of the extracellular milieu: Physiological measurements (reviewed by Attwell et al., 1993) have demonstrated that the intracellular glutamate concentration is a powerful factor in determining the lowest level of extracellular glutamate that can be maintained by glutamate uptake. This is evident from the kinetics of high affinity, sodium dependent glutamate uptake, which depends on the glutamate gradient across the membrane. This gradient is four orders of magnitude, the concentrations being approx. 1µM extracellularly and 10mM intracellularly. Without instantaneous intracellular removal of glutamate taken up, glutamate transporter kinetics (Attwell et al., 1993) predict a rise in extracellular glutamate that is directly proportional to the intracellular rise:

$$[Glu]_{ext} \sim [Glu]_{int}$$

$[Glu]_{ext}$ being the lowest extracellular concentration that can be maintained by the transporter.

GLAST mediated uptake has a value Km=11µM or 77µM for glutamate (depending on cell type; Storck et al., 1992, Klöckner et al.,1993) so that it could efficiently control low extracellular glutamate concentrations of around 10µM. Amongst the glutamate metabolizing enzymes (Kugler, 1993), GS appears well-suited to control intracellular, cytoplasmatic levels of glutamate: Its Km-value for glutamate is 2,5mM (Meister, 1974), and it is localized exclusively in glial cells (brain: Martinez-Hernandez et al., 1977; retina: Riepe and Norenberg, 1977), in the direct vicinity of glutamatergic synapses, and even in the lamellar, organelle-free processes (brain: Derouiche and Frotscher, 1991; Norenberg, 1979; Norenberg and Martinez-Hernandez, 1979; retina: present findings).

Seen in this light, GS would enhance glutamate clearance by GLAST, which is even dependent on subsequent glutamate degradation. Thus, the anatomical colocalization in the same (glial) cells of GS and a glutamate transporter would be a functional requirement. In the retina, the transporter providing the substrate for GS is GLAST but in the brain this may be GLT-1, a glia-bound glutamate transporter (Rothstein et al., 1994).

ACKNOWLEDGEMENTS

The author is greatly indebted to Dr. B. Hamprecht and Dr. W. Stoffel for the anti-GS antiserum and the anti-GLAST antiserum, respectively. He also thanks M. Ehms-Sommer, who did most of the photographic work.

REFERENCES

Attwell D, Barbour B and Szatkowski, M, Nonvesicular release of neurotransmitter. Neuron 11:401-407 (1993)

Balazs R, Patel AJ and Richter D, Metabolic compartments in the brain: their properties and relation to morphological structures. In: Balazs R and Cremer (eds.) Metabolic Compartmentation in the Brain, Macmillan, NY, 1972, pp 167-184

Barbour B, Keller BU, Llano I and Marty A, Prolonged presence of glutamate during excitatory synaptic transmission to cerebellar Purkinje cells. Neuron 12:1331-1343 (1994)

Barbour B, Magnus C, Szatkowski M, Gray PTA, Attwell D, Changes in NAD(P)H fluorescence and membrane current produced by glutamate uptake into salamander Müller cells. J Physiol 466:573-597 (1993)

Berl S and Clarke DD, Metabolic Compartmentation of the Glutamate-Glutamine System: Glial Contribution. In: Fonnum F (ed), Amino Acids as Chemical Transmitters. Plenum NY 691-708 (1978)

Chee PY, Dahl JL and Fahien LA, The purification and properties of rat brain glutamate dehydrogenase. J Neurochem 33:53-60 (1979)

Conti F, Rustioni A, Petrusz P and Towle AC, Glutamate-positive neurons in the somatic sensory cortex of rats and monkeys. J Neurosci 7:1887-1901 (1987)

Danbolt NC, The high affinity uptake system for excitatory amino acids in the brain. Prog Neurobiol 44:377-396 (1994)

Derouiche A, Härtig, W, Brauer K and Brückner G, Spatial relationship of lectin-labelled extracellular matrix and glutamine synthetase-immunoreactive astrocytes in rat cortical forebrain regions. J Anat (1996in press)

Derouiche A and Frotscher M, Astroglial processes around identified glutamateric synapses contain glutamine synthetase: evidence for transmitter degradation. Brain Res 552:346-350 (1991)

Derouiche A and Rauen T, Coincidence of Glutamate-Aspartate-Transporter-(GLAST) and Glutamine Synthetase- (GS) Immunoreactions in Retinal Glia: Evidence for Coupling of GLAST and GS in Transmitter Clearance. J Neurosci Res 42:131-143 (1995)

Ehinger B and Falck B, Autoradiography of some suspected neurotransmitter sbstances: GABA, glycine, glutamic acid, histamine, dopamine, and L-DOPA. Brain Res 33:157-172 (1971)

Fairman WA, Vandenberg RJ, Arriza JL, Kavanaugh, MP and Amara SG, An excitatory amino-acid transporter with properties of a ligand-gated chloride channel. Nature 375:599-603 (1995)

Garcia Y, Ibarra C and Jaffe EH, NMDA- and non-NMDA receptor mediated release of [^{3}H]glutamate from granule cell dendrites of rat olfactory bulb. J Neurochem 64:662-669 (1995)

Gebhard R, Histochemical demonstration of glutamate dehydrogenase and phosphate-activated glutaminase activities in semithin sections of the rat retina. Histochem 97:101-103 (1992)

Halász N and Shepherd GM, Neurochemistry of the vertebrate olfactory bulb. Neurosci 10 3:579-619 (1983)

Hallermayer K and Hamprecht B, Cellular heterogeneity in primary cultures of brain cells revealed by immunocytochemical localisation of glutamine synthetase. Brain Res 295:1-11 (1984)

Klöckner U, Storck T, Conradt M and Stoffel W, Electrogenic L-glutamate uptake in xenopus leavis oocytes expressing a cloned rat brain L-glutamate/L-aspartate transporter (GLAST-1). J Biol Chem 268:14594-14596 (1993)

Kollegger H, McBean GJ and Tipton KF, The inhibition of glutamine synthetase in rat corpus striatum in vitro by methionine sulfoximine increases the neurotoxic effects of kainate and N-methyl-D-aspartate. Neurosci Lett 130:95-98 (1991)

König JFR and Klippel RA, The rat brain. A stereotaxic atlas of the forebrain and lower parts of the brain-stem. The Williams and Wilkins Comp., Baltimore, 1963

Kugler P, Enzymes involved in glutamatergic and GABAergic neurotransmission. Int Rev Cytol 147:285-336 (1993)

Laming PR, Cosby SL and O'Neill JK, Seizures in the mongolian gerbil are related to a deficiency in cerebral glutamine synthetase. Comp Biochem Physiol 94C, 2:399-404 (1989)

Logan WJ and Snyder SH, High affinity uptake systems for glycine, glutamic and aspartic acids in synaptosomes of rat central nervous tissues. Brain Res 42:413-431 (1972)

Maragos WF, Penny JB and Young AB, Anatomic correlation of NMDA and ^{3}H-TCP-labelled receptors in rat brain. J Neurosci 8:493-501 (1988)

Martinez-Hernandez A, Bell KP and Norenberg MD, Glutamine Synthetase: Glial Localization in Brain. Science 195:1356-1358 (1977)

Massey SC, Cell types using glutamate as a neurotransmitter in the vertebrate retina. In: Osborne N and Chader G (eds): Prog Retin Res, Vol 9, Oxford: Pergamon Press, pp 399-425 (1990)

Massey SC, Redburn DA, Transmitter circuits in the vertebrate retina. Prog Neurobiol 28:55-96 (1987)

Meister A, Glutamine Synthetase of Mammals. The Enzymes 10:699-754 (1974)

Monaghan DT, Holets VR, Toy TW and Cotman CW, Anatomical distributions of four pharmacologically distinct ^{3}H-L-glutamate binding sites. Nature 306:176-179 (1983)

Norenberg MD, The distribution of glutamine sythetase in the rat central nervous system. J Histochem Cytochem 27:756-762 (1979)

Norenberg MD and Martinez-Hernandez A, Fine structural localization of glutamine synthetase in astrocytes of rat brain. Brain Res 161:303-310 (1979)

Oliver C, Starke-Reed P, Stadtman E, Liu G, Carney JFR, Oxidative damage to brain proteins, loss of glutamine synthetase activity, and production of free radicals during ischemia/reperfusion-induced injury to gerbil brain. Proc Natl Sci USA 87:5144-5147 (1990)

Ottersen OP and Storm-Mathisen J, Glutamate- and GABA- containing neurons in the mouse and rat brain, as demonstrated with a new immunocytochemical technique. J Comp Neurol 229:374-392 (1984)

Palay S and Chan-Palay V, Cerebellar cortex: Cytology and organisation. Heidelberg: Springer Verlag, (1974)

Petito CK, Chung MC, Verkhovsky LM and Cooper AJL, Brain glutamine synthetase increases following cerebral ischemia in the rat. Brain Res 569:275-280 (1992)

Pin J-P, Bockaert J, Recasens M, The Ca^{2+}/Cl^{-} dependent L-[^{3}H]glutamate binding: a new receptor or a particular transport process? FEBS Lett 175:31-36 (1984)

Pow DV and Robinson SR, Glutamate in some retinal neurons is derived solely from glia. Neurosci 60:355-36 (1994)

Price JL, An autoradiographic study of complementary laminar patterns of termination of afferent fibres to the olfactory cortex. J Comp Neurol 150:87-108 (1973)

Rauen T and Kanner BI, Localization of the glutamate transporter GLT-1 in rat and macaque monkey retina. Neurosci Lett 169:137-140 (1994)

Reichelt W, Stabel J, Pannicke T, Heinemann U, Glutathione synthesis is limited by the velocity of the Na^{+}/ glutamate transporter in Müller glial cells. Society for Neuroscience Meeting, Miami, Abstract No 267.11. (1994)

Riepe RE and Norenberg MD, Müller cell localization of glutamine synthetase in rat retina. Nature 286: 654-655 (1977)

Rothstein JD, Martin L, Levey AI, Dykes-Hoberg M, Jin L, Wu D, Nash N and Kuncl RW, Localization of neuronal and glial glutamate transporters. Neuron 13:713-725 (1994)

Schor NF, Inactivation of mammalian brain glutamine synthetase by oxygen radicals. Brain Res 456:17-21 (1988)

Schultz K, Stell W, Immunocytochemical localization of the high-affinity glutamate transporter, EAAC1, in the retina of representative vertebrate species. Neurosci Lett 211:191-194 (1996)

Shao Y, Enkvist MOK, McCarthy KD, Glutamate blocks astroglial stellation: effect of glutamate uptake and volume changes. Glia 11:1-10 (1994)

Sher PK and Hu S, Increased glutamate uptake and glutamine synthetase activity in neuronal cell cultures surviving chronic hypoxia. Glia 3:350-357 (1990)

Steward O, Topographic organization of the projections from the entorhinal area to the hippocampal formation of the rat. J Comp Neurol 167:285:314 (1976)

Storck T, Schulte S, Hofmann K and Stoffel W, Structure, expression and functional analysis of a Na^{+}-dependent glutamate/aspartate transporter from rat brain. Proc Natl Acad Sci USA 89:10955-10959 (1992)

Storm-Mathisen J, Leknes AK, Bore AT, Vaaland JL, Edminson P, Haug F-M S and Ottersen OP, First visualization of glutamate and GABA in neurones by immunocytochemistry. Nature 301:517-520 (1983)

Tanaka H, Araki M and Masuzawa T, Reaction of astrocytes in the gerbil hippocampus following transient ischemia: immunohistochemical observations with antibodies against Glial Fibrillary Acidic Protein, glutamine synthetase, and S-100 protein. Exp Neurol 116:264-274 (1992)

Tiffany-Castiglioni EC, Peterson SL and Castiglioni AJ, Alterations in glutamine synthetase activity by $FeCl_2$-induced focal and kindled amygdaloid seizures. J Neurosci Res 25:223-228 (1990)

Torp R, Danbolt NC, Babaie E, Bjoras M, Seeberg E, Storm-Mathisen J and Ottersen OP, Differential expression of two glial glutamate transporters in the rat brain: an *in situ* hybridization study. Eur J Neurosci 6:936-942 (1994)

Wässle H and Boycott B, Functional architecture of the mammalian retina. Physiol Rev 71:447-480 (1991)

Westrum LE and Bakay RAE, Plasticity in the rat olfactory cortex. J Comp Neurol 243:195-206 (1986)

Witter MP, Griffioen AW, Jorritsma-Byham B and Krijnen JLM, Entorhinal projections to the hippocampal CA1 region in the rat: an underestimated pathway. Neurosci Lett 85:193-198 (1988)

Wolff JR, Quantitative aspects of astroglia. Proceedings of the 6th International Congress of Neuropathology, Masson&Cie., Paris, pp 327-352 (1970)

Modulation of Neurotransmitter Uptake

Şakire Pöğün

Ege University Center for Brain Research and School of Medicine, Department of Physiology, Bornova, 35100, Izmir, Turkey

Neurotransmitter transporters or reuptake carriers, terminate the action of released neurotransmitter by reuptake; they are the initial sites of action for many drugs including psychostimulants and antidepressants. Neurotransmitter transporters have some general properties which can be summarized as follows: Sodium, chloride and temperature dependency, "High affinity" with K_m in the μM range or lower, pharmacologic selectivity, and selective localization to nerve terminals and glia (Kuhar, 1993).

A new family of neurotransporters have been cloned which include γ-amino-butyric acid (GABA) (Guestella et al., 1990; Nelson et al., 1990), norepinephrine (NE) (Pacholczyk et al., 1991), dopamine (DA) (Shimada et al., 1991; Kilty et al., 1991; Usdin et al., 1991; Giros et al., 1992), serotonin (5-HT) (Hoffman et al., 1991; Blakely et al., 1991) and betaine transporters. The betaine transporter is from kidney and not restricted to brain and neuronal function (Yamauchi et al., 1992). This new protein family has 619-693 amino acids, 12 transmembrane regions, consensus glycosylation and phosphorylation sites, sodium, chloride and temperature-dependency and close homology. Transporters for monoamines (DA, 5-HT, NE) as well as GABA, proline, glycine, betaine, creatine are coupled to inward cotransport of sodium ions. The ionic stoichiometry, however, is different: One Na^+ and one Cl^- for 5-HT and NE transporters, but two Na^+ and one Cl^- for DA transporter (DAT). Involvement of other ions, including Cl^- and K^+ were also demostrated.

Additional transporter families have also been characterized including the synaptic vesicle transporter family in which the driving force is a proton gradient and the glutamate transporter family with 8 transmembrane regions, and sodium and potassium dependency. There is yet another multigene family including neutral amino acid transporters for alanine, serine and cysteine (Uhl and Hartig, 1992).

1 Methods for Studying Transporters

The choice of preparation for studying neurotransmitter uptake range from in-vivo, ex-vivo or in-vitro (tisue slices, synaptosomes) preparations to primary cell cultures, cell lines, computer simulations and imaging.

Uptake assays reflect transporter function, a combined process of binding and transporter translocation, and are employed to measure maximal velocity (V_{max}), substrate affinity (K_m) and affinity for an inhibitor (K_i). Ligand binding assays, on the other hand, measure initial binding and provide information on the number of

NATO ASI Series, Vol. H 100
Neurotransmitter Release and Uptake
Edited by Şakire Pöğün

transporters (B_{max}), binding affinity of a radioligand (K_D) and binding affinity of a compound (K_i).

In-vivo microdialysis and voltammetry are also employed to study neurotransmitter uptake.

Short term modulation involves direct modulation of the transporter while long term modulation requires changes in transporter gene expression. Neurotransmitter receptor activation can alter gene expression and currently gene expression studies have provided invaluable information in the understanding of uptake regulation. The relatively new technological approaches including cloning and transfection are being widely employed to understand neurotransmitter transporter function. Site-directed mutants are used to find out which domains are responsible for particular properties, such as determining maximal transport rate.

Depolarization may be another modulatory factor on neurotransmitter uptake since neuronal transporters for choline and glutamate can be regulated by neuronal activity and nerve terminal depolarization (Murrin and Kuhar, 1976; Murrin et al., 1978).

Finally, transporters can be visualized by histochemical techniques using electron- and light-microscopy, as well as by PET and SPECT imaging in the living brain.

2 Second Messenger Pathways Involved in Uptake Regulation

Consensus sites on the intracellular domains of transporters for various protein kinases suggest receptor-mediated regulation by intracellular signals. Therefore second messenger pathways involved in uptake regulation can be studied by utilizing assays of the phosphoinositides, adenylate cyclase (AC), guanylate cyclase (GC) or phospholipase A_2 (PLA_2). In the central nervous system some molecules are regulated by post-translational modifications including phosphorylation by protein kinases (PKs). Since most of the neurotransmitter transporters show dependency on ion gradients, study of ion channels and ion pumps is yet another tool in transporter research.

2.1 The Serotonin Transporter

5-HT transporter has been cloned (Hoffman et al., 1991; Blakely et al., 1991) and found to belong to a superfamily of Na^+/Cl^- dependent monoamine and amino acid neurotransmitter transporters, including those of DA and NE (Uhl, 1992; Amara and Kuhar, 1993; Browstein and Hoffman, 1993).

5-HT transport inhibitors alter brain function and behavior through increasing the duration of action and concentration of 5-HT in the synapse.

Serotonin transporters in neurons, platelets and endothelium have been studied (Sneddon, 1973; Ross, 1982; Davis, 1984; Graham and Langer, 1988; Marcusson and Ross, 1990). The maternal-facing brush-border membrane of the human placental syncytiotrophoblast has a high affinity transporter for serotonin as well, with similar properties to those of neuronal and platelet serotonin transporters. JAR human placental choriocarcinoma cells, expressing the plasma membrane serotonin transporter were used in two studies. In one study, a cell line derived from human choriocarcinoma (JAR) was evaluated as a model to probe the regulation of several transport systems (Cool et al., 1991). Serotonin transporter was shown to be modulated by cAMP. Transport was shown to be energized by a NaCl gradient, inhibited by antidepressants but not by reserpine or 5-hydroxytriptophan. Ouabain, gramicidin and nigericin abolished Na gradient and inhibited transport. Phlordizin, which interferes with glucose transport, or incubation in a glucose-free medium also inhibited transport. Incubation of this cell line with cholera toxin (CTX) stimulated transport. CTX ADP-ribosylates the α subunit of the Gs protein which results in the stimulation of adenylate cyclase and increased levels of cAMP; cAMP potentiates cAMP dependent protein kinases. Incubating cells with increasing concentrations of CTX is reported to cause a three fold stimulation of serotonin uptake. Effects of other agents elevating intracellular cAMP levels were also studied: Forskolin, IBMX and dibutyryl cAMP. Forskolin is an activator of adenylate cyclase, IBMX inhibits cyclic nucleotide phosphodiesterase and thus inhibits cAMP degredation, and dibutyryl cAMP is a membrane permeable cAMP derivative. All these agents were found to stimulate serotonin transport, indicating a role of cAMP. H-9, a PK inhibitor, inhibited CTX induced stimulation. Stimulation of serotonin uptake by CTX or forskolin started after at least 8 hours and increased up to 24 hours. The effects of CTX on 5-HT uptake appears to be selective as it does not effect taurin uptake. CTX increased the maximal velocity of 5-HT transport as well as transporter affinity. The authors propose that the observed effects can not be due to the phosphorylation of the transporter but are due to cAMP induced de novo synthesis of transporter since cAMP levels increase in 2 hours but stimulation of transport is not observed until 8 hours (Cool et al, 1991). Since PKC has been shown to inhibit serotonin transport, it appears as if there is a reciprocal control by PKC and PKA. After cloning the cDNA for the human placental serotonin transporter, Northern blot analysis showed that mRNAs, which hybridize to the serotonin transporter cDNA, are elevated by treatment with CTX. This effect is specific and is not observed with the taurin transporter.

The second study investigated the cellular mechanisms responsible for the stimulation of serotonin transport by cAMP, induced by CTX and forskolin (Ramamoorty et al, 1993). In this study, effects of CTX and forskolin on serotonin transport activity were studied in the presence and absence of cycloheximide, an inhibitor of protein synthesis at the level of translation, with the aim of determining the involvement of cellular protein synthesis. CTX and forskolin were found to stimulate serotonin synthesis in the absence but not in the presence of cycloheximide. Lack of leucine incorporation showed that protein synthesis was blocked. Actinomycin D gave similar results.

These results indicate that cAMP-dependent enhancement of serotonin transport was through *de novo* synthesis of the transporter and subsequent increase in the density of the transporter in the plasma membrane. Serotonin transporter density was measured using RTI-55 and CTX treated cells had higher B_{max} compared to controls.

Phosphorylation of the serotonin transporter may underlie the inhibition of transport in platelets by PKC since platelets are nonnucleated cells and the transporter numbers do not change following inhibition by PKC. Therefore, the mechanism of PKC inhibition and cAMP regulation must be different.

The 5-HT plasma membrane transporter in platelets and nerve endings is driven by transmembrane gradients of sodium, potassium and chloride, has an affinity (K_m) for 5-HT of approximately 1 μM and can be blocked by re-uptake inhibitors including tricyclic antidepressants such as imipramine (Andersen and Horne, 1992).

Anderson and Horne (1992) treated platelets with activators of PKC and observed a reduction in 5-HT transport. β-TPA (4--β-12-*O*-tetradecanoylphorbol 13-acetate) had an inhibitory effect that was stereospecific (no inhibition with α-TPA). The decrease in the V_{max} of the transporter by TPA could not be attributed to a reduction of the transporter number on the platelet surface since imipramine binding was not affected (Imipramine labels the transporter selectively). Mezerin, a synthetic phorbol ester analogue, had a similar effect to TPA, and staurosporine, a PKC inhibitor, blocked TPA and mezerin effects. PKC inhibitors had little effect on release of 5-HT or on the number of transporters on the platelet surface. The effects of dibutyril cAMP and calcium ionophore A23187 on serotonin uptake were also tested in the same study, and no significant effects were observed following short term (5 min) preincubation. Minimizing ion gradients had little effect on transport, as well. The results of Andreson and Horn (1992) imply that PKC activators may down-regulate the activity of the serotonin transporter in platelets. The same phenomena is observed in endothelial cells and therefore, could be operating in neurons.

Both tricyclic and non-tricyclic antidepressants have inhibitory effects on high affinity serotonin transporter. Role of serotonergic synapse as a causative factor in depression is less certain. Even though there is some controversy, there are reports of decreased number of 5-HT transporters and reduced serotonin transport (V_{max} and antidepressant binding B_{max}) in depressed patients. Therapeutic effect of antidepressant treatment is delayed (days-weeks) indicating an indirect compensatory or regulatory mechanism. Antidepressant therapy is known to downregulate the number of 5-HT transporters (King et al, 1992). In a study employing JAR choriocarcinoma and PC12 pheochromocytoma cells, CTX has been shown to effect the imipramine-sensitive component of 5-HT uptake. Uptake was increased in JAR cells 3-fold but was inhibited in PC12 cells. Uptake was also inhibited in the C33-14-B1 fibroblast cell line established through gene transfer. Other cAMP enhancing agents (forskolin, IBMX) gave similar results (King et al, 1992). Stimulatory effects of CTX on JAR cells were due mainly to a V_{max} increase. In PC12 cells, there were high and low affinity components and possibly CTX decreased the affinity of both, with little effect on V_{max}. CTX effect, in these studies, became evident after a long delay of 15-24 hours, which indicates that the observed effect is not parallel to cAMP production. Authors' explanation for the observed effects are that the CTX effect does not involve direct phosphorylation of the transporter, and that the delayed effect of CTX could be through the modulation of transporter biosynthesis by cAMP, as opposed to rapid regulation such as recruitment of previously synthesized transporter to the cell surface. Rapid indirect phosphorylation leads to slow increase in V_{max} by stabilizing transporter against degradation. If rate of transporter biosynthesis is constant, the t½ of V_{max} shift would equal new t½ of phosphorylated transporter.

2.2 The Dopamine Transporter

cDNAs for the DAT have been cloned and reveal consensus sites for phosphorylation by several PKs, including PKC, in putative intracellular portions of the molecule. PKC activity is regulated by several neurotransmitter systems known to modulate the function of the Dopaminergic system (Kitayama et al., 1994).

Rat DAT expressed in COS cells was studied with regard to phosphorylation/dephosphorylation events, including those mediated by PKC. It has been shown that transfected COS cells (with pcD-NADAT1) display robust uptake of DA and high affinity binding of ^{3}H-WIN 35428 (Shimada et al, 1991; Kitayama et al, 1992). In a study where transfected COS cells were incubated with ^{3}H-DA and the phorbol ester PMA (Phorbol 12-myristrate 13-acetate), which activates PKC, PMA treatment reduced DA uptake significantly in a time-dependent manner. The effect was due to a reduction in transport velocity (V_{max}) with little effect on affinity (K_m). ^{3}H-WIN 35428 binding, a measure of specific DAT binding, was also reduced: There was no difference in B_{max} but in K_D, indicating that transporter number was not reduced but affinity was decreased. The phorbol ester antagonist, staurosporine reversed the PMA effect. A non-phorbol PKC receptor agonist, OAG (1-oleoyl-2-acetyl-sn-glycerol), on the other hand, mimicked the effect. OAG effect was reversed by H-9 in a reversible fashion. H-9 is structurally different from staurosporine but is a biological antagonist of PKC activation. The effects of phorbol esters were not through an interaction with the affinity of the transporter for Na^+. This study shows that DAT transport function can be altered by neurotransmitter systems that alter PKC function. Since B_{max} of the DAT did not change in the study mentioned above, PKC does not seem to alter DAT expresion. The authors point to the fact that dephosphorylation through the calcineurin family of phosphatases or other phosphatases could also regulate DAT function and emphasize that there may be similarities between the action of PKC activation and cocaine.

Metabolites of arachidonic acid, specificaly 12-lipoxygenase metabolites, have been implicated as second messengers mediating presynaptic inhibition in Aplysia sensory neurons (Piomelli et al, 1987a,b; Buttner et al., 1989). Arachidonic acid is released from membrane phospholipids by the action of phospholipase A_2 and is metabolized further by lipoxygenases and cyclooxigenase into several lipid mediators.

The role of arachidonic acid and metabolites in presynaptic inhibition in the striatum was examined by determining the effects of key enzymes in the arachidonic acid cascade (Piomelli et al., 1987b; Williams and Bliss, 1988) as well as determining the effects of arachidonic acid itself. Cass et al. (1991) used male

Sprague Dawley (SD) rats to study the involvement of arachidonic acid metabolites in dopamine release and uptake. p-Bromophenacyl bromide (BPB) and U-73122 were used to inhibit phospholipase A_2. Nordihydroguaiaretic acid (NDGA) was used to inhibit lipoxygenase, and indomethacin (INDO) and aspirin were used to inhibit cyclooxygenase. BPB increased electrically evoked overflow of endogenous DA from rat striatal slices and appeared to block the modulatory effects of D2 DA receptor agonist N-0437 on this release. BPB also increased spontaneous DA outflow in a dose dependent manner. U-73122 decreased evoked outflow of DA, but did not effect the action of N-0437 and increased spontaneous outflow of DA. Arachidonic acid, on the other hand, produced no effects. In low concentrations (μM) BPB, U-73122 and NDGA inhibited ^{3}H-DA uptake into synaptosomes and binding of ^{3}H-mazindol, but aspirin and indomethacin did not. These findings show that inhibitors of arachidonic acid metabolism are relatively potent DA uptake blockers and that this action is unrelated to their inhibition of arachidonic acid cascade (Cass et al., 1991).

Tuberoinfundibular dopaminergic (TIDA) neurons with cell bodies in the arcuate and periventricular nuclei and short axons terminating in the vicinity of the hypophyseal portal vasculature are considered to inhibit prolactin release. These neurons are thought to take up dopamine as well as release it from nerve terminals. DA uptake requires energy and depends on Na^+ in the external medium. It also is dependent on Na^+,K^+-ATPase because it is inhibited by ouabain. The involvement of Ca^+ or cAMP in uptake regulation are not clear. Kadowaki et al. (1990) prepared cell cultures from fragments of arcuate-periventricular nuclei-median eminence (containing TIDA neurons) taken from newborn male and female HLA-Wistar rats. Dibutyryl cAMP enhanced DA uptake in a dose dependent manner after an incubation period of 4 hours with ^{3}H-DA and preincubation for 15 mins with Dibutyryl cAMP. Forskolin, activator of AC, had a similar effect. Metabolites of dibutyryl cAMP, n-Butyrate and adenosine, did not enhance uptake of DA.

3 Modulation of Uptake Systems by Drugs

Modulation of uptake systems by drugs and the understanding of the mechanisms of actions of these drugs provide information on physiological processes as well as possible therapeutic interventions. Antidepressants (fluoxetine, sertraline, paroxetine, citalopram), cocaine, amphetamine and methylphenidate act on the serotonin transporter; the tricyclic antidepressant desipramine, cocaine, amphetamine and methylphenidate modulate the norepinephrine transporter; and cocaine, amphetamine and methylphenidate interact with dopamine uptake.

3.1 The Serotonin Transporter

Studies on human and animal platelets of 5-HT transport indicate that age (Langer et al., 1980; Severson et al., 1985; Marazziti et al., 1989), membrane fluidity (Block et al., 1987), hypoxia (Lee and Fanburg, 1986), physical stress (Thakur et al., 1990), oxidant stress (Bosin, 1989), and time of day (Wirz-Justice and Richter, 1979; Rausch et al., 1982; Arora et al., 1984; Humphries et al., 1985) change uptake rates rapidly, showing that 5-HT transporter might be regulated. Serotonin uptake is also altered in disease states such as hypertension, cirrhosis of the liver, and depression (Ahtee et al., 1981; Kamal et al., 1984; Briley et al., 1980). Chronic treatment with antidepressants either upregulate (Menini et al., 1987; Fattacini et al., 1990) or down regulate (Hridina, 1987; Lesch et al., 1993) the transporter, depending on the antidepressant used.

Serotonin transporter is the site of action for antidepressants such as fluoxetine (Prozac), appetite suppressant drugs such as fenfluoramine, and drugs of abuse such as 3,4-methylenedioxymethamphetamine and cocaine (Fuller and Wong, 1990; Douglas and Munro, 1982; Rowland and Carlton, 1986; Rudnick and Wall, 1992; Kuhar et al., 1991)

There is controversy regarding regulation of 5-HT-transporter following long-term antidepressant treatment: Antidepressants are reported to downregulate ^{3}H-Imipramine binding sites (Plenge and Mellerup, 1982; Brunello et al., 1987) but not ^{3}H-Paroxetine binding sites (Graham et al, 1987; Cheetham et al, 1991; Foy et al, 1991). Paroxetine binds to a single population of sites selectively localized on 5-HT neurites, whereas imipramine labels heterogenous sites in brain.

Pineyro et al. (1994) used male SD rats to study the effects of Paroxetine, delivered by osmotic mini pumps. The firing rate of CA_3 pyramidal neurons in the dorsal hippocampus after microiontophoretic application of ^{3}H-5-HT was taken as the index of in-vivo reuptake activity. After 21 days of paroxetine treatment and 2 days washout, there was a 50-60% reduction of in vitro uptake and 60-70% reduction in B_{max} of paroxetine binding.

3.2. The Norepinephrine Transporter

Chronic treatment by tricyclic antidepressants down regulate α2-adrenergic and β-adrenergic receptors in addition to serotonergic receptors in brain and supress the firing rate of neurons in the locus coeruleus (LC), the major source of neurons providing noradrenergic innervation, of rat brain.

Following in-situ hybridization for norepinephrine transporter (NET) mRNA in LC in animals receiving short (2 days) and long (4 weeks) term desipramine treatment, significant increase in labeling was observed in both groups (Szot et al, 1993). Since tricyclics block NE uptake, increased mRNA can not be due to up-regulation of transporter protein. Increased turnover due to chronic blockade by desipramine is a possible explanation. Similar antidepressants, on the other hand, have been shown to decrease serotonin transporter mRNA (Lesch et al., 1993)

3.3. The Dopamine Transporter

Dopamine transporters (DATs) terminate the action of released DA by reuptake and are the initial sites of action for many drugs including psychostimulants and antidepressants.

Cocaine binds to the DAT and inhibits DA uptake. This site of action of cocaine is believed to play a major role in its reinforcing properties (Ritz et al., 1987). Dopaminergic neurons that originate in the ventral tegmental area and terminate in the forebrain, particularly in the n. accumbens, are thought to play an important role in the reinforcing effects of cocaine. Treatment with dopaminergic agonists reduces and with dopaminergic antagonists increases cocaine self administration. Therefore, the reinforcing effects of cocaine are presumed to be related to enhanced dopaminergic activity.

Cocaine administration alters DA synthesis and release, firing pattern of dopaminergic cells, DA receptor density and affinity and DAT binding and function (for review, Ng et al, 1991):

Cocaine inhibits neuronal monoamine transporters and blocks voltage-gated Na^+ channes (local anaesthetic effects). Cocaine inhibits uptake of DA, 5-HT and NE transporters in high nanomolar to low micromolar range (for review, Zahniser et al., 1995).

Chronic cocaine administration produces behavioral sensitization and/or tolerance. Repeated administration of a constant dose results in the augmentation of observed behaviors: A long lasting behavioral sensitization during withdrawal from chronic cocaine and without a challenge dose, basal extracellular DA concentration is depressed, but there is an augmentation of extracellular DA in response to cocaine challenge. This implies that there is regulation of DAT function as a response to repeated cocaine administration, which may also underlie the persistence of behavioral changes. The ability of cocaine to antagonize the DAT has been altered with repeated exposure (for review, Zahniser et al., 1995).

Repeated intermittent cocaine administration causes a transient increase in DA uptake and DAT binding shortly after cessation of treatment but after relatively long withdrawal periods, DA uptake and DA binding are reduced in limbic areas. Parallel results are obtained in in-vivo microdialysis studies. Fewer transporters to inhibit may underlie behavioral sensitization. Another explanation is increased affinity of the transporter (for review, Zahniser et al., 1995).

Chronic cocaine administration, unlike intermittent, leads to behavioral tolerance (cocaine induced locomotor activity and stereotypy); binding, but not uptake of the DAT is altered (for review, Zahniser et al., 1995).

In postmortem brain samples of people with a history of cocaine abuse, ^{3}H-WIN 35,428 binding sites are increased in caudate, putamen and n. accumbens (Little et al, 1993) while increased binding in the putamen is observed in postmortem brain samples of cocaine abusers who died from cocaine overdose (Staley et al., 1993). These studies report increased transporter binding following chronic exposure and no withdrawal. However, there is one study where decreased binding is reported, under same conditions, and measurable cocaine levels at time of death (Hurd and Herkenham, 1993).

DAT mRNA in the rat substantia nigra was visualized by in situ hybridization and quantified by nuclease protection assay with a human DAT cDNA clone that is highly homologous with the rat DAT. DAT mRNA was detected in rat and human substantia nigra but not in striatum (Xia Y et al., 1992). Male SD rats received either acute or repeated cocaine treatments followed by a withdrawal period. Repeated cocaine administration resulted in a 40% decrease in rat nigral DAT mRNA levels. Acute cocaine, 4 hrs before death, did not alter DAT mRNA. After 72 days of withdrawal, DAT mRNA levels were the same as control levels (Xia Y et al., 1992). This study shows that repeated cocaine administration may down-regulate DAT gene expression in DA neurons.

Withdrawal of repeated intravenous infusions of cocaine persistently reduces binding to dopamine transporters in the nucleus accumbens of Lewis rats. Ten days after withdrawal of chronic intermittent cocaine, reduced mazindol binding (Sharpe et al,1991) and decreased dopamine uptake was observed in the n. accumbens (Isenwasser and Cox, 1990). Concordantly, decreased extracellular DA concentrations in the N. accumbens is reported after cocaine withdrawal (Parsons et al., 1991; Robertson et al, 1991; Rosetti et al., 1992). As mentioned above, postmortem binding to DAT in the striatum is reduced in humans who have abused cocaine (Hurd and Herkenham, 1993). Acute cocaine injections, on the other hand, increase extracellular DA but chronic application does not have this effect. Acute and chronic effects, and effects after the termination of cocaine application are not the same.

Cocaine also interacts with glucose utilization. Acute cocaine injections increase use of glucose (Porrino et al, 1988a) whereas abstinence from cocaine decrease glucose metabolism in N. Accumbens (Clow and Hammer, 1991).

Pilotte et al. (1994) treated Male Lewis rats, behaviorally more sensitive to cocaine than other strains, with cocaine or saline i.v. in an intermittent fashion-mimicking the self-administration pattern and assayed ^{3}H-WIN 35,428 binding in c. striatum and n.accumbens. Rats were treated 5-20 days with cocaine and were decapitated 0-60 days after the termination of treatment. Significant reductions in binding in the n.accumbens but not in the c.striatum was observed in cocaine treated rats relative to the saline controls, and the reduction in binding was associated with a change in B_{max} rather than K_d . These findings show that withdrawal of cocaine may lead to persistent changes in dopamine neurons.

Using the same treatment regimen Cerutti et al. (1994) performed in-situ hybridization studies on rats withdrawn for 10 days and found that the decreases in DAT mRNA were restricted to cells of the interfascicular and caudal linear nuclei (These dopaminergic neurons in the ventral tegmentum, project to n. accumbens and other limbic areas.) Other cell groups did not show a decrease in mRNA. This study shows that withdrawal from cocaine can alter DAT expression and provides an example of transporter gene regulation.

4 Modulation of Transporter Function by a G-Protein Coupled Membrane Receptor

Rat basophilic leukemia cell line (RBL 2H3) provides a model system with cellular compartments similar to those involved in synaptic neurotransmission. Since both 5-HT transporter and A_3 adenosine receptors are present on the RBL 2H3 cells, Miller and Hoffman (1994) used these cells to determine the possible regulation of 5-HT transporter function by a membrane-bound G-protein coupled receptor.

Incubation of rat basophilic leukemia (RBL) cells with the adenosine receptor agonist 5'-N-carboxamidoadenosine (NECA) led to an increase in ^{3}H-5-HT uptake, due to an increase in V_{max} and K_t. Another agonist (N^6-cyclopentlyadenosine) and 8-Bromo-cGMP had similar effects and the adenosine receptor antagonist xanthine amine congener (XAC) totally blocked this effect. The maximal effect of NECA was observed at 10 minutes.

Adenosine receptor is coupled to an elevation in IP_3 and Ca^+ via pertusus toxin sensitive G-protein. Therefore elevations in Ca^{2+} could act through protein kinases including PKC. Direct stimulation of PKC by PMA decreased uptake (Miller and Hoffman, 1994). This shows that the 5-HT transporter can be differentially regulated by intracellular second messengers.

Adenosine receptor in RBL does not effect cAMP and PKC has an opposing effect. Miller and Hoffman (1994) examined the possible mediation by cGMP in the same study. 8-Bromo-cGMP had a similar effect to that of NECA and cyclic nucleotide dependent PK inhibitor H-8 blocked the stimulation. Thus it is proposed that the increase in 5-HT transport by NECA is mediated by cGMP. Nitric oxide synthase (NOS) inhibitors blocked the effect of NECA and nitric oxide (NO) generators stimulated uptake in cultured cells. Calmidozolium blocks calmodulin and therefore would prevent NOS activation. Methylene blue scavanges NO and LY-83,583 inhibits guanylate cyclase. Therefore the effects of these compounds on uptake were examined and all compounds were found to inhibit uptake alone or with NECA at the same level.

Direct measurement of cGMP after A_3 receptor stimulation with NECA show a 170% increase while paroxetine binding was not affected either by PMA or NECA. The agents used, did not affect alanine uptake showing that the observed effects were not generalizable and further experiments showed that they were not due to ion gradient alterations but to a change in the activity of the transporter (Miller and Hoffman, 1994).

Glutamate transport can be regulated by α_1-adrenergic receptor activation in primary astrocyte cultures (Hansson and Ronnback, 1989 and 1991). PKC activation results in an increase of glutamate uptake in primary cultures of glia cells (Casado et al, 1991). Phosphorylation of the glutamate transporter by PKC has been demonstrated (Casado et al, 1993).

Although it is not known if adenosine receptors are present in neurons possessing 5-HT transporters, the studies by Miller and Hoffman (1994) provide evidence for the modulation of transporter function by a membrane-bound G-protein coupled receptor.

5 Modulation of Neurotransmitter Uptake by Nitric Oxide

NO is a novel, gaseous biological messenger molecule in the CNS which is enzymatically formed from L-arginine by nitric oxide synthase, a Ca^{2+}-calmodulin requiring enzyme; citrulline is the co-product. Activation of calmodulin is caused by calcium influx, at least partly derived from the stimulation of glutamate receptors. NO has been proposed to mediate in the neurotoxic actions of glutamate, to promote long term potentiation and long term depression and to modulate neurotransmitter release. Inhibition of transporters for some neurotransmitters in brain by NO may be a new type of transynaptic retrograde regulation mechanism for uptake regulation (Pöğün and Kuhar, 1994).

In synaptosomes prepared from male SD rat brain, NO has been shown to inhibit uptake of DA, glutamate and 5-HT in a dose-, time- and temperature dependent fashion, but was without effect on NE uptake (Pogun et al, 1994a and 1994b). The NOS inhibitor N^{ϖ}-Nitroarginine increased DA uptake after systemic administration.

NO has been implicated in drug tolerance, dependence and withdrawal (Trujillo and Akil, 1991). Inhibitors of NOS attenuate opioid withdrawal (Kimes et al., 1993). An abused nitrite, amyl nitrite, is also thought to release NO (Ignarro et al., 1981). The inhibition of DA uptake by NO may be involved in the rewarding effects of drug (Pöğün et al., 1994).

Glutamate is a major excitatory amino acid neurotransmitter in the brain and glutamate uptake is thought to be a mechanism of inactivation of released glutamate. Strengthening of glutamatergic neurotransmission is characteristic of long term potentiation (LTP) and therefore glutamate plays a key role in learning and memory. It has also been suggested that NO is the retrograde mediator of hippocampal LTP. It seems possible that glutamate uptake inhibition could modulate learning and memory by increasing the availability of glutamate in the synaptic cleft. The finding that the uptake inhibition in-vitro is only partially reversible after repeated washing of the synaptosomes could be compatible with this role since by definition, LTP is long lasting (Pöğün et al., 1994a).

While glutamate is a neurotransmitter, it is also a well known neurotoxin in certain situations such as in the ischemic or post-ischemic brain and therefore, regulatory or protective mechanisms must fail. For years it has been known that excitotoxicity must involve more than just an increase in glutamate release since glutamate transporters, by reuptake, can save neurons from excessive stimulation and damage. If the processes that lead to neurotoxicity produce substances that inhibit uptake as well as those that stimulate release of glutamate, the conditions leading to cell damage and finally cell death will be augmented. NO is thought to be a mediator of glutamate's neurotoxicity because nitric oxide synthase (NOS) inhibitors prevent glutamate's neurotoxicity. The results shown here suggest that glutamate uptake inhibition by NO, if it occurs in-vivo, could play a role in toxicity of glutamate.

In the studies above, saturation analyses indicated that the reduction in ^{3}H-glutamate and ^{3}H-dopamine uptake was due to a change in both V_{max} and K_m in the presence of 300 μM sodium nitroprusside (SNP). ^{3}H-Serotonin uptake was also inhibited by NO. The inhibition of uptake observed in all cases was prevented when reduced hemoglobin (Hb), a substance that binds NO, was added to the medium. SNP did not effect ^{3}H-WIN 35428 binding to the dopamine transporter suggesting that transporter function rather than substrate binding is involved. Calcium free medium augmented the inhibitory effect of NO on ^{3}H-dopamine uptake. Potassium ferricyanide and ferrocyanide, reagents similar to SNP but do not produce NO, did not effect uptake. Norepinephrine uptake in the frontal cortex, however was insensitive to NO modulation and surprisingly, was fascilitated by cyanide exposure.

Using an approach similar to the studies mentioned above, Lonart and Johnson (1994), studied the effects of NO on dopamine and glutamate uptake in synaptosomes prepared from rat brain. S-Nitroso-L-Cysteine (NO-CYS) was used as NO donor. 10 min. superfusion with 1 nM NO-CYS caused a time dependent increase in DA efflux which was blocked by reduced hemoglobin. NO-CYS and NO caused a dose dependent inhibition of ^{3}H-DA uptake: 73% inhibition was observed at 1 mM. NO-CYS inhibited ^{3}H-Glu uptake in a concentration dependent manner as well, but NO is found to be a less potent inhibitor of Glu uptake compared to DA uptake. Scatchard analyses showed that NO inhibits uptake through reducing capacity and not by affecting affinity. NO-CYS had no effect on mazindol binding.

On the other hand, Miller and Hoffman (1994) observed a completely opposite effect of NO on serotonin uptake in RBL cells (Please refer to Section 4). This discrepency may be reflecting variance in nitric oxide effects in different cell types.

6 Modulation of the Dopamine Transporter by Gonadal Steroid Hormones

Modulation of presynaptic dopamine (DA) activity (e.g. Becker et al, 1984; Dluzen and Ramirez, 1990; Morissette et al, 1990; Becker, 1990; Pilotte et al., 1984) and postsynaptic DA binding (Levesque et al, 1989; Levesque and Di-Paolo, 1990; Di Paolo et al, 1988; Hruska and Silbergeld, 1980; Pasqualini et al ,1984; Pilotte et al, 1989) by ovarian hormones have been well documented in rodents. DA neural activity and a variety of non-reproductive behaviors involving the central dopaminergic system are also known to show estrus cycle dependendent variations (Becker and Cha, 1989; Barraclough and Wise, 1982; Fernandez-Ruiz et al., 1991).

Estrogen enhances behavioral sensitization to cocaine (Glick and Hinds, 1984; Peris et al., 1991) and influences cocaine self-administration (Dalton, Vickers and Roberts, 1986; Roberts, Dalton and Vickers, 1987). In a cocaine self-administration study, female rats were found to reach higher breaking points during the estrus phase of the estrus cycle on a progressive ratio, but not on a fixed ratio scale, implying influences of the estrus cycle on motivation (Roberts et al., 1989).

Sex and estrus cycle dependent variations of the DA uptake sites have been reported in the rat striatum with no change in affinity (Morisette and Di Paolo, 1993a). In ovariectomized (OVX) rats, estradiol and progesterone, used alone or in combination, caused an increase in DA metabolites in the striatum. DA levels, however, were increased only after progesterone alone or in combination with estradiol (Morisette et al, 1990a). A single injection of estradiol, but not progesterone, to OVX rats caused a slight increase in the maximum density of striatal DA uptake sites measured with [^{3}H] GBR-12935 binding. (Morisette et al., 1990b). When applied chronically, however, both estradiol and progesterone increased DAT density in the striatum (Morisette and Di Paolo, 1993b).

Pöğün et al. (1996) assessed the effects of ovariectomy and hormone replacement on DAT binding using [^{3}H]WIN 35428. Female SD rats were used and the hormone replacement regimen provided low physiological concentrations of estradiol (E2) and progesterone (P4), and simulated hormonal profiles of estrus animals. DAT density was lowest in ovariectomized rats and progesterone replacement caused an increase in transporter B_{max}. Estrogen treatment, on the other hand, was ineffective when used alone. No change in affinity was observed between the experimental groups. These findings point to the significant role of progesterone in the regulation of the DAT density in the rat striatum.

In a study by Morisette et al. (1990b), a single injection of estradiol (100 ng, SC) to OVX rats caused a 24% and 18% increase (15 and 30 min after the injection, respectively) in the maximum density of striatal DA uptake sites measured with [^{3}H] GBR-12935 binding. One injection of progesterone (110 μg, SC), on the other hand, did not cause any changes in binding up to 120 minutes after the injection (Morisette et al., 1990b). However, when OVX rats were treated twice daily with E2 (10μg) or P (0.72 μg) or their combination for two weeks, the highest densities of the DAT were observed in P treated animals although E2 alone also caused a significant increase in B_{max} as compared to the OVX group (Morisette and Di Paolo, 1993b).

The comparison of these two studies underscore the importance of the duration and regimen of hormone replacement to godadectomized animals as well as providing a means to compare acute and chronic effects of ovarian steroid hormones.

Summary

Uptake of neurotransmitters by nerve endings is a mechanism to terminate the synaptic actions of released neurotransmitter. Neurotransmitter transporters are also targets for many drugs including antidepressants and psychostimulants. Drug administrations have long-term regulatory influences on transport, possibly through gene expression, resulting in behavioral modifications. While the regulation of these uptake transporters are not fully understood, factors such as depolarization, inhibition by endogenous substances, phosphorylation and changes in transcription have been suggested as possibilities. Within this scope, intracellular signal pathways, drugs, membrane receptors, sex hormones and nitric oxide have been evaluated as possible modulators of neurotransmitter uptake.

References

Ahtee L, Briley M, Raisman R, Lebrec D and Langer SZ (1981) Life Sci. 29:2323

Amara SG and Kuhar M (1993) Annu. Rev. Neurosci. 16:73

Anderson GM and Horne WC (1992) Biochim Biophys Acta. 1137:331

Arora RC, Kreegel L, and Melttzer HY (1984) Biol. Psychiatry 19:1579

Barbour, B., Szatkowski, M. Ingledew, N. & Atwell, D. (1989) Nature 342:918

Barraclough CA and Wise PM (1982) Endocr. Rev. 3:91

Becker JB and Cha JH (1989) Behav. Brain Res. 35: 117

Becker JB; Beer ME and Robinson, TE (1984) Brain. Res. 311:157

Becker JB (1990) Synapse 5: 157

Blakely RD, Berson HE, Fremeay RT, Caron MG, Peek MM, Prince HK and Bradley CC (1991) Nature 354: 66

Block ERand Edwards DA (1987) Am. J. Physiol. 253:C672

Bosin TR (1989) J. Pharm. Exp. Ther. 248: 67

Briley MS, Langer SZ, Raisman R, Sechter D and Zarifian E (1980) Science 209: 303

Browstein M and Hoffman BJ (1993) Recent Prog. Hormone Res. 49:27

Brunello N, Riva M, Volterra A, Racagni G (1987) Fundam Clin Pharmacol 1: 327

Buttner N, Siegelbaum SA and Volterra A (1989) Nature (Lond.) 342: 553

Casado M, Zafra F, Aragon C and Gimenez C (1991) J. Neurochem. 57:1185

Casado M, Bendahan A, Zafra F. Danbold NC, Aragon C, Gimenez C and Kanner BI (1993) J. Biol. Chem. 268:27313

Cass WA, Larosn G, Fitzpatrick, FA, Zahniser NR (1991) J Pharmacol. Exp. Ther. 257:990

Ceruti C, Pilotte NS, Uhl G, Kuhar MJ (1994) Mol Brain Res 22:132

Cheetham SC, Viggers JA, Bucket WR (1991) Br. J. Pharmacol 102:242P

Clow DW and Hammer RP (1991) Neuropsychopharmacology 4:71

Cool DR, Leibach FH, Bhalla VK, Manesh VB and Ganapathy V (1991) J. Biol. Chem. 266:15750

Dalton JC, Vickers GJ and Roberts DC (1986) Pharmacol. Biochem. Behav. 25:497

Davis A (1984) Experientia 40:782

Di Paolo T, Falardeau P and Morissette M, (1988) Life-Sci. 43 :665

Dluzen DE and Ramirez VD (1990) Brain. Res. 536:163

Douglas JG and Munro JF (1982) Pharmacol. & Ther. 18:351

Fattacini CM, Bolanos-Jimenez F, Gozlan H and Hamon M (1990) Neuropharmacology 29:1

Fernandez-Ruiz JJ, Hernandez ML de-Miguel R, Ramos JA (1991) J. Neural. Transm. Gen. Sect. 85:223

Foy B, Hepner E and Hrdina PD (1991) In: Presynaptic receptors and neuronal transporters (Eds. SZ Langer, AM Galzin, J Cosentin) pp. 231, Oxford:Pergamon

Fuller RW and Wong DT (1990) New York Academy of Sciences 600:68

Giros B, Mestikawy SE, Godinot N, Zheng K, Han Y, Yang-Feng T and Caron MG (1992) Mol. Pharm. 42:383

Glick SD and Hinds, PA (1984) Eur. J. Pharmacol. 99:119

Graham D and Langer SZ (1988) in Neuronal Serotonin (Ossborne, N. N. & Hamon, M..eds) 367 ohn Wiley & Sons. New York

Graham D, Tahraoui L and Langer SZ (1987) Neuropharmacology 26:1087

Guestella J, Nelson N, Nelson H., Czyzyk L, Keynan S, Miedel MC, Davidson N, Lester HA and Kanner BI (1990) Science 249:1303

Hansson E and Ronnback L (1989) Life Sci. 44:27

Hansson E and Ronnback L (1991) Brain Res. 548:215

Hoffman BJ, Mezey E and Brownstein MJ (1991) Science 254:579

Hridina PD (1987) Eur. J. Pharmacol. 138:159

Hruska RE, Silbergeld EK (1980) Eur. J. Pharmacol. 61:397

Humphries LL Shirley P, Allen M, Codd EE and Walker RF (1985) Biol. Psychiaatry 20:1073

Hurd YL and Herkenham M (1993) Synapse 13:357

Ignarro LJ, Lippton H, Edwards JC, Baricos WH, Hyman AL, Kadowitz PJ, Gruetter CA (1981) J. Pharmacol. Exp. Ther. 218:739

Isenwasser S and Cox BM (1990) Brain Res. 531:338

Kadowaki K, Hirota K, Koike K, Ohmichi M, Kiyama H, Miyake A and Tanizawa O (1990).Neuroendocrinology 52:256

Kamal LA, Raisman R, Meyer P and Langer SZ (1984) Life Sci. 34:2083

Kilty J, Lorang D and Amara SG (1991) Science 254:578

Kimes AS, Vaupel DB, London ED (1993) Psychopharmacology-Berl. 112:521

King S, Tiller A, Chang A-S, Lam D-K (1992) Biochem Biophys Res Commun. 183:487

Kitayama S, Dohi T., Uhl GR (1994) Eur. J. Pharmacol. 268:115

Kitayama SK, Shimada S, Xu H, Markham L, Donovan DM and Uhl GR (1992) Proc. Natl. Acad. Sci. USA 89:7782

Kuhar MJ, Ritz C and Boja JW (1991) Trends Neurosci. 14:299

Kuhar, M.J. (1993) The pharmacologist 35:28

Langer SZ, Briley MS, Raisman K, Henry JF and Morselli PL (1980) Naunnyn-Schmiedeberg's Arch. Pharmacol. 313:189

Lee SL and Fanburg BL (1986) Am. J. Physiol. 250:C766

Lesch, KP, Aulakh, CS, Wolozin, BL, Tolliver, TJ, Hill, JL and Murphy, DL (1993) Mol. Brain Res. 17:31

Levesque D and Di Paolo, T (1990) Brain-Res-Bull. 24:281

Levesque D, Gagnon S, Di Paolo T (1989) Neurosci.Lett. 98:345

Little KY, Kirkman JA, Carroll FI, Clark TB and Duncan GE (1993) Brain Res. 628:17

Lonart G, Johnson KM (1994) J. Neurochem. 63:2108

Marazziti D, Falcone MF, Rotondo A, and Castrogiovanni P (1989) Naunyn-Schmiedeberg's Arch. Pharmacol. 340:593

Marcusson JO and Ross SB (1990) Psychopharmacology 102:145

Menini T, Mocaer E and Garattini S (1987) Naunyn-Schmiedeeberg's Arch. Pharmacol. 336:478
Miller KJ and Hoffman BJ (1994) J. Biol. Chem. 269:27351
Morisette M and Di Paolo T (1993a) Neuroendocrinology 58:16
Morisette M and Di Paolo T (1993b) J. Neurochem., 60:1876
Morissette M, Levesque D, Belanger A and Di Paolo T (1990a) Can. J. Physiol. Pharmacol., 68:1520
Morissette M, Biron D and Di Paolo T (1990b) Brain-Res-Bull. 25:419
Murrin LC and Kuhar MJ (1976) Mol. Pharmacol. 12:1082
Murrin LC., Lewis MS. & Kuhar MJ (1978) Life Sci. 22:2009
Nelson H, Mandiyan S and Nelson N (1990) Fed. Eur. Biochem. Soc. 269:181
Ng JP, Hubert GW and Justice JB (1991) J. Neurochem. 56:1485
Pacholczyk T, Blakely RD and Amara SG (1991) Nature 350:350
Parsons LH, Smith AD and Justice JB (1991) Synapse 9:60
Pasqualini C, Lenoir V, El-Abed A, Kerdelhue B (1984) Neuroendocrinology. 38:39
Peris J, Decambre N, Coleman-Hardee ML, Simpkins JW (1991) Brain-Res. 566:255
Pilotte NS, Burt DR and Barraclough CA (1984) Endocrinology 114:2306
Pilotte NS, Burt DR and Barraclough CA (1989) Endocrinology 124:805
Pilotte NS, Sharpe LG, Kuhar MJ (1994) J Pharmacol Exp Ther. 269:963
Pineyro G, Blier P, Dennis T, de Montigny C (1994) J Neurosci. 14:3036
Piomelli D, Shapiro E, Feinmark SJ and Schwartz, JH (1987a) J. Neurosci. 7:3675
Piomelli D, Volterra A, Dale N, Siegelbaum SA, Kandel ER, Schwart, JH and Belardtti F (1987b) Nature (Lond.) 328:38
Plenge P and Mellerup ET (1982) Psychopharmacology 77:94
Pöğün, Ş. and Kuhar, MJ (1994) In: Neurobiology of NO and OH (Eds CC Chiueh, Gilbert, DL. and Colton, CA.) Ann. New York Acad. Sci., Vol. 738. pp. 316, New York
Pöğün, Ş., Baumann M and Kuhar MJ (1994b) Brain Res. 641:83
Pöğün, Ş., Dawson, V and Kuhar MJ. .(1994a) Synapse18:21
Pöğün, Ş., Pilotte N, Kuhar MJ (1996) In preparation
Porrino LJ, Domer FR, Crane AM and Sokoloff L (1988a) Neuropsychopharmacology 1:109
Ramamoorty S, Cool DR, Mahesh VB, Leibach FH, Melikian HE, Blakely RD and Ganapathy V (1993) J. Biol. Chem. 268:21626
Rausch JL, Shah NS, Burch EA and Donald AG (1982) Biol. Psych. 17:121
Ritz MC, Lamb RJ, Goldberg SR and Kuhar MJ (1987) Science 237:1219
Roberts DCS, Bennet SAL and Vickers GJ (1989) Psychopharmacology 98:408
Roberts DCS, Dalton JCH and Vickers, GJ (1987) Pharmacol. Biochem. Behav., 26:37
Robertson MW, Leslie CA and Bennett JP (1991) Brain Res. 538:337
Rosetti ZL, Hmaidan J and Gressa GL (1992) Eur. J. Pharmacol. 221:227
Ross, SB (1982) In: Biology of Serotonergic Transmission (Osborne, N. N., ed) John Wilev & Sons. New York 159
Rowland NE and Carlton J (1986) Prog. Neurobiol. 27:13
Rudnick G and Wall SC (1992) Proc. Natl. Acad. Sci. U.S.A. 89:1817

Severson JA, Marcusson JO, Osterburg HH, Finch CE and Winblad B (1985) J. Neurochem. 45:1382
Sharpe LG, Pilotte NS, Mitchell WM, Dax EM and De Souza EB (1991) Eur. J. Pharmacol. 203:141
Shimada , S, Kitayama, SK, Lin, C-L, Patel, A, Nanthakumar, E, Gregor, P, Kuhar, MJ and Uhl, G (1991) Science, 254:576
Sneddon JM (1973) Prog. Neurobiol. 1:151
Staley J, Flynn DD, Stitt F, Wetli CV and Mash DC (1993) Soc. Neurosci. Abstr. 19:1843
Szot P, Ashliegh E, Kohen R, Petrie E, Dorsa D, Veith R (1993) Brain Res. 618:308
Thakur R, Verma M, Narain VS, Puri M, Hasam M, and Shanker K (1990) In: Cardiovascular Syste, Hypertension, and Serotonin Antagonists (Pasletti, R. et al., eds.) Kluwer, Boston, MA
Trujillo KA, Akil H (1991) Science 251:85
Uhl GR and Hartig PR (1992) TIPS. 13:421
Uhl, GR (1992) Trends Neurosci. 15:265
Usdin, TB, Mezey R, Chen C, Brownstein MJ. and Hoffman BJ (1991) Proc. Nat'l. Acad Sci. 88:11168
Williams JH and Bliss TVP (1988) Neurosci. Lett. 88:81
Wirz-Justice A and Richter R (1979) Psychiatry Res. 1:53
Xia Y, Goebel DJ, Kapatos G and Bannon MJ (1992) J Neurochem. 59:1179
Yamauchi A, Uchida S, Kwon H.M, Preston AS, Robey RB, Garcia-Perez A, Burg MB and Handler JS (1992) J. Biol., Chem. 649
Zahniser N, Gerhardt G, Cass W (1995) In "The Neurobiology of Cocaine:cellular and molecular mechanisms" Editor:Hamer R, Boca Raton, pp.181, Florida, CRC Press

Catalytic and Non-Catalytic Activities of Acetylcholinesterase Implied from Transgenic ACHE Expression in Vertebrates

Meira Sternfeld, Shlomo Seidman, Rachel Beeri and Hermona Soreq*

Dept. of Biological Chemistry, Institute of Life Sciences, The Hebrew University of Jerusalem, Jerusalem 91904, Israel

*To whom correspondence should be addressed

Abstract

Transgenic overexpression of human acetylcholinesterase (AChE) was employed as an experimental tool for creating a subtle cholinergic imbalance in vertebrates. In Xenopus laevis tadpoles, transgenic AChE accelerated the development of neuromuscular junctions. In stably transgenic mice, overexpression of AChE was associated with deterioration in both cognitive and neuromotor functions, suggesting requirement for balanced cholinergic transmission for stable maintenance of these functions. However, transfections of cultured glioma cells demonstrated a morphogenic activity for AChE which was apparently non-catalytic. This suggested that putatively distinct feedback mechanisms are activated in response to the catalytic and non-catalytic activities of excess AChE, which causes the observed symptoms.

1 Introduction

Cholinergic neurotransmission involves the coordinated and well-balanced interplay between presynaptic and post-synaptic processes. Pre-synaptic events involve the production, synaptic vesicle loading, storage and release of the neurotransmitter acetylcholine (ACh). Post synaptic events conduct the signal by the subsequent interaction of ACh with its receptors and their activation. Finally, rapid hydrolysis of ACh terminates the neurotransmission process.

Many different functions in vertebrates, such as learning and memory, therrnoregulation, voluntary and involuntary muscle control are controlled through cholinergic neurotransmission. When balanced cholinergic neurotransmission is perturbed, such functions are impaired, which leads to variable disease states. For example, accelerated death of ACh producing neurons is associated with the cognitive decline characteristic of Alzheimer's disease (Davies and Maloney, 1976). Similarly, the requirement for precise cholinergic neuromuscular communication is not fulfilled in myasthenia gravis patients, where autoimmune antibodies to the nicotinic muscle ACh receptor block post-synaptic responses (Drachman et al., 1982). In addition, many autonomous functions are subject to central cholinergic

NATO ASI Series, Vol. H 100
Neurotransmitter Release and Uptake
Edited by Şakire Pöğün

control, with the failure of such control leading to different syndromes. However, in none of these correlations is it clear whether the apparent impairment in these important functions is merely the outcome of the disease state or whether this impaired cholinergic function participates in promoting the pathology. One approach to this key question is to create transgenic organisms in which cholinergic communication is deregulated by excess expression of an introduced transgene. Such manipulations, and follow-up of embryonic development and subsequent adult functioning in the transgenic organisms should reveal the answer to the above question.

If indeed imbalanced cholinergic communication is causally involved in creating disease symptoms, transgenic organisms expressing excess of certain key cholinergic proteins should develop cognitive, neuromotor or autonomous impairments. Alternatively, the cholinergic imbalance could be causally involved with specific symptoms but not others. A more complex possibility involves the option that the transgenic organisms, being live entities, will compensate for the introduced imbalance by suppressing the production of the elements in excess or overproducing other elements in the imbalanced system. This will prevent the development of apparent malfunction(s), at least for a while, in spite of the fact that the cholinergic imbalance would in principle cause it. In that case, we might expect immediate changes at the level of gene expression shortly after the inserted transgene has been turned on (i.e., early in embryonic development); however, disease symptoms might only develop long after this time, in the adult organism, or not develop at all in the chosen species. This latter possibility is in line with the fact that most of the diseases associated with cholinergic imbalance are late onset, progressive ones. Last, but not least, one should always consider the possibility that the protein product of an introduced transgene has more than one activity. In this yet more complicated situation one should perceive interference with additional functions, causing unpredictable perturbations. Control experiments should be designed to account for each of these possibilities and pinpoint the right answer.

2 Acetylcholinesterase overexpression as an experimental tool for cholinergic imbalance

A subtle option for manipulating cholinergic neurotransmission in transgenic vertebrates is by accelerating the rate of termination of this process. This can be performed by transgenic overexpression of the ACh hydrolysing enzyme acetylcholinesterase (AChE; EC3.1.1.7). The elevated levels of AChE in the brain of transgenic mice suffice for hydrolyssing more ACh than normal, which will shorten the duration of the post-synaptic input but will not stop the process altogether. This should prevent drastic alterations in embryonic development and create organisms in which cholinergic neurotransmission is mildly suppressed but not totally blocked: in essence, a mimic of a number of disease states in humans (e.g. Alzheimer's disease, myasthenia gravis and others) and hence an experimentally useful model for biomedical research. To reach this goal, we have isolated the human ACHE gene (Soreq et al; 1990, Soreq and Zakut, 1993) and characterized its 3 major alternative splicing mRNA products (Karpel et al., 1994, 1996). The next step was to search for appropriate organisms in which consequent perturbations of embryonic development and adult functioning can be studied.

3 Accelerated synaptic development in ACHE transgenic *Xenopus laevis* tadpoles

The organism we selected first was the South African frog *Xenopus laevis*. It lays eggs throughout the year, is easily subject to hormonal stimulation and the eggs are conveniently amenable for *in vitro* fertilization (Seidman and Soreq, 1996). Being very large (> 1 mm diameter), hundreds of fertilized eggs may easily be injected in each experiment; the resultant embryos develop within 3 days to freely swimming tadpoles which can be watched, dissected and molecularly and structurally analysed in statistically significant numbers. The versatility and utility of this system thus offer important advantages to the experimentalist, especially for interdisciplinary studies combining different methodologies.

Our initial experiments involved the construction and introduction of two alternatively spliced ACHEcDNA species into *Xenopus laevis* embryos. These constructs included the potent cytomegalovirus (CMV) promoter and the coding sequence from the primary human brain ACHEmRNA species, 3'-terminated with exon No. 6 (E_6ACHE; Ben Aziz-Aloya et al., 1993); or the alternative ACHEmRNA subtype, expressed in hemopoietic cells and 3'-terminated with pseudointron No. 4 and exon No. 5 (I_4/E_5ACHE; Seidman et al., 1995). Using biochemical and electron microscopy methods, these experiments revealed that E_6ACHE codes in *Xenopus* tadpoles for catalytically active AChE which accumulates in neuromuscular junctions (NMJs) and accelerates their ultrastructural development (Shapira et al., 1994). In contrast, I_4/E_5ACHE was found to produce a similarly active enzyme which surprisingly found its way to epidermal cells, but not NMJs (Seidman et al., 1995). These studies hence led to the conclusion that AChE acquires distinct properties that are dependent on the 3'-exon included in the corresponding ACHEmRNA and/or the resultant C-terminal peptide (ca. 40 amino acid residues out of 600 in the entire protein). In particular, this distinction referred to the tissue or cell type in which the active protein accumulated (cholinergic synapses for E_6ACHE, epidermis for I_4/E_5ACHE) and to the morphogenic effect it exerted on such cells: a 160% increase in the area of NMJs and an enlargement of the embryonic synaptic cleft, as compared with no apparent difference in the epidermal cells secreting I_4/E_5 AChE.

4 Multisystem deterioration observed in ACHE transgenic mice

To create transgenic mice, we constructed a DNA vector carrying part of the human ACHE promoter, the 1st intron and the E_6ACHE coding sequence. When microinjected into mouse eggs, two copies of this DNA integrated into the genome of a transgenic mouse pedigree. In adult mice, this transgene was actively expressed but only in central nervous system neurons. This new and unique experimental model has thereafter served for investigating the effects of E_6ACHE overexpression on neuromotor, autonomous and cognitive functions.

Neuromuscular junctions in adult E_6ACHE transgenic mice were enlarged similar to those of *Xenopus* embryos transgenic for this transcript (Andres et al., 1996a). This suggested a universal morphogenic effect for this transgene on vertebrate NMJs. Transgenic mice displayed deficient grip strength which deteriorated with age, abnormal electromyographic properties and structurally altered NMJs. NMJ abnormalities resembled those in myasthenia gravis patients, where ACh receptors in NMJs are blocked (Drachman et al., 1982), and in those suffering from the Eaton-Lambert syndrome, (Eaton and Lambert, 1957) with inefficient presynaptic release of ACh. In both cases, the symptoms were hence apparently correlated with cholinergic deficits.

Further analyses indicated an inability of adult E_6ACHE transgenic mice to improve their performance following training in the Morris water maze test for learning and memory (Beeri et al., 1995). These studies revealed a progressive deterioration in the cognitive properties of the transgenic mice, suggesting that in this function as well, impaired cholinergic balance may contribute toward neuropathological disease symptoms. Autonomous functions were affected too: E_6ACHE transgenic mice were more resistant to the hypothermic effects of organophosphate AChE inhibitors and agonists of muscarinic and nicotinic ACh receptors, although their capacity to regulate body temperature in response to external changes in ambient temperature was sustained (Andres et al., 1996b). Altogether, our two distinct transgenic animal models therefore supported the notion that excess AChE may lead to progressive deterioration in neuromotor, cognitive and autonomous functions in vertebrates by promoting chronic deregulation of balanced cholinergic neurotransmission.

5 Cell culture transfections demonstrate a non-catalytic morphogenic activity for AChE

To further examine the morphogenic activities of E_6AChE and I4/E_5AChE we employed cell culture transfections. When transfected into rat C6 glioma cells, E_6ACHE induced rapid process extension and cell body enlargement. In contrast, I_4/E_5ACHE caused a totally different phenotype of small, round cells. E_6ACHE was partially associated with cell membranes but I_4/E_5ACHE was completely secretory (Karpel et al., 1996). However, C_6 glioma cells neither possess ACh receptors nor synthesize the enzyme choline acetyl transferase (ChAT) which produces ACh. In line with the previous reports of Layer et al. (1993) and Small et al. (1995), the distinct morphogenic functions of AChE subtypes in C_6 glioma therefore suggested that AChE might have additional, non-catalytic functions. Moreover, our findings indicated that the morphogenic activities of AChE susbtypes differ in a manner dependent on the variable C-terminal peptides generated by alternative splicing.

How, then, would we expect ACHE overexpression to affect neurons in the mammalian brain? As a preliminary approach to this question, we suppressed AChE production in primary mouse brain neurons by synthetic antisense oligodeoxynucleotides targetted at the ACHE gene (Grifman et al., 1996). These caused the retraction of neuronal processes, within 24 hr and with 0.1μM oligonucleotide. Our conclusion was, therefore, that we could expect morphogenic effects of AChE in mammalian glia as well, and that these morphogenic effects might depend on catalytic and non-catalytic activities.

6 Putative feedback Mechanisms

The transgenic animal models and the transfected cell culture experiments together indicate both catalytic and non-catalytic activities for overexpressed AChE. Yet, expression of the ACHE transgene was not lethal, and the pathological symptoms detected in the ACHE transgenic organisms were relatively limited. This suggests the existence of compensatory feedback mechanisms which enabled survival of the AChE-overexpressing organisms. The questions therefore arise - what were these feedback mechanisms and did they depend on catalytic, non-catalytic activities or both?

An organism may be expected to compensate for cholinergic deficits by upregulating production of key cholinergic elements (Wurtman, 1992), such as ChAT and the high affinity choline transporter that translocates choline from the synaptic cleft into the presynaptic terminal (Simon et al., 1976). Recent evidence suggests enhanced choline uptake in the Alzheimer's disease brain, demonstrating an example for such compensation attempts in a disease which involves ACh deficits (Slotkin et al., 1994). Both ACh synthesis and high affinity choline transport take place in the presynaptic cell. Our studies indicate that such compensation occurred in our mice and could be triggered by transgenic AChE through at least three possible different pathways.

(1) Overexpressed AChE would cause a decrease in ACh levels which should be "sensed" by both the pre and the post-synaptic ACh receptors. Signals reporting reduced ACh levels would be transduced through cAMP (for nicotinic ACh receptors) or G proteins (for muscarinic ACh receptors). Such signal transduction pathways can subsequently induce transcriptional and post-transcriptional responses compensating for the initial imbalance.

(2) Overexpressed AChE, through its catalytic activity would lead to elevated choline levels. This, in turn, may trigger enhancement of the high affinity choline uptake mechanism. That such enhancement occurs in Alzheimer's disease (Slotkin et al., 1994) was interpreted as implying that the membrane elements of decomposing dead cholinergic neurons can increase choline levels and enhance its uptake. Although it is not yet clear whether choline's interaction with the yet unidentified high affinity transporter transduces a signal to the nucleus, changes in the rate of this transport process may cause an imbalance in choline metabolism. This imbalance, in turn, is suspected to induce neuronal membrane vulnerability and enhance further neuronal damage (Wurtman, 1992).

(3) The apparent distinct morphogenic functions of the equally active proteins coded by E_6ACHE and I_4/E_5ACHE in *Xenopus* embryos and C_6 glioma cells, suggest morphogenic function(s) for AChE variants which are unrelated to their catalytic properties. Therefore, specific phenomena observed in the transgenic animals could reflect changes caused by the non-catalytic activities of AChE. In this latter state of events, cholinergic balance is not necessarily directly involved. A possible explanation to the non-catalytic morphogenic activity of AChE relates to the fact that the amino acid sequence of ACHE shares considerable sequence homologies with a group of nervous system proteins that bind the family of neurexins, neuronal receptors involved in presynaptic release (De la Escalera et al., 1990, Hortsch et al., 1990, Olson et al., 1990, Barthalay et al., 1992, Auld et al., 1995). Competition of AChE with these neurexin ligands could perhaps activate the neurexins and enhance presynaptic release of ACh. Such activation would not be caused by the perturbation in cholinergic communication; rather, it would only depend on the structural properties of the relevant AChE subtype. However, it could nevertheless correct the cholinergic imbalance. Therefore, subsequent events caused by this enhancement in pre-synaptic release would be similar to those discussed above.

To dissociate the catalytic from the non-catalytic activities of AChE, we have more recently constructed ACHE vectors encoding a mutation-inactivated enzyme and a C-terminally truncated one (Sela et al., 1995). Both these enzymes caused embryonic malformations in *Xenopus* tadpoles (Sternfeld et al., 1995). Transgenic mice carrying these DNAs should be most useful for delineating the effects of cholinergic imbalance as distinct from those correlated with the non-catalytic properties of AChE and for attributing specific role(s) to the different C-terminal peptides present in its alternative variants.

Acknowledgements

This research was supported by the Israel Science Fund at the Academy of Sciences and Humanities, by the USA Army Medical Research and Development Command (grant DAMD 17-94-C-4031) and by the U.S.-Israel Binational Science Foundation (to H.S.).

References

Andres, C., Beeri, R., Friedman, A., Lev-Lehman, E., Timberg, R., Shani, M. and Soreq, H. (1996a). Transgenic ACHE induces Neuromuscular deterioration in mice. Submitted for publication.

Andres, C., Beeri, R., Huberman, T., Shani, M. and Soreq, H. (1996b) Cholinergic drug resistance and impaired spatial learning in transgenic mice overexpressing human brain acetylcholinesterase. K. Loffelholz, Ed. Progress in Brain Res., in press.

Auld, V.J., Fetter, R.D., Broadie, K. and Goodman, C.S. (1995) Gliotactin, a novel transmembrane protein on peripheral glia, is required to form the blood-nerve barrier in *Drosophila*, Cell 81, 757-767.

Barthalay, Y., Hipeau-Jacquotte, R., Escalera, S., Jimenez, F., and Piovant, M. (1990) *Drosophila* neurotactin mediates heterophilic cell adhesion, EMBO J., 9, 3603-3609.

Beeri, R., Andres, C., Lev-Lehman, E., Timberg, R., Huberman, T., Shani, M. and Soreq, H. (1995) Transgenic expression of Human acetylcholinesterase induces progressive cognitive deterioration in mice. Current Biology, 5, 1063-1071.

Ben Aziz-Aloya, R., Seidman, S., Timberg, R., Sternfeld, M., Zakut, H. and Soreq, H. (1993) Expression of a human acetylcholinesterase promoter-reporter construct in developing neuromuscular junctions of Xenopus embryos. Proc. Natl. Acad. Sci. U.S.A., 90, 2471-2475.

Davies, P. and Maloney, A.J. (1976). Selective loss of central cholinergic neurons in Alzheimer's disease. Lancet 2, 1403.

De la Escalera, S., Bockamp, E.O., Hoya, F., Riovant, M. and Jimenez, F. (1990) Characterization and gene cloning of neurotactin, a *Drosophila* transmembrane protein related to cholinesterases. EMBO J. 9, 3593-3601.

Drachman, D.B., Adams, R.N. and Josifek, L.F. (1982). Functional activity of autoantibodies to acetylcholine receptor and the clinical severity of myasthenia gravis. N. Eng. J. Med. 307, 769-775.

Eaton, L.M. and Lambert, E.H. (1957). Electromyography and electric stimulation of nerves in diseases of motor unit: observations on the myasthenic syndrome associated with malignant tumors. J. Am. Med. Assoc. 163, 1117-1124.

Grifman, M., Lev-Lehman, E., Ginzberg, D., Eckstein, E., Zakut, H. and Soreq, H. (1996) Potential antisense oligonucleotide therapies for neurodegenerative diseases. M. Strauss and J.A. Barranger, Eds. Concepts in Gene Therapy, in press. Walter de Gruyter & Co., Berlin.

Hortsch, M., Patel, N.H., Bieber, A.J., Traquina, Z.R. and Goodman, C.S. (1990). *Drosophila* neurotactin, a surface glycoprotein with homology to serine esterases, is dynamically expressed during embryogenesis. Development 110, 1327-1340.

Karpel, R., Ben Aziz-Aloya, R., Sternfeld, M., Ehrlich, G., Ginzberg, D., Tarroni, P., Clementi, F., Zakut, H. and Soreq, H. (1994) Expression of three alternative acetylcholinesterase messenger RNAs in human tumor cell lines of different tissue origins. Exptl. Cell Res. 210, 268-277.

Karpel, R., Sternfeld, M., Ginzberg, D., Guhl, E., Graessman, A. and Soreq H. (1996) Overexpression of alternative human acetylcholinesterase forms modulates process extensions in cultured glioma cells. J. Neurochem., 66, 114-123.

Layer P.G., Weikert T. and Alber R. (1993) Cholinesterases regulate neurite growth of chick nerve cells *in vitro* by means of a non-enzymatic mechanism. Cell Tissue Res. 273, 219-226.

Olson, P.F., Fessler, L.I., Nelson, R.E., Sterne, R.E., Campbell, A.G. and Fessler, J.H. (1990). Glutactin, a novel *Drosophila* basement membrane related glycoprotein with sequence similarity to serine esterases. EMBO J. 9, 1219-1227.

Seidman S. and Soreq, H. (1996) Transgenic *Xenopus* Microinjection Methods and Developmental Neurobiology. Humana Press, Neuromethods series, in press. A. Boulton, Ed.

Seidman, S., Sternfeld, M., Ben Aziz-Aloya, R., Timberg, R., Kaufer, D. and Soreq, H. (1995) Synaptic versus epidermal accumulation of human acetylcholinesterase is encoded by alternative 3'-terminal exons. Mol. Cell Biol., 15, 2993-3002.

Sela, K., Sternfeld, M. and Soreq, H. (1995) Genetic engineering of modular human acetylcholinesterase variants. Israel J. of Medical Sci., 31, 755.

Shapira, M., Seidman, S., Sternfeld, M., Timberg, R., Kaufer, D., Patrick, J.W. and Soreq, H. (1994) Transgenic engineering of neuromuscular junctions in Xenopus laevis embryos transiently overexpressing key cholinergic proteins. Proc. Natl. Acad. Sci. USA, 91, 9072-9076.

Simon, J.R., Atweh, S. and Kuhar, M.J. (1976) Sodium-dependent high affinity choline uptake: a regulatory step in the synthesis of acetylcholine. J. Neurochem. 26, 909-922.

Slotkin, T.A., Nemeroff, C.B., Bissette, G. and Seidler, F.J. (1994). Overexpression of the high affinity choline transporter in cortical regions affected by Alzheimer's disease. J. Clin. Invest. 94,696-702.

Small, D.H., Reed G., Whitefield, B. and Nurcombe V. (1995) Cholinergic regulation of neurite outgrowth from isolated chick sympathetic neurons in culture. J. Neurosci. 15, 144-151.

Soreq, H. and Zakut, H. (1993) Human Cholinesterases and anticholinesterases. Academic Press, San Diego.

Soreq, H., Ben-Aziz, R., Prody, C.A., Gnatt, A., Neville, L., Lieman-Hurwitz, J., Lev-Lehman, E., Ginzberg, D., Seidman, S. Lapidot-Lifson, Y. and Zakut, H. (1990). Molecular cloning and construction of the coding region for human acetylcholinesterase reveals a G,C-rich attenuating structure. Proc. Natl. Acad. Sci. USA, 87, 9688-9692.

Sternfeld, M., Sela, K., Timberg, R. and Soreq H. (1995) Molecular dissection of protein domains directing the tissue targeting of acetylcholinesterase isoforms. Israel J. of Medical Sci., 31, 755.

Wurtman, R.J. (1992) Choline metabolism as a basis for the selective vulnerability of cholinergic neurons. Trends Neurosci. 15, 117-122.

Molecular Dissection of Protein Domains Directing the Tissue Targeting of Acetylcholinesterase Isoforms

Meira Sternfeld, Keren Sela, Shlomo Seidman, Rina Timberg and Hermona Soreq

Dept. of Biological Chemistry, The Institute of Life Sciences, The Hebrew University, Jerusalem, Israel, 91904.

1 Expression pattern of human acetylcholinesterase alternative forms

Acetylcholinesterase (AChE) accumulates in neuromuscular junctions (NMJs) where it terminates synaptic transmission by hydrolyzing the neurotransmitter acetylcholine (ACh) (Salpeter, 1967). In addition, AChE is expressed in non-cholinergic tissues such as the hematopoietic system (Paulus et al., 1981), germ cells (Gundersen and Miledi,1983), embryonic tissues at developmental stages before the onset of cholinergic transmission (Layer, 1991) and malignant tumors (Karpel et al., 1994). The role of AChE in these non-cholinergic tissues is unknown and is sometimes speculated to be non-catalytical. ACHE, the human gene encoding AChE, is located on the long arm of chromosome 7 (7q22) (Getman et al., 1992; Ehrlich et al., 1992), and spans a region of 7 Kb. Three 3'- alternative splicing options yield 3 mature AChE mRNAs whose catalytically active protein products differ at their C-termini (Ben Aziz-Aloya et al., 1993; Karpel et al., 1994).

Figure 1 depicts, on its upper left side, a representation of the human ACHE gene and its RNA transcripts. ACHE contains 6 exons, 3 introns and one pseudointron (I4). The major AChE mRNA form in brain and muscle is the form in which the common domain, including exons 1-4, is spliced to exon 6. A second alternative mRNA which is abundant in erythrocytes is the one in which the common domain is spliced to exon 5, and the third splicing option leaves pseudointron I4 to yield a readthrough mRNA found in human cell lines of tumorigenic origin (Karpel et al., 1994). A model of the three dimensional structure of human AChE based on the structure of the *Torpedo* enzyme (Sussman et al., 1991) is represented on the lower left side of figure 1. The 3' splice site connecting the common domain to the alternative sequences is marked by an arrow.

2 Human brain and muscle AChE accumulates in frog neuromuscular junctions

We have previously demonstrated that microinjection of a DNA construct encoding the brain and muscle AChE form (ACHE-E6) to developing *Xenopus laevis* embryos results in overexpression of the enzyme in myotomes (Fig. 1, upper right corner) and NMJs (Fig. 1, lower right part) of two- and three-day-old embryos (Seidman et al.,

NATO ASI Series, Vol. H 100
Neurotransmitter Release and Uptake
Edited by Şakire Pöğün

1994; 1995). Overexpression was assessed by AChE activity staining using the reaction of Karnovsky and Roots (1964), the products of which appear as dark crystals, which can be visualized by light or electron microscopy. In contrast, although the AChE forms encoded by the three alternatively spliced mRNAs differ only in their 40 C-terminal amino acids, when a cDNA construct encoding the erythrocyte and/or readthrough forms of AChE mRNA (ACHE-I4/E5) was microinjected to *Xenopus* embryos, the recombinant enzyme was overexpressed in ciliated and secretory epidermal cells (Seidman et al., 1995). In addition, when either of these constructs were microinjected to cultured glioma cells, and when the cells were stably transfected by ACHE-E6, overexpression caused cell body enlargement and process extensions. On the other hand, stable transfection with ACHE-I4/E5 caused the appearance of small round cells, demonstrating the different morphogenic effects of the enzymes encoded by these constructs (Karpel et al., 1995).

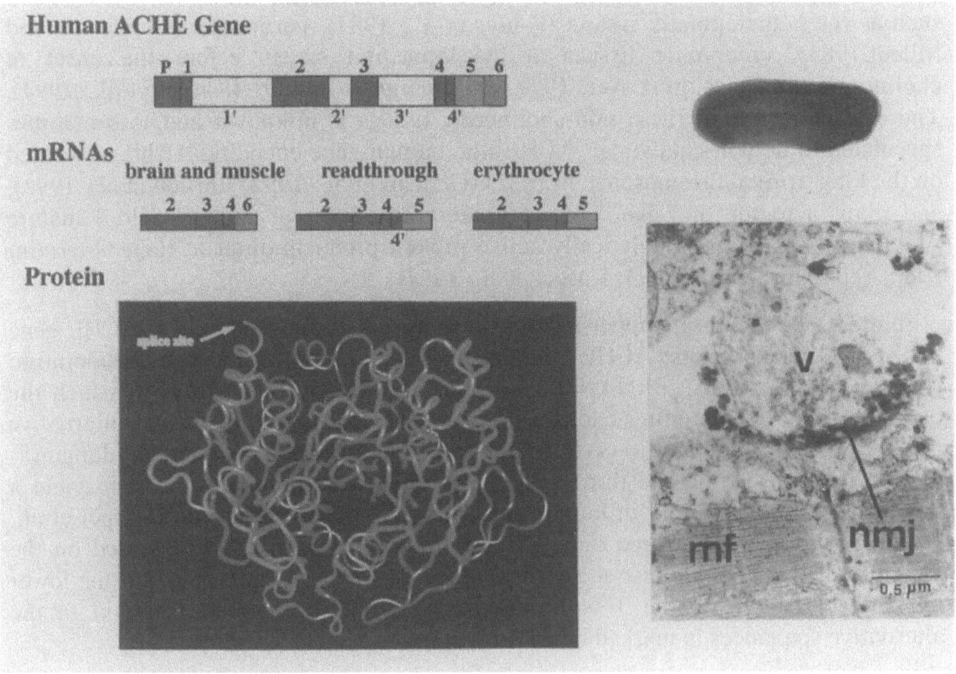

Fig 1. Human AChE pre-mRNA is alternatively spliced. The brain and muscle mRNA form encodes for a protein that accumulates in *Xenopus laevis* myotomes and NMJs (see text for details). The promoter (P) is designated by a dark box, common exons are numbered above and appear as dark gray boxes, introns are numbered below and are designated by empty boxes, and alternative sequences are presented by light gray boxes. mf- myofibrils, v- vesicle.

These findings called for investigating the mechanism(s) by which the NMJ-localization takes place, and the role the C-terminus of the enzyme plays in this mechanism(s). In addition, they raised the question of whether or not AChE catalytic activity is necessary for NMJ targeting. To this end, we employed PCR engineering to construct a truncated form of the enzyme (ACHE-E4) containing that part of the coding sequence common to all alternative forms but missing a natural C-terminus. In addition, a disrupted ACHE coding sequence was constructed by inserting an in-frame tag sequence of 7 amino acids, two residues downstream of the active site serine. The accuracy and integrity of these constructs was validated by DNA sequencing and their biochemical properties were tested in microinjected *Xenopus oocytes*. These analyses showed that the two natural variants and the truncated enzyme were all active, whereas the enzyme with the disrupted sequence displayed almost no catalytic activity, and the construct encoding it was thus named Inactive-ACHE.

3 Differential expression of human ACHE constructs in microinjected *Xenopus* embryos

To further delineate protein domains determining the tissue targeting patterns characteristic of the alternative forms of AChE, we expressed these engineered vectors in microinjected embryos of *Xenopus laevis*. *Xenopus* eggs were fertilized *in vitro* and microinjected within the first two cleavage cycles. Two day old embryos were stained for catalytically active AChE and visualized as wholemounts (Fig. 2), or transversally sectioned and then stained for AChE activity (Fig. 3). In the lower right corner of each panel in Figure 2 is a schematic presentation of the cDNA construct injected to the embryo shown. In control embryos, dark staining was apparent within myotomes, along the spinal cord and in the brain. Weak staining could be also seen in a few of the epidermal cells. In E6-ACHE injected embryos, prominent black horizontal staining was evident in the lower part of the myotomes and in brain, indicating overexpression. Furthermore, in these embryos there was a slight increase in the number of pigmented cells. Inactive-ACHE injected embryos demonstrated a high variability in staining. Some embryos, such as the one presented in figure 2, showed a slight reduction in myotomal staining as compared to control embryos especially in the mid trunk region. Limited staining is apparent in the posterior somites in these embryos. Microinjection of E4-ACHE resulted in prominent myotomal staining similar to E6-ACHE injected embryos. However, a larger percentage of these embryos seemed deformed as compared to that observed among control embryos and embryos injected with other constructs (note the apparent spine curvature). Here too, the number of pigmented cells in the ventral part of the embryo was increased. In addition, the area in the anterior ventral part of the embryo which represents the heart-anlage and the blood islands which contain the progenitor cells of the blood (Hausen and Reibesell, 1991; Zon, 1995) was darkly stained. I4/E5-ACHE injected embryos did not display increased myotomal staining. However, epidermal cells were stained for active AChE, as previously reported (staining was not very obvious in the shown focal depth, but was apparent *in vitro*). Moreover, pigmented cell numbers were significantly larger in this group of embryos. Therefore, the C-terminus of AChE affects the tissue localization of the

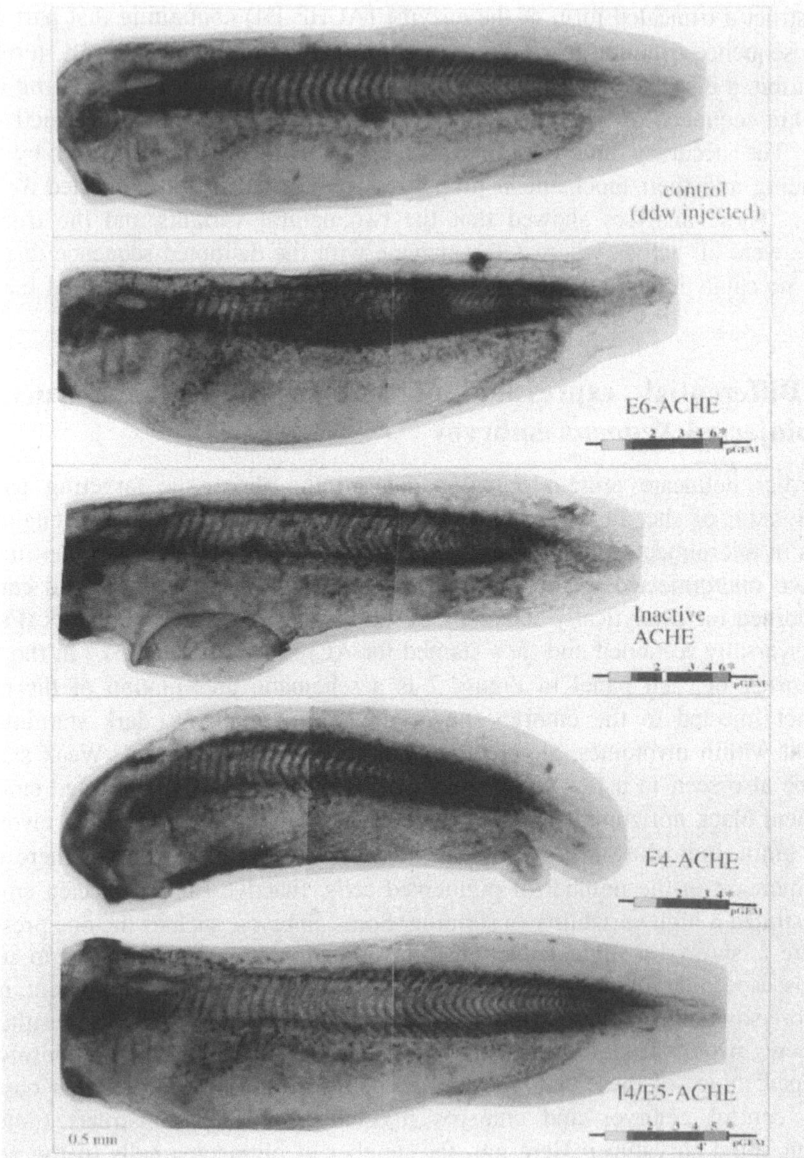

Fig 2. Human AChE variants are differentially expressed in *Xenopus* embryos as shown by staining of microinjected *Xenopus* embryos for catalytically active AChE in wholemount (see text for details). Exons and introns are designated as above, the cytomegalovirus promoter is shown as a light gray box, the AChE signal peptide is presented by a darker box and an empty box designates the insert in the active site coding sequence of ACHE used to create Inactive AChE.

Figure 3 shows sections from the microinjected embryo stained for catalytically active AChE. Transversal sections (50 μm) from head and trunk areas were prepared from *Xenopus* embryos microinjected with the different human ACHEcDNA constructs, stained for AChE activity and viewed under (100 magnification. Top panels show head sections and bottom ones show trunk sections. Above are schematic presentations of the cDNAs injected, as in figure 2. Head sections of control embryo sections showed AChE activity staining in the eye, brain and epidermis tissues. Myotomes, spinal cord and epidermis were stained in the trunk sections of these embryos. E6-ACHE injection significantly increased staining in all the above tissues except for epidermis. Overall, Inactive-ACHE injected embryos showed a slight increase in staining as compared to control ones, although individual embryos displayed more variability than those injected with other DNAs. E4-ACHE injected embryos demonstrated staining in myotomes and spinal cord at least as prominent as the staining caused by E6-ACHE microinjection, but head sections were not stained. I4/E5-ACHE microinjection elevated staining in spinal cord and in ventral epidermis in the trunk section. To summarize, different AChE isoforms were overexpressed in different tissues and E6-ACHE, particularly, caused a wider variety of overexpression patterns than E4-ACHE.

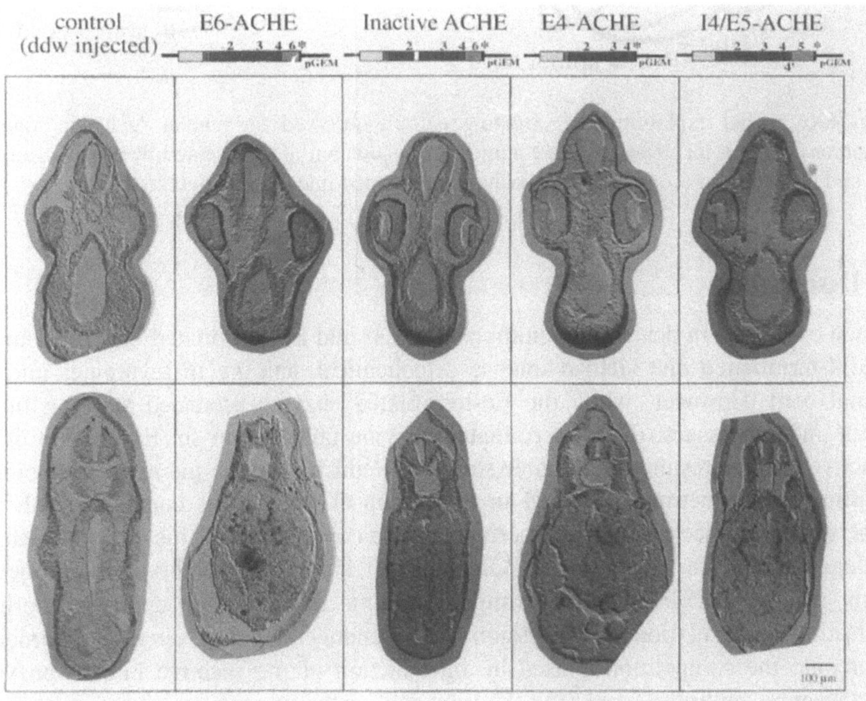

Fig 3. Differential expression of human ACHE constructs as shown by AChE activity staining of transversal sections of microinjected *Xenopus* embryos (see text for details). ACHE constructs are designated as above.

In turn, E4-ACHE was more widely overexpressed than I4/E5-ACHE. Additionally, all AChE isoforms caused increase in pigmentation, especially in head sections.

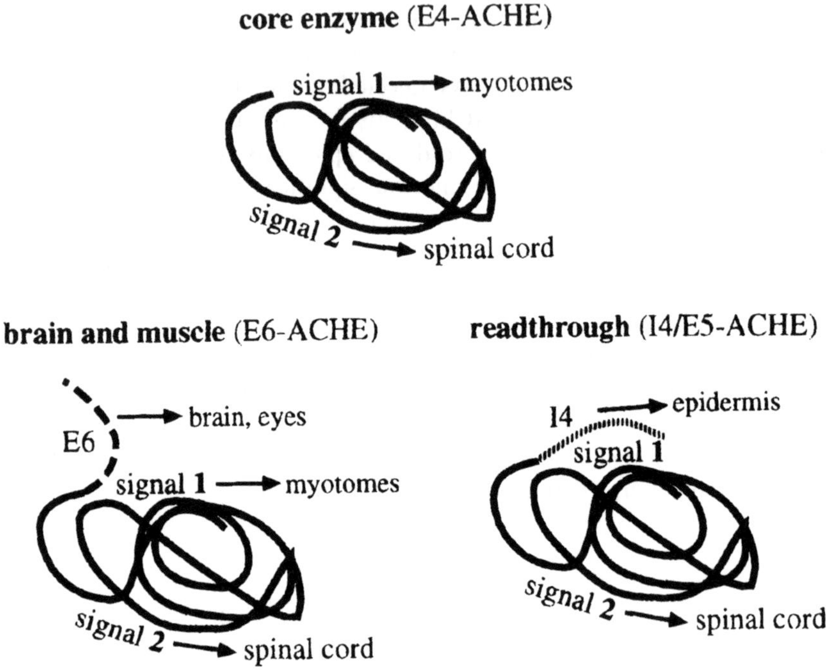

Fig. 4. A model explaining the targeting pattern observed for human AChE in *Xenopus* embryos (see text for details). The common AChE domain is represented by a black line, the E6 and I4 C-termini are depicted by dashed and striped lines, respectively.

4. Discussion

When expressed in developing embryos, the E4- and E6-terminated enzymes but not the I4-terminated one yielded intense cytochemical staining in myotomes and the spinal cord. However, while the E6-terminated enzyme produced staining in the brain and eye tissues, the E4-terminated enzyme failed to do so. Expression of the inactive enzyme resulted in variable staining. Reduction in the intensity of myotomal staining was sometimes observed in this group of embryos as compared with that observed in uninjected embryos, perhaps due to competition of this enzyme with the endogenous one for the substrate. On the other hand, other embryos microinjected with Inactive-ACHE showed stronger staining than control embryos, perhaps reflecting up-regulation of expression of the endogenous frog enzyme in order to overcome the competition caused by introduction of the inactive human enzyme. Furthermore, embryos expressing the truncated or the inactive enzymes were subject to developmental defects more often than control embryos or embryos injected with other ACHE cDNA constructs.

The model depicted in figure 4 could explain our findings: the core domain of AChE (demonstrated by E4-ACHE) could contain a signal for targeting to the myotomes (signal 1) and another one directing the enzyme to the spinal cord (signal 2). The E6 encoded C-terminus is positioned in such a way that both these signals are exposed and functional. In addition, this C-terminus could contain a signal directing the enzyme to the brain and eye tissues. According to this model, the I4 C-terminus would be positioned such that it masks the myotome-signal, thus preventing targeting to this tissue, but leaving the signal to the spinal cord exposed. Additionally, the I4 C-terminal peptide should contain a signal directing the enzyme to epidermal cells. Thus, our findings suggest that the tissue targeting of AChE is directed by a mechanism involving the core domain of AChE, and both C-termini. Furthermore, preliminary results suggest that this mechanism is independent of the catalytic activity of AChE. In addition, these findings demonstrate that excess amounts of either highly active or inactive AChE, both of which may cause cholinergic imbalance, can hamper with normal embryogenesis in vertebrates.

Acknowledgments:

This study was supported by the U.S. Army medical Research and Development Command (DAMD 17-94-4031), the U.S.-Israel Binational Science Foundation (93-307) and the Israeli Science Foundation.

References:

Ben Aziz-Aloya, R., Seidman, S., Timberg, R., Sternfeld, M., Zakut, H., and Soreq, H. 1993. Expression of a human acetylcholinesterase promoter-reporter construct in developing neuromuscular junctions of *Xenopus* embryos. Proc. Natl. Acad. Sci. USA 90: 2471-2475.

Ehrlich, G., Viegas-Pequignot, E., Ginzberg, D., Sindel, L., Soreq, H., and Zakut, H. 1992. Mapping the human acetylcholinesterase gene to chromosome 7q22 by fluorescent in situ hybridization coupled with selective PCR amplification from a somatic hybrid cell panel and chromosome-sorted DNA libraries. Genomics 13: 42-47.

Getman, D.K., Eubanks, J.H., Camp, S., Evans, G.A. and Taylor, P. 1992. The human gene encoding acetylcholinesterase is located on the long arm of chromosome 7. Am. J. Hum. Genet. 51:170-177.

Gundersen, C.B. and Miledi, R. 1983. Acetylcholinesterase activity in *Xenopus laevis* oocytes. Neurosci. 10: 1487-1495.

Hausen, P. and Riebesell, M. 1991. The early development of *Xenopus laevis*. An atlas of the histology. Eds. Springer-Verlag. Berlin, Heidelberg, New York.

Higuchi, R. 1990. Recombinant PCR. In: PCR Protocols. Academic Press, San Diego. Michael A. Innis. David H. Gellfand, John J. Sninsky, Thomas J. White, Eds. pp 177-183.

Karnovski, M.J. and Roots, L., 1964. A direct coloring method for cholinesterases. J. Histochem. Cytochem. 12: 219-221.

Karpel, R., Ben Aziz-Aloya, R., Sternfeld, M., Ehrlich, G., Ginzberg, D., Tarroni, P., Clementi, F., Zakut, H., and Soreq, H. 1994. Expression of three alternative acetylcholinesterase messenger RNAs in human tumor cell lines of different tissue origins. Experimental Cell Research 210: 268-277.

Karpel, R., Sternfeld, M., Ginzberg, D., Guhl, E., Graessman, A. and Soreq, H. 1995. Overexpression of alternative human acetylcholinesterase forms modulates process extensions in cultured glioma cells. J. Neurochem. 66: 114-123.

Layer, P.G. 1991. Cholinesterases during development of the avian nervous system. Cell. Mol. Neurobiol. 11: 7-33.

Paulus, J.M., Maigen, J. and Keyhani, E. 1981. Mouse megakaryocytes secrete acetylcholinesterase. Blood 58: 1100-1106.

Salpeter, M., 1967. Electron microscope radioautography as a quantitative tool in enzyme cytochemistry. I. The distribution of acetylcholinesterase at motor endplates of a vertebrate twitch muscle. J. Cell. Biol. 32: 379-389.

Seidman, S., Ben Aziz-Aloya, R., Timberg, R., Loewenstein, Y., Velan, B., Shafferman, A., Liao, J., Norgaard-Pedersen, Brodbeck, U. and Soreq, H. 1994. Overexpressed monomeric human acetylcholinesterase induces subtle ultrastructural modifications in developing neuromuscular junctions of *Xenopus laevis* embryos. J. Neurochem. 62: 1670-1681.

Seidman, S., Sternfeld, M., Ben Aziz-Aloya, R., Timberg, R., Kaufer-Nahum, D. and Soreq, H. 1995. Synaptic and epidermal accumulation of human acetylcholinesterase are encoded by alternative 3'-terminal exons. Mol.Cel. Biol. 15: 2993-3002.

Sussman,J.L., Harel, M., Frolow, F., Oefner, C., Goldman, A., Toker, L. and Silman, I. 1991. Atomic structure of acetylcholinesterase from Torpedo californica: a prototypic acetylcholine-binding protein. Science 253: 872-879.

Zon, L.I. 1995. Developmental biology of hematopoiesis. Blood. 86: 2876-2891.

Interdisciplinary Interactions Across the Biobehavioral Divide: A Pre-Socratic Perspective

John J. Furedy

Department of Psychology, University of Toronto, Toronto, Ontario, M5S 3G3, Canada
furedy@psych.utoronto.ca (E-mail)

As an atypical participant at the conference on which this book is based, I am aware that I have to make a special case for the relevance of this chapter. I begin by elaborating on the concept of genuine interdisciplinaty interactions, and then consider the nature of the Pre-Socratic, phenomenon-focussed perspective. This is followed by a brief account of one example of an interdisciplinary approach applied to a biobehavioral problem. The example indicates that attention to measurable differences in psychological functions can illuminate our understanding of the phenomena. Finally, I conclude with three opinions that, as an ignorant outsider to the area of research covered in the conference, I developed from my "super student" perspective as I listened to the papers, and looked at the posters, that were presented.

1 Elaboration of Interdisciplinary Interactions

Much of scientific activity is not interdisciplinary at all. Many investigations are confined not only to one discipline, but to quite narrow areas of specialization. Indeed, as an outsider, I was struck by the fact that even within what I think of as an already highly specialized field, neurotransmitter functions, there is a further marked distinction between release and uptake specialists, and many investigations are appropriately confined to one or the other sub-speciality.

In addition, although there are important differences in methods of investigation between specialities within a discipline (e.g., the release/uptake methodological differences), these methodological (and perhaps conceptual) differences are vastly increased as one moves along the disciplinary continuum from, say, single-cell biochemistry to social psychology. It is here that the biobehavioral divide can become a veritable chasm. Yet there are phenomena like human drug addiction and its cure that require reaching across that divide, if there is to be any hope of progressing towards an understanding of the phenomena and solutions of the practical problems that they engender. Interdisciplinary research on these sorts of issues does not mean that one tries to ignore or minimize the vast differences in methodology and conceptualization that exist. Nor should the integrity of each relevant discipline be compromised in such research. Indeed, it is necessary to treat the phenomena at all levels of analysis with the same level of conceptual rigor, at least in relation to the issues being investigated. This

NATO ASI Series, Vol. H 100
Neurotransmitter Release and Uptake
Edited by Şakire Pöğün

is because, although social psychological phenomena are more difficult to quantify and analyse than single-cell biological ones, they must both be viewed as being part of the same spatio-temporal realm that is the object of scientific analysis and technological manipulation. The approach that I am advocating here stems from a Pre-Socratic perspective, on which I elaborate in the next section.

2 Elaboration of the Pre-Socratic, Disinterested Perspective: "Thinking About the World in the Greek Way"

Although disinterestedness is often confused, even in academic writings, with lack of interest, the distinction is clearly part of our current intellectual and even political heritage. A board of inquiry has, in principle, to be independent or detached as well as expert (or at least interested) in the issue under discussion. That this idea is understood by the population at large rather than just by intellectuals, and that attacks on such a board's independence are politically as well as logically effective, indicate that the concept of disinterestedness is woven into our culture.

Nevertheless, disinterestedness is both a relative newcomer to civilization and a notion that is constantly opposed by other influences. The notion of disinterested inquiry, or considering *x* for its own sake, first arose among a group of Ionian or Anatolian philosophers who are generally known as the Pre-Socratics. They were the first to demonstrate "thinking about the world in the Greek way"[1] (Burnet, 1930, p.v), and laid the foundation for a problem- rather than person-oriented approach to both scientific and literary fields of inquiry.

In terms of scientific fields, it is just this phenomenon-centered aspect that was missing from an otherwise technically advanced civilization like the one in Babylon, which had a well-developed discipline of nonscientific astrology but no genuine science of astronomy. To look at the heavens in the Babylonian way is to engage in observation and quantified data-gathering activities, but these observations are made from an interested perspective. That is, the movements of the heavenly bodies are not viewed astronomically as

[1] Experts on Ephesus (toured by the conference participants) and other Anatolian cities of this period like Mehlika Seval have argued that the term "Greek" is at least anachronistic if not worse, as most of these Pre- Socratics preceded the union of the Greek cities in the face of the Persian threat, and that, in any case, there was no long-lasting political Greek "empire" in the sense that the Romans, the Ottomans, and the British had empires. Nevertheless, in terms of culture and intellectual civilization, the (mostly Athenian) "glory that was Greece" is, I suggest, a meaningful concept, and that those who contributed to the possibility of its development include Leonidas and his 300 Spartans, despite the fact that it was the later wars between Sparta and Athens that played a significant role in preventing a political Greek empire from being established. So I follow the philosopher-historian John Burnet in referring to a "Greek way of thinking about the world", and include the Pre-Socratic Ionian philosophers in that mode of thought.

phenomena of interest in their own right. They are rather seen astrologically as being related to individual and (in the ancient world before an important battle--even the technically, politically, and militarily advanced Romans engaged in this practice) groups of human lives. In contrast, the Greeks, who amassed far less information about those movements than the Babylonians, nevertheless developed astronomy, because they treated those movements as problems to be considered for their own sake rather than in relation to individual or social concerns. Moreover, in the long run, such advanced technical applications as space flight are based on astronomy rather than astrology, no matter how sophisticated and quantified those Babylonian astrological systems may have been.

For biology also, I believe that the phenomenon-centered, disinterested approach of the Pre-Socratics is of current relevance. The principle of "saving the appearances" embodies the notion that theoretical explanations must strive to account for *all* observations, rather than only those that happen to fit with the explanation. Moreover, I suggest that the words of a particular Pre-Socratic, the Ephesian Heraclitus, are especially relevant. As a group, the Pre-Socratics were concerned with the general question of what was common to all things, and typically their answers involved relatively static substances like water (Thales). Heraclitus, however, asserted that change and interaction is what is common to all things. He put forward the fire metaphor, because fire, unlike water, is obviously ever changing, and is dramatically both influenced by, and is a dramatic influencer of (in other words, interacts with), the environment. It is this Heraclitian, interactive approach that seems most suitable for the biological sciences, and those interactions occur not only between the organism and its environment, but also among the various functions that occur within the organism. And, I suggest, only an interdisciplinary approach to both research and practice can provide the path towards progress in understanding and controlling biobehavioral phenomena. As an illustration (though not a full justification) of what I have been suggesting, I now turn to an example of an interdisciplinary approach to a biobehavioral problem.

3 Example of an Interdisciplinary Approach to a Biobehavioral Problem: Effects of Sex and Nicotine on Levels of Cognitive Functions

Although cognitive functions are clearly more varied and sophisticated in humans. experimental psychologists have studied them in rats for at least fifty years. It is, in fact, the learning of *place* that is the clearest mark of cognitive functioning in rats, and the maze has been the most common apparatus for studying these functions (e.g., Tolman, Ritchie, & Kalish, 1946). Still, learning in these early mazes was still subject to a non-cognitive, stimulus-response interpretation. The development of the Morris Water Maze (MWM) by Morris (1981) allowed a more unambiguous cognitive interpretation. In the MWM, rats learn to locate a hidden platform by swimming to it from several different starting points around the perimeter of a circular pool, and so provide a relatively unequivocal demonstration of cognitive *place* learning, i.e., the learning of the proposition that the platform is in a specific place.

When the platform is hidden, the only cues that can be utilized are *spatial* ones. However, if the platform is raised above water level to be visible, then the place-learning problem can be solved by employing *visual* cues. Both sets of cues lead to a cognitive, propositional form of learning, but in terms of "levels" of cognitive functioning, it is clear that using spatial cues involves a higher level of functioning than using visual cues. One might say that spatial-cue functioning is more "reflective" (in Rodin's thinker sense of the term) than visual-cue functioning. Finally, an evolutionary psychologist or sociobiologist might even associate the spatial/visual cue distinction[2] with the sex-roles-connected hunter/gatherer distinction.

The experiment that I would like to briefly report will, by the time that the present book is published, be available in abstract form in the 1996 Proceedings of the Society for Neuroscience (Kanit et al., 1996). Sex and the administration of nicotine were varied between subjects in a 2 x 2 factorial design with 7 Sprague-Dawley rats in each cell. The training phase comprised a 12-day period in the MWM apparatus with the platform hidden (requiring spatial-cue-based cognitive functioning) on the 4th, 8th, and 12th days. The task itself was rendered a clearly place-learning one, because while the platform was always in the same place, animals were started from (four) different positions on the perimeter of the circular pool. The results showed classical learning curves accelerating negatively towards asymptotic performance of approx. 4 and 2 s., respectively, for the hidden- and visible-platform conditions. The latter value is close to the limit set by the rat's swimming speed, and hence represents animals swimming directly to the target, or knowing exactly where the target is. There were no clear effects

---

[2]The spatial/visual cue distinction has been criticized because the "spatial" cues employed (relationally) to learn where the platform is when it is hidden are also visual, these cues being various objects hung from the ceiling in the room containing the circular pool. Because the MWM models the problems faced by human sailors on the high seas, an alternative to the spatial/visual distinction that is feasible is navigation versus sighting. It will be noted that navigation involves a clearly deeper level of cognizing than sighting, even though, depending on the situation, it may be the shallower level of cognitive functioning that can be more adaptive. These insights were gained in the course of a sometimes heated discussion with my wife, Dr. Christine Furedy, whose disciplinary specialities (history and anthropology) are even further from the biological end of the biobehavioral continuum than mine. Initially I was sceptical whether a historian-anthropologist could help an experimental psychologist clarify a terminological problem in his own field, but now I see that this scepticism was misplaced, and probably fuelled by the irrational territorial behavior manifested by many specialists.

due to sex or nicotine, suggesting that, at least in this context[3], sex and nicotine did not affect cognitive *abilities*.

An opportunity for assessing differential *preference* for the two levels of cognitive function was provided by *shifting* the position of the platform on the 13th day, and making that platform visible. This allows the spatial and visual cues to compete. Only the females without the nicotine showed complete visual-cue preference, by swimming straight to the moved (but visible) platform in the mean asymptotic time of about 2 s. The mean times of the other three groups ranged between 16 and 18 s., did not differ significantly from each other, and hence showed a preference for the more "reflective" (and, in this case, unadaptive) spatial-cue level of cognitive functioning. Another measure of preference for the deeper, "reflective" cognitive function is time spent in the quadrant where the platform used to be during the first 12 days. On this measure the females without the nicotine again differentiated themselves (spending about 0.75 s.) from the other three groups (for nicotinized females and both male groups, respectively, the mean scores were 6.75 and 10 s.).

Following the 13th day, animals were sacrificed and the pattern and amount of dopamine uptake was assessed. These pharmaco-neuronal functions are known to be associated with the psychological reward/pleasure functions, and, as expected, the usual group sex differences emerged. However, the nicotine manipulation had no main or interactive effect. In summary then, this study yielded differences among two cognitive functions and a hedonic one. Specifically, sex and nicotine did not affect cognitive ability with respect to two different levels of cognitive functioning; sex and nicotine interacted in affecting preference for a deeper, more "reflective" level of cognitive functioning, with maleness and nicotine appearing to be equivalent ways of producing such a preference for spatial- cue-based functioning (even when that preference was unadaptive in the context of the task); the hedonic reward system showed the usual sexual dimorphism effect, but was unaffected by nicotine.

At a more speculative level it is interesting to note that one of the attractive features of cigarettes reported by smokers is that smoking "helps one think". Again, at the level of evolutionary psychology or sociobiology, the parallel between the visual and spatial cues and the contrasting requirements for female gatherers and male hunters in primitive society is obvious enough. Of course in rats the hunter/gatherer distinction is not as obvious. But perhaps the preference sex difference is hard-wired in vertebrates, and develops, through cultural, environmental, and even evolutionary factors into ability differences in humans. We are presently engaged in hormonal manipulations in the

---

[3]In another context that was an interdisciplinary examination of verbal cognitive ability in humans and employed psychophysiological as well as behavioral measures (Algan et al., 1995), sex and nicotine yielded an interactive effect on cognitive ability, with females being superior to males, but that superiority being eliminated by (smoking-administered) nicotine.

MWM context to assess the stage at which the male preference for spatial cues is "laid down".

Another issue concerns the *mechanism* through which nicotine works to produce the reflective preference in females. The apparent preference may not be for the reflective cognitive function *per se*, but (as noted at the conference presentation) could be due to a change of attitude to the water. That is, nicotine may reduce anxiety, cause relaxation, and hence decrease motivation to reach the platform as soon as possible. More generally, although the MWM is primarily designed to examine cognitive functions, it must not be forgotten that living organisms (unlike computers) are affected by motivational ones as well. Varying water temperature and administering "questionnaires" to measure individual rat motivation to avoid the water (done by raising and then lowering each of two platforms, and measuring swimming speed from the lowered to the raised platform) should allow the relevant information to be gained regarding these motivational influences.

There are also some extensions of our MWM preparation for other interdisciplinary problems that examine different biological variables. Just from this conference alone, there are at least two possibilities that appear to be promising. One such extension arises from the transgenic mice studies reported by Hermona Soreq, which employed a standard MWM behavioral preparation. Employing the preparation used here, which allows a distinction to be made between cognitive abilities, as well as between ability and preference, could provide additional information on the psychological effects of her physiological manipulations. Another such possible extension arises from the anandamidic rats described by Julius Axelrod, where cannabis rather than nicotine was the pharmacological manipulation. The effects of cannabis on changes in cognitive abilities as well as preference could be examined in the present MWM preparation. In addition, effects on mood could be examined either in the "questionnaire" preparation, or even in the more conventional behavioral open field test.

I hope that this example of interdisciplinary research has indicated that attention to psychological differences such as that between different levels of cognitive function, ability versus preference, and cognitive versus non-cognitive (e.g., motivational) functions, can illuminate our understanding of biobehavioral phenomena. In addition, I would also like to note that in the example of research that I have given, I have used no jargon or terms that are intelligible only to specialists in specific psychological areas of current interest. Rather, I have used terms to make psychological distinctions that are intelligible to any intelligent reader who is unacquainted with the quasi talmudic complexities of much of current psychological terminology.[4] The language, moreover, has been that of phenomena rather than of prominent scientists' names. Finally, I have referred constantly to theories and hypotheses (to be tested in terms of their truth) rather than to models (to be tested in terms of their usefulness or fruitfulness). This phenomenon-centered, realist approach to science is one that I have tried to adopt long before I attended the present NATO conference. In the concluding section I shall outline three sets of opinions that, as a predisposed realist, I developed as a result of my participation in the conference.

4 A "Super Student" Perspective on Neurotransmitter Release and Uptake Research

At the conference there was a distinction made between "students" and "lecturers" as a function of degree of expertise in the field. My claim to being a "super student" in these terms is based on an almost total lack of expertise, stemming from my extreme behavioral position on the biobehavioral disciplinary continuum. Sometimes, however, the opinions of a rank outsider can be useful, and I offer three of these here.

My first opinion concerns the place of models in the conference's area of research. Contrary to the impression I have given above, I am not against the use of models. They are clearly useful for originating hypotheses. They can even be useful for testing hypotheses, provided that *competing* models are considered in what Platt (1964) has called the strong- inference mode. This mode requires specifying what the essential differences are between competing models, and then specifying how the data to be gathered will bear on these differences.

It is also necessary for strong-inference testing that positions be formulated in such a way that they are falsifiable in Popper's (1959) sense of that term. The Freudian libido hypothesis of psychoanalysis has often been soundly criticized on the grounds of being unfalsifiable, meaning that it was compatible with *all* behavioral observations. As I

---

[4] So, for example, even though the originators of the modified MWM preparation have described their results as indicating "parallel information processing" that constitutes evidence for "independent memory systems" (McDonald & White, 1994), I have eschewed this terminology. The former expression is based on computer models of behavior that "treat the mind (and the brain) as an information processor" (Furedy, 1994, p. 177), as if living organisms functioned like computers (see also Searle, 1992 for more systematic philosophical critique of the adoption of the computer analogy by most current cognitive psychologists). The latter expression involves an assumption that comes down on one side of an issue which may well not be empirically resoluble: whether there is a single memory system operating at different levels, or whether there are separate and different memory systems. Experts in memory have been arguing about this one for at least two decades, and it looks at least to me that the argument may be more of a theological rather than an empirical debate.

listened to how the calcium hypothesis was employed by many of the release specialists at the conference, I could not help but wonder whether the way in which this hypothesis was formulated suffered from the shortcomings of the libido hypothesis--that it, with suitable auxiliary assumptions, was broad enough to predict all outcomes, and hence explained none of them adequately.

More generally, for strong-inference testing I think one needs, in all scientific disciplines, to move from a functional or teleological analysis to one that seeks necessary and sufficient conditions, and provides deterministic explanations. A functional analysis, like modelling, can be useful for suggesting hypotheses and theories to be tested, but is no substitute for them. Those hypotheses and theories, moreover, need to be formulated in a deterministic, necessary-and-sufficient mode. The problem with using functional analysis as an explanation is that it tends to be circular, because one can invent as many functions as there are phenomena, just as a hundred years ago faculty psychologists invented as many faculties as there were abilities to explain.

An adequate scientific account should, as the Pre-Socratics put, "save the appearances", or should take into consideration all classes of observations. What I take this to mean is that all classes of observations, including psychological ones, need to be the subject of serious enquiry, though not of "proof". In empirical science, no matter how "hard", there is never proof, as all empirical generalizations are open to error. Even in astronomy, there is the story of the deductive "proof" of Vulcan to remind us of this. Briefly, the planets outside Jupiter (visible without a telescope) were all discovered by a cycle of observation (of perturbations of observed planets not accounted for by the planets and the sun), hypothesis (that an outer planet exists), and confirmation/"proof" (observation of the outer planet predicted by the hypothesis based on the perturbations). By the time astronomers had "proofed" the existence of Neptune by this, they had got so cocky that when they observed unexplained perturbations in the innermost planet, Mercury, they proceeded not only to predict the existence of an additional planet between Mercury and the sun, but also named it Vulcan (no relation to Mr. Spock's home). In fact, several astronomers even reported observing Vulcan, but later work indicated that these observations were false, and according to current views, Mercury's perturbations are accounted for by Einstein's special theory of relativity. I could not help thinking of the Vulcan tale from what is quite possibly our "hardest" science when I hear several speakers at the conference talk of "proof" and even "conclusive proof" when referring to views that clearly enjoy widespread current expert support, but that may turn out to be false on the basis of future observations and theorizing.

My second opinion is about interdisciplinary, biobehavioral research, and is relevant to the extent that a significant proportion of the papers presented referred to behavioral implications (e.g., implications of cocaine uptake processes for understanding drug addiction and improving its treatment). The approach here has to be concerned both with scientific and applied purposes. A critical point, in my view, is that there be common definitions of basic terms, so that, for example, the term "plasticity" has to be distinguished from the term "learning", even though plasticity may be one of the mechanisms that underlie learning.

A related point is that all phenomena, whether they are single-cell electrical or behavioral, have to be treated with equal seriousness, and as part of the same scientifically studiable and technologically evaluatable spatio-temporal realm. So, as I hope I have shown in the previous section, cognitive abilities are indeed evaluatable, even if these evaluations may be embarrassing to some (as in the case of sex differences--but recall that the astronomical heliocentric hypothesis was also "uncomfortable" to many at one time). Furthermore the applied standards for treatment evaluation must be equally rigorous in the logical sense across the biobehavioral realm. For example, just as we require controlled, double-blind clinical trials for drugs like aspirin, so we must demand the same level of rigor for evaluating biofeedback (a learning, instrumental-conditioning phenomenon) or even psychotherapy that purports, say, to cure addiction.

My third and final opinion is directed at those "students" at the conference who are relatively young in professional age.[5] At this conference and else where in the literature as well as in grant applications, there are examples of promissory notes being issued that refer to relations between single cell work and behavior. However, the relations are stated in such a vague way that the note is never even potentially cashed. I predict that this practice will cease to be good enough in the next couple of decades. So while senior and well-established researchers can afford to be unserious about stepping across the biobehavioral divide, you cannot. Serious interdisciplinary work is increasingly going to become essential for survival.

That, however, is not the main reason for engaging in interdisciplinary research. The main reason, rather, is that it provides a far better way for both science and technology. For science, it provides an understanding of the phenomena that is genuinely progressive (as in the case of astronomy), although there is never any "proof" (remember the Vulcan tale). For technology or applications, interdisciplinary work provides better control, though that too is never complete. Only God is omnipotent, and s/he is not part of science or technology.

Just as the Pre-Socratics were interested not only in enquiry, but were also people with considerable practical interests, so to-day's young researchers must be prepared to focus

[5] In general I agreed with the conference's organizer, Sakire Pogun, when she suggested at the outset that we drop the student/lecturer distinction, because there were many situations where a "student" would be more expert in the topic under discussion than a "lecturer", and because even if this were not so, fruitful scientific discussion is not facilitated by these sorts of distinctions. However, as there was at least a correlation (with some exceptions) between age and the student/lecturer distinction, I have used this distinction here.

on considering phenomena for their own sakes, on critically evaluating competing accounts of those phenomena, and on being prepared to seek solutions that can be shown to *really* work in the sense of having a demonstrable positive influence. Interdisciplinary research can do this, and so I invite at least some of you take the plunge.

References

Algan, O., Furedy, J.J., Demirgoren, S., Vincent, A., & Pogun, S. (1995). Effects of tobacco smoking on interhemispheric processing and psychophysiological responses during verbal and spatial tasks: gender differences. *Society for Neurosciences Proceedings*(Abstract).

Burnet, J. (1930). *Early Greek philosophy.* London: Adam & Charles Black.

Furedy, J.J. (1994). Review of J. Searle's *The rediscovery of mind.* Cambridge, MA: MIT press.

Kanit, L., McDonald, R., Taskiran, D., Kulali, B., Furedy, J.J., & Pogun, S. (1996). Morris Water Maze reveals interactions between effects of nicotine and sexually diomorhicpreferences for different cognitive strategies in rats. *Society for Neuroscience Proceedings* (Abstract), in press.

McDonald, R.J., & White, N.M. (1994). Parallel information processing in the water maze: Evidence for independent memory systems involving dorsal striatum and hippocampus. *Behavioral and Neural Biology, 61*, 260-270.

Morris, R.G.M. (1981). Spatial localization does not require the presence of local cues.*Learning and Motivation, 12*, 239-260.

Platt, J.R. (1964). Strong inference. *Science, 146*, 347-352.

Popper, K.R. (1959). *The logic of scientific discovery.* New York: Basic Books.

Searle, J. (1992). *The rediscovery of mind.* Cambridge, MA: MIT press.

Tolman, E.C., Ritchie, B.F., & Kalish, D. (1946). Studies in spatial learning: II. Placelearning versus response learning.*Journal of Experimental Psychology, 35*, 221-29.

Subject Index

E

F

G

H

I

K

L

M

N

T

V

W

Author Index

NATO ASI Series H

Vol. 1: **Biology and Molecular Biology of Plant-Pathogen Interactions.** Edited by J.A. Bailey. 415 pages. 1986.

Vol. 2: **Glial-Neuronal Communication in Development and Regeneration.** Edited by H.H. Althaus and W. Seifert. 865 pages. 1987.

Vol. 3: **Nicotinic Acetylcholine Receptor: Structure and Function.** Edited by A. Maelicke. 489 pages. 1986.

Vol. 4: **Recognition in Microbe-Plant Symbiotic and Pathogenic Interactions.** Edited by B. Lugtenberg. 449 pages. 1986.

Vol. 5: **Mesenchymal-Epithelial Interactions in Neural Development.** Edited by J. R. Wolff, J. Sievers, and M. Berry. 428 pages. 1987.

Vol. 6: **Molecular Mechanisms of Desensitization to Signal Molecules.** Edited by T M. Konijn, P J. M. Van Haastert, H. Van der Starre, H. Van der Wel, and M.D. Houslay. 336 pages. 1987.

Vol. 7: **Gangliosides and Modulation of Neuronal Functions.** Edited by H. Rahmann. 647 pages. 1987.

Vol. 8: **Molecular and Cellular Aspects of Erythropoietin and Erythropoiesis.** Edited by I.N. Rich. 460 pages. 1987.

Vol. 9: **Modification of Cell to Cell Signals During Normal and Pathological Aging.** Edited by S. Govoni and F. Battaini. 297 pages. 1987.

Vol. 10: **Plant Hormone Receptors.** Edited by D. Klämbt. 319 pages. 1987.

Vol. 11: **Host-Parasite Cellular and Molecular Interactions in Protozoal Infections.** Edited by K.-P. Chang and D. Snary. 425 pages. 1987.

Vol. 12: **The Cell Surface in Signal Transduction.** Edited by E. Wagner, H. Greppin, and B. Millet. 243 pages. 1987.

Vol. 13: **Toxicology of Pesticides: Experimental, Clinical and Regulatory Perspectives.** Edited by L.G. Costa, C.L. Galli, and S.D. Murphy. 320 pages. 1987.

Vol. 14: **Genetics of Translation. New Approaches.** Edited by M.F. Tuite, M. Picard, and M. Bolotin-Fukuhara. 524 pages. 1988.

Vol. 15: **Photosensitisation. Molecular, Cellular and Medical Aspects.** Edited by G. Moreno, R. H. Pottier, and T. G. Truscott. 521 pages. 1988.

Vol. 16: **Membrane Biogenesis.** Edited by J.A.F Op den Kamp. 477 pages. 1988.

Vol. 17: **Cell to Cell Signals in Plant, Animal and Microbial Symbiosis.** Edited by S. Scannerini, D. Smith, P. Bonfante-Fasolo, and V. Gianinazzi-Pearson. 414 pages. 1988.

Vol. 18: **Plant Cell Biotechnology.** Edited by M.S.S. Pais, F. Mavituna, and J. M. Novais. 500 pages. 1988.

NATO ASI Series H

Vol. 19: **Modulation of Synaptic Transmission and Plasticity in Nervous Sys tems.**
Edited by G. Hertting and H.-C. Spatz. 457 pages. 1988.

Vol. 20: **Amino Acid Availability and Brain Function in Health and Disease.**
Edited by G. Huether. 487 pages. 1988.

Vol. 21: **Cellular and Molecular Basis of Synaptic Transmission.**
Edited by H. Zimmermann. 547 pages. 1988.

Vol. 22: **Neural Development and Regeneration. Cellular and Molecular Aspects.**
Edited by A. Gorio, J. R. Perez-Polo, J. de Vellis, and B. Haber. 711 pages. 1988.

Vol. 23: **The Semiotics of Cellular Communication in the Immune System.**
Edited by E.E. Sercarz, F. Celada, N.A. Mitchison, and T. Tada. 326 pages. 1988.

Vol. 24: **Bacteria, Complement and the Phagocytic Cell.**
Edited by F. C. Cabello und C. Pruzzo. 372 pages. 1988.

Vol. 25: **Nicotinic Acetylcholine Receptors in the Nervous System.**
Edited by F. Clementi, C. Gotti, and E. Sher. 424 pages. 1988.

Vol. 26: **Cell to Cell Signals in Mammalian Development.**
Edited by S.W. de Laat, J.G. Bluemink, and C.L. Mummery. 322 pages. 1989.

Vol. 27: **Phytotoxins and Plant Pathogenesis.**
Edited by A. Graniti, R. D. Durbin, and A. Ballio. 508 pages. 1989.

Vol. 28: **Vascular Wilt Diseases of Plants. Basic Studies and Control.**
Edited by E. C. Tjamos and C. H. Beckman. 590 pages. 1989.

Vol. 29: **Receptors, Membrane Transport and Signal Transduction.**
Edited by A. E. Evangelopoulos, J. P. Changeux, L. Packer, T. G. Sotiroudis, and K.W.A. Wirtz. 387 pages. 1989.

Vol. 30: **Effects of Mineral Dusts on Cells.**
Edited by B.T. Mossman and R.O. Begin. 470 pages. 1989.

Vol. 31: **Neurobiology of the Inner Retina.**
Edited by R. Weiler and N.N. Osborne. 529 pages. 1989.

Vol. 32: **Molecular Biology of Neuroreceptors and Ion Channels.**
Edited by A. Maelicke. 675 pages. 1989.

Vol. 33: **Regulatory Mechanisms of Neuron to Vessel Communication in Brain.**
Edited by F. Battaini, S. Govoni, M.S. Magnoni, and M. Trabucchi. 416 pages. 1989.

Vol. 34: **Vectors asTools for the Study of Normal and Abnormal Growth and Differentiation.**
Edited by H. Lother, R. Dernick, and W. Ostertag. 477 pages. 1989.

Vol. 35: **Cell Separation in Plants: Physiology, Biochemistry and Molecular Biology.** Edited by D. J. Osborne and M. B. Jackson. 449 pages. 1989.

NATO ASI Series H

Vol. 36: **Signal Molecules in Plants and Plant-Microbe Interactions.** Edited by B.J.J. Lugtenberg. 425 pages. 1989.

Vol. 37: **Tin-Based Antitumour Drugs.** Edited by M. Gielen. 226 pages. 1990.

Vol. 38: **The Molecular Biology of Autoimmune Disease.** Edited by A.G. Demaine, J-P. Banga, and A.M. McGregor. 404 pages. 1990.

Vol. 39: **Chemosensory Information Processing.** Edited by D. Schild. 403 pages. 1990.

Vol. 40: **Dynamics and Biogenesis of Membranes.** Edited by J. A. F. Op den Kamp. 367 pages. 1990.

Vol. 41: **Recognition and Response in Plant-Virus Interactions.** Edited by R. S. S. Fraser. 467 pages. 1990.

Vol. 42: **Biomechanics of Active Movement and Deformation of Cells.** Edited by N. Akkas. 524 pages. 1990.

Vol. 43: **Cellular and Molecular Biology of Myelination.** Edited by G. Jeserich, H. H. Althaus, and T. V. Waehneldt. 565 pages. 1990.

Vol. 44: **Activation and Desensitization of Transducing Pathways.** Edited by T. M. Konijn, M. D. Houslay, and P. J. M. Van Haastert. 336 pages. 1990.

Vol. 45: **Mechanism of Fertilization: Plants to Humans.** Edited by B. Dale. 710 pages. 1990.

Vol .46: **Parallels in Cell to Cell Junctions in Plants and Animals.** Edited by A. W Robards, W. J . Lucas, J . D. Pitts, H . J . Jongsma, and D. C. Spray. 296 pages. 1990.

Vol. 47: **Signal Perception and Transduction in Higher Plants.** Edited by R. Ranjeva and A. M. Boudet. 357 pages. 1990.

Vol. 48: **Calcium Transport and Intracellular Calcium Homeostasis.** Edited by D. Pansu and F. Bronner. 456 pages. 1990.

Vol. 49: **Post-Transcriptional Control of Gene Expression.** Edited by J. E. G. McCarthy and M. F. Tuite. 671 pages. 1990.

Vol. 50: **Phytochrome Properties and Biological Action.** Edited by B. Thomas and C. B. Johnson. 337 pages. 1991.

Vol. 51: **Cell to Cell Signals in Plants and Animals.** Edited by V. Neuhoff and J. Friend. 404 pages. 1991.

Vol. 52: **Biological Signal Transduction.** Edited by E. M . Ross and K . W. A. Wirtz. 560 pages. 1991.

Vol. 53: **Fungal Cell Wall and Immune Response.** Edited by J. P. Latge and D. Boucias. 472 pages. 1991.

NATO ASI Series H

Vol. 54: **The Early Effects of Radiation on DNA.**
Edited by E. M. Fielden and P. O'Neill. 448 pages. 1991.

Vol. 55: **The Translational Apparatus of Photosynthetic Organelles.**
Edited by R. Mache, E. Stutz, and A. R. Subramanian. 260 pages. 1991.

Vol. 56: **Cellular Regulation by Protein Phosphorylation.**
Edited by L. M. G. Heilmeyer, Jr. 520 pages. 1991.

Vol. 57: **Molecular Techniques in Taxonomy.**
Edited by G . M . Hewitt, A. W. B. Johnston, and J. P. W. Young .
420 pages. 1991.

Vol. 58: **Neurocytochemical Methods.**
Edited by A. Calas and D. Eugene. 352 pages. 1991.

Vol. 59: **Molecular Evolution of the Major Histocompatibility Complex.**
Edited by J. Klein and D. Klein. 522 pages. 1991.

Vol. 60: **Intracellular Regulation of Ion Channels.**
Edited by M. Morad and Z. Agus. 261 pages. 1992.

Vol. 61: **Prader-Willi Syndrome and Other Chromosome 15q Deletion Disorders.**
Edited by S. B. Cassidy. 277 pages. 1992.

Vol. 62: **Endocytosis. From Cell Biology to Health, Disease and Therapie.**
Edited by P. J. Courtoy. 547 pages. 1992.

Vol. 63: **Dynamics of Membrane Assembly.**
Edited by J. A. F. Op den Kamp. 402 pages. 1992.

Vol. 64: **Mechanics of Swelling. From Clays to Living Cells and Tissues.**
Edited by T. K. Karalis. 802 pages. 1992.

Vol. 65: **Bacteriocins, Microcins and Lantibiotics.**
Edited by R. James, C. Lazdunski, and F. Pattus. 530 pages. 1992.

Vol. 66: **Theoretical and Experimental Insights into Immunology.**
Edited by A. S. Perelson and G. Weisbuch. 497 pages. 1992.

Vol. 67: **Flow Cytometry. New Developments.**
Edited by A. Jacquemin-Sablon. 1993.

Vol. 68: **Biomarkers. Research and Application in the Assessment of Environmental Health.** Edited by D. B. Peakall and L. R. Shugart.
138 pages. 1993.

Vol. 69: **Molecular Biology and its Application to Medical Mycology.**
Edited by B. Maresca, G. S. Kobayashi, and H. Yamaguchi. 271 pages.
1993.

Vol. 70: **Phospholipids and Signal Transmission.**
Edited by R. Massarelli, L. A. Horrocks, J. N. Kanfer, and K. Löffelholz.
448 pages. 1993.

NATO ASI Series H

Vol. 71: **Protein Synthesis and Targeting in Yeast.**
Edited by A. J. P. Brown, M. F. Tuite, and J. E. G. McCarthy. 425 pages. 1993.

Vol. 72: **Chromosome Segregation and Aneuploidy.**
Edited by B. K. Vig. 425 pages. 1993.

Vol. 73: **Human Apolipoprotein Mutants III. In Diagnosis and Treatment.**
Edited by C. R. Sirtori, G. Franceschini, B. H. Brewer Jr. 302 pages. 1993.

Vol. 74: **Molecular Mechanisms of Membrane Traffic.**
Edited by D. J. Morré, K. E. Howell, and J. J. M. Bergeron. 429 pages. 1993.

Vol. 75: **Cancer Therapy. Differentiation, Immunomodulation and Angiogenesis.**
Edited by N. D'Alessandro, E. Mihich, L. Rausa, H. Tapiero, and T. R. Tritton. 299 pages. 1993.

Vol. 76: **Tyrosine Phosphorylation/Dephosphorylation and Downstream Signalling.**
Edited by L. M. G. Heilmeyer Jr. 388 pages. 1993.

Vol. 77: **Ataxia-Telangiectasia.** Edited by R. A. Gatti, R. B. Painter. 306 pages. 1993.

Vol. 78: **Toxoplasmosis.** Edited by J. E. Smith. 272 pages. 1993.

Vol. 79: **Cellular Mechanisms of Sensory Processing. The Somatosensory System.**
Edited by L. Urban. 514 pages. 1994.

Vol. 80: **Autoimmunity: Experimental Aspects.**
Edited by M. Zouali. 318 pages. 1994.

Vol. 81: **Plant Molecular Biology. Molecular Genetic Analysis of Plant Development and Metabolism.**
Edited by G. Coruzzi, P. Puigdomènech. 579 pages. 1994.

Vol. 82: **Biological Membranes: Structure, Biogenesis and Dynamics.**
Edited by Jos A. F. Op den Kamp. 367 pages. 1994.

Vol. 83: **Molecular Biology of Mitochondrial Transport Systems.**
Edited by M. Forte, M. Colombini. 420 pages. 1994.

Vol. 84: **Biomechanics of Active Movement and Division of Cells.**
Edited by N. Akkaş. 587 pages. 1994.

Vol. 85: **Cellular and Molecular Effects of Mineral and Synthetic Dusts and Fibres.**
Edited by J. M. G. Davis, M.-C. Jaurand. 448 pages. 1994.

Vol. 86: **Biochemical and Cellular Mechanisms of Stress Tolerance in Plants.**
Edited by J. H. Cherry. 616 pages. 1994.

NATO ASI Series H

Vol. 87: **NMR of Biological Macromolecules.**
Edited by C. I. Stassinopoulou. 616 pages. 1994.

Vol. 88: **Liver Carcinogenesis. The Molecular Pathways.**
Edited by G. G. Skouteris. 502 pages. 1994.

Vol. 89: **Molecular and Cellular Mechanisms of H^+ Transport.**
Edited by B. H. Hirst. 504 pages. 1994.

Vol. 90: **Molecular Aspects of Oxidative Drug Metabolizing Enzymes: Their Significance in Environmental Toxicology, Chemical Carcinogenesis and Health.**
Edited by E. Arınç, J. B. Schenkman, E. Hodgson. 623 pages. 1994.

Vol. 91: **Trafficking of Intracellular Membranes: From Molecular Sorting to Membrane Fusion.**
Edited by M. C. Pedroso de Lima, N. Düzgüneş D. Hoekstra. 371 pages.1995.

Vol. 92: **Signalling Mechanisms – from Transcription Factors to Oxidative Stress.**
Edited by L. Packer, K. Wirtz. 466 pages.1995.

Vol. 93: **Modulation of Cellular Responses in Toxicity.**
Edited by C. L. Galli, A. M. Goldberg, M. Marinovich. 379 pages. 1995.

Vol. 94: **Gene Technology.**
Edited by A. R. Zander, W. Ostertag, B. V. Afanasiev, F. Grosveld. 556 pages. 1996.

Vol. 95: **Flow and Image Cytometry.**
Edited by A. Jacquemin-Sablon. 241 pages. 1996.

Vol. 96: **Molecular Dynamics of Biomembranes.**
Edited by J. A. F. Op den Kamp. 424 pages. 1996.

Vol. 97: **Post-transcriptional Control of Gene Expression.**
Edited by O. Resnekov, A. von Gabain. 283 pages. 1996.

Vol. 98: **Lactic Acid Bacteria: Current Advances in Metabolism, Genetics and Applications.**
Edited by T. F. Bozoğlu, B. Ray. 412 pages. 1996.

Vol. 99: **Tumor Biology Regulation of Cell Growth, Differentiation and Genetics in Cancer.**
Edited by A. S. Tsiftsoglou, A. C. Sartorelli, D. E. Housman, T. M. Dexter. 342 pages. 1996.

Vol. 100: **Neurotransmitter Release and Uptake.**
Edited by Ş. Pöğün. 344 pages. 1997.

MIX
Papier aus verantwortungsvollen Quellen
Paper from responsible sources
FSC® C105338

If you have any concerns about our products,
you can contact us on
ProductSafety@springernature.com

In case Publisher is established outside the EU,
the EU authorized representative is:
Springer Nature Customer Service Center GmbH
Europaplatz 3, 69115 Heidelberg, Germany

Printed by Libri Plureos GmbH
in Hamburg, Germany